TEUBNER-TEXTE zur Mathematik   Band 135

S. G. Mikhlin / N. F. Morozov /
M. V. Paukshto

The Integral Equations
of the Theory of Elasticity

TEUBNER-TEXTE zur Mathematik

Herausgegeben von
Prof. Dr. Jochen Brüning, Augsburg
Prof. Dr. Herbert Gajewski, Berlin
Prof. Dr. Herbert Kurke, Berlin
Prof. Dr. Hans Triebel, Jena

Die Reihe soll ein Forum für Beiträge zu aktuellen Problemstellungen der Mathematik sein. Besonderes Anliegen ist die Veröffentlichung von Darstellungen unterschiedlicher methodischer Ansätze, die das Wechselspiel zwischen Theorie und Anwendungen sowie zwischen Lehre und Forschung reflektieren. Thematische Schwerpunkte sind Analysis, Geometrie und Algebra.

In den Texten sollen sich sowohl Lebendigkeit und Originalität von Spezialvorlesungen und Seminaren als auch Diskussionsergebnisse aus Arbeitsgruppen widerspiegeln.

TEUBNER-TEXTE erscheinen in deutscher oder englischer Sprache.

The Integral Equations of the Theory of Elasticity

By Prof. Dr. Solomon G. Mikhlin

Prof. Dr. Nikita F. Morozov
University St. Petersburg

Prof. Dr. Michael V. Paukshto
Marine Technical University St. Petersburg

Translated from the Russian by Prof. Dr. Rainer Radok
Salaya

Edited by

Prof. Dr. Herbert Gajewski
Weierstraß Institute Berlin

Springer Fachmedien Wiesbaden GmbH 1995

Prof. Dr. Solomon G. Mikhlin

Born in 1908 in Kholmetch (Belorussia). In 1929 he finished his studies at the Leningrad University as a Bachelor of Science in Mathematics, where among others N. M. Günter and W. I. Smirnov were his teachers. He achieved in 1935 the scientific degree "Doctor of the Physical-Mathematical Sciences", and in 1937 he was appointed a professorship. Since 1944 until his death in 1990 he has been working as a professor at the Leningrad University. His scientific achievements were highly recognized – especially abroad. This was proved among others by the award of the degree of Honorary Doctor of the Technische Hochschule Karl-Marx-Stadt (Chemnitz) in 1968, his election to a Member of the Deutsche Akademie der Naturforscher Leopoldina in 1970 and to the Italian Accademia Nazionale dei Lincei in 1982.

Prof. Dr. Nikita F. Morozov

Born in 1932 in Leningrad (St. Petersburg). Studied mathematics and mechanics in St. Petersburg University from 1949 to 1954. Received Dr. rer. nat. in 1958 and Dr. sc. nat. in 1967 from St. Petersburg University. Head of Department of Elasticity at St. Petersburg University since 1976. Corresponding Member of the Russian Academy of Science since 1994. Fields of interest: Nonlinear elasticity and mathematical problems of fracture.

Prof. Dr. Michael V. Paukshto

Born in 1952 in Yoshkar-Ola. Studied mathematics in Novosibirsk from 1970 to 1972 and in Leningrad (St. Petersburg) from 1973 to 1975. Received Dr. rer. nat. in 1978 and Dr. sc. nat. in 1991 from St. Petersburg University. Since 1994 Professor of Mathematics at Marine Technical University.
Fields of interest: Mathematical problems of fracture mechanics, molecular mechanics and software support.

Die Deutsche Bibliothek – CIP-Einheitsaufnahme

Michlin, Solomon G.:
The integral equations of the theory of elasticity / by Solomon G. Mikhlin,
Nikita F. Morozov ; Michael V. Paukshto.
Transl. from the Russ. by Rainer Radok. Ed. by Herbert Gajewski. –
Stuttgart ; Leipzig : Teubner, 1995
  (Teubner-Texte zur Mathematik ; Bd. 135)
   ISBN 978-3-663-11627-1   ISBN 978-3-663-11626-4 (eBook)
   DOI 10.1007/978-3-663-11626-4

NE: Morozov, Nikita F.:; Paukshto, Michail V.:; GT

© Springer Fachmedien Wiesbaden 1995
Ursprünglich erschienen bei B.G Teubner Verlagsgesellschaft Leipzig 1995

Umschlaggestaltung: E. Kretschmer, Leipzig

PREFACE

It was the last book the outstanding mathematician, mechanician and lecturer S.G. Mikhlin took an active part in writing. Having been completed during his lifetime, this book could not be published in Russia due to well-know difficulties.

Since that time new results in integral equations of elasticity theory have appeared. The works of W. Wendland and his school on numerical methods of solving boundary integral equations, the works of I. Chudinovich on investigation of non–stationary integral equations, the works of S. Kuznetsov connected with the construction of the fundamental solutions for anisotropic media and others deserve special mentioning. The authors recognize that though the book is devoted to integral equations of elasticity theory, its contents do not cover all possible directions in this field. So the book does not contain the investigations of pseudo-differential equations of three–dimensional problems of elasticity theory, connected with the works of R. Goldstein, I. Klein, G. Eskin; the questions of solving by integral transformations (I. Ufland, L. Slepian, B. Budaev); the theory of symbols of pseudo–differential operators on non–smooth surfaces developed in the works of B. Plamenevski et al. and the new methods of numerical solution of pseudo–differential equations as developed by a school of V. Mazya.

The present book gives the classical methods of potential theory in elasticity and their development and also the solution of a number of problems which here are published in English for the first time. The book contains the work of S.G. Mikhlin on the Cosserat spectrum, the results of which are now applied effectively.

The authors would like to express their sincere gratitude to the editor of the book H. Gajewski, to the translator R. Radok, and to S. Proessdorf and A. Koshelev for their help and assistance.

Our special gratitude is to T. Efimova and Ch. Huber who prepared this book for edition.

St. Petersburg, August 1994

N. Morozov

M. Paukshto

CONTENTS

PART I
INTEGRAL EQUATIONS

CHAPTER 1
GENERAL RESULTS ON LINEAR INTEGRAL EQUATIONS

We will assume that the reader is familiar with the theory of Fredholm integral equations (for example, as presented in one of the books [24, 33, 39, 41]). We also require the reader to know the simplest concepts of Functional Analysis such as Banach Spaces (in particular, the spaces C and L_2), separable spaces, linear functionals and operators (bounded and unbounded), dual spaces and operators, precompactness and compactness of manifolds (a space is M-precompact, if $\bar{M}$ is compact). The reader will find an adequate introduction to these topics, for example, in [10, 12, 40].

§1. Compact operators and Fredholm's theorems

1^0. Equations with compact operators form a natural class (which is wider than the class of Fredholm equations) to which the Fredholm Theory extends. Let X and Y be Banach spaces which, in general, are complex, and T the linear operator which is defined on the entire space X and maps X into Space Y; as is customary, we will write: $T \in (X \to Y)$. The operator T is said to be compact, if it transforms any manifold, bounded in X, into a manifold, precompact in Y.

The simplest compact operator has finite dimension; it is of the form

$$Tx = \sum_{k=1}^{n} \varphi_k \, f_k(x) \tag{1.1.1}$$

where n is a natural number, φ_k are given elements of Y, f_k linear functionals, defined and bounded throughout X. In fact, let M be a bounded manifold in $X: \forall x \in M$, $\|x\| \leq C = const$. By the well known Weierstrass theorem, we may separate from M sequence $\{x_j\}$ so that all sequences $\{f_k(x_j)\}$, $k = 1, 2, ..., n$ converge, when in the norm of Y the sequence $\{Tx\} = \{\sum_{k=1}^{n} \varphi_k \, f_k(x)\}$ will converge. As a consequence, the operator T in (1.1.1) will be compact.

2^0. We will now study certain theorems on compact operators which will be used widely in the sequel.

Theorem 1.1. Every compact operator is bounded.

If an operator T is unbounded, then there exists some sequence of elements $x_n \in X$, $n = 1, 2, ...$ so that $\|x_n\| = 1$ and $\|Tx_n\|_Y \longrightarrow \infty$ as $n \to \infty$. However, then the manifold $\{Tx_n\}$ is not precompact, although the manifold $\{x_n\}$ is bounded, hence follows that the operator T is not compact.

Theorem 1.2. A sum of two compact operators is compact. The product of two operators is compact, if one of the factors is bounded, the other is compact.

The assertion on a sum of compact operators is obvious. We consider the product: Let $A \in (Z \to Y)$ be bounded, $T \in (X \to Z)$ be a compact operator, X, Y, Z Banach spaces and M a bounded manifold in X. We will prove that the operator AT transforms M into a manifold, precompact in Y. The compact operator T transforms the bounded manifold $M \subset X$ into the precompact manifold $M' = T \subset Z$, which is transformed by the bounded operator A into a manifold, precompact in Y.

In a similar manner, the compactness of the operator TA may be established, if $A \in (X \to Z)$ is bounded and $T \in (Z \to Y)$ is compact.

Theorem 1.3. Let T and T^* be two abjoint operators. If one of them is compact, then likewise the second operator is compact.

We will not give here the proof of this theorem which may be found in the monographs [10, 12].

Theorem 1.4. Let $T_n \in (X \to Y)$, $n = 1, 2, ...$, a sequence of compact operators, converge in a norm to some operator T_n so that $\|T_n - T\| \longrightarrow 0$ as $n \to \infty$. Then T is a compact operator.

Let $M \subset X$ be a bounded manifold: $\forall x \in M$, $\|x\| \leq c$, $c = const$. Separate from M a sequence $\{x_{1n}\}$ such that $T_1 x_k$ converges for $k \to \infty$ to some element $y_1 \in Y$. Separate from this sequence a sequence $\{x_{2k}\}$, such that $T_2 x_{2k}$ tends for $k \to \infty$ to some element $y_2 \in Y$, etc. In this manner, we obtain a sequence of sequences $\{x_{nk}\}$ such that every one of them is a part of all preceding sequences and for any n there exists $\lim\limits_{k \to \infty} T_n x_{nk} = y_n \in Y$. We select diagonally the sequence $\{x_{kk}\}$. It has the property that it is contained with the exception of some finite number of elements in the sequence $\{x_{nk}\}$

for any fixed n, hence $\forall n \; \lim\limits_{k\to\infty} T_n x_{kk} = y_n$.

We shall prove next that the sequence $\{T x_{kk}\}$ converges, and hence theorem 1.4. is valied. We have

$$\|T\,x_{kk} - T\,x_{jj}\| \le \|T_n(x_{kk} - x_{jj})\| + \|(T - T_n)(x_{kk} - x_{jj})\| \,.$$

Given a small number $\varepsilon > 0$, we pick out and fix the number n such that $\|T - T_n\| < \varepsilon/4c$ when

$$\|(T - T_n)(x_{kk} - x_{jj})\| < \varepsilon/2 \,.$$

Using the fact that for the fixed n the sequence $\{T_n\,x_{jj}\}$ converges, we select a number I so large that for $j,\,k \ge I$ the inequality $\|T_n(x_{jj} - x_{kk})\| < \varepsilon/2$ holds true. Hence we find for $j,\,k \ge I$ the inequality $\|T\,x_{kk} - T\,x_{jj}\| < \varepsilon$ and the sequence $\{T\,x_{jj}\}$ converges.

3^0. Theorem 1.4 permits us to present certain new and important classes of compact operators two of which will now be considered.

a) Let S be a region in Euclidean space R^m of dimension m, bounded or unbounded (possibly, in particular, $S = R^m$) and let dS be an element of measure in S. We will denote by $L_2(S)$ the space of functions which are given almost everywhere and quadratically symmetric in S. Consider the integral operator

$$(K\,x)(t) = \int_S K(t,\tau)\,x(\tau)d_\tau S \,, \qquad (1.1.2)$$

where t and τ are points in S, $x \in L_2(S)$ and $K(t,\tau)$ satisfies the inequality

$$\int_S \int_S |\,K(t,\tau)\,|^2 \, d_t S \, d_\tau S = C_0^2 < \infty \,. \qquad (1.1.3)$$

The operator (1.1.2) will be called a Fredholm operator, the function $K(t,\tau)$ the kernel of this operator. We will show that the Fredholm operator is compact in $L_2(S)$. First of all, it is bounded: Integrating the Cauchy-Bunyakovski inequality

$$|\,(K\,x)\,|^2 = |\int_S K(t,\tau)\,x(\tau)d_\tau S\,|^2 \le \|x\|^2 \int_S |\,K(t,\tau)\,|^2 \, d_\tau S \,;$$

with respect to t, we find $\|K\,x\|^2 \le C_0^2 \|x\|^2$ or

$$\|K\| \le C_0 \,. \qquad (1.1.4)$$

The separable Hilbert space $L_2(S)$ has an enumerable, ortho-normalized base. Let $\{\varphi_n(t)\}$ be such a base. The kernel $K(t,\tau)$ may be expanded in the, convergent in the mean, double series

$$K(t,\tau) = \sum_{j,k=1}^{\infty} A_{jk}\varphi_j(t)\varphi_k(\tau), \tag{1.1.5}$$

where $A_{jk} = \int_S \int_S K(t,\tau)\,\varphi_j(t)\,\varphi_k(\tau)\,d_tS\,d_\tau S$.

$$K^{(n)}(t,\tau) = \sum_{j,k=1}^{n} A_{jk}\varphi_j(t)\varphi_k(\tau). \tag{1.1.6}$$

Let $K^{(n)}$ be the Fredholm operator with kernel $K^{(n)}(t,\tau)$. This operator has finite dimension, and, consequently, is compact. Furthermore, the sequence (1.1.5) converges in the mean, hence

$$\int_S \int_S \mid K(t,\tau) - K^{(n)}(t,\tau) \mid^2 d_tS\,d_\tau S \longrightarrow 0, n \to \infty.$$

By (1.1.4), $\|K - K^{(n)}\| \longrightarrow 0$, $n \to \infty$, and, by theorem 1.4, the operator K is compact.

In particular, the operator (1.1.2) is compact in $L_2(S)$, if the region S is finite and the kernel $K(t,\tau)$ is bounded.

b) Let S be a finite region of the space R^m. We consider the integral operator with a weak singularity

$$(K\,x)(t) = \int_S \frac{A(t,\tau)}{r^\alpha}\,x(\tau)\,d_\tau S\,, \quad K(t,\tau) = \frac{A(t,\tau)}{r^\alpha}\,, \tag{1.1.7}$$

where t and τ are, as before, points of S, $A(t,\tau)$ is a bounded, measurable function, α a constant, $0 < \alpha < m$, and r is the distance between the points t and τ. We will prove that the operator K, defined by (1.1.7), is compact in $L_2(S)$.

We select a sequence of numbers $\eta_n \longrightarrow 0$ and let

$$K^{(n)}(t,\tau) = \begin{cases} K(t,\tau), r \geq \eta_n, \\ \qquad 0, r < \eta_n; \end{cases} \tag{1.1.8}$$

where $K^{(n)}$ is a Fredholm operator will kernel (1.1.8). We will obtain the norm of the difference $K - K_{(n)}$. In order to simplify the calculations, we will continue the function $x(t)$ into the outside of space S, setting it there equal to zero. Then, if $\mid A(t, \tau) \mid \leq C_0$, then

$$\mid ((K - K^{(n)})x)(t) \mid = \mid \int\limits_{r < \eta_n} \frac{A(t, \tau)}{r^\alpha} x(\tau)\, d_\tau S \mid \leq C_0 \int\limits_{r < \eta_n} \frac{\mid x(\tau) \mid}{r^{\alpha/2}} \frac{1}{r^{\alpha/2}}\, d_\tau S ,$$

and, by the Cauchy-Bunyakovski inequality,

$$\mid ((K - K^{(n)})x)(t) \mid^2 \leq C_0^2 \int\limits_{r < \eta_n} \frac{\mid x(\tau) \mid^2}{r^\alpha}\, d_\tau S \int\limits_{r < \eta_n} \frac{d_\tau S}{r^\alpha} . \tag{1.1.9}$$

The value of the second integral on the right hand side, readily evaluated by transition to spherical coordinates, is $2\pi^{m/2} \cdot \eta_n^{m-\alpha}/(m - \alpha)\Gamma(m/2)$. Its substitution into (1.1.9) and integration over S yields

$$\|K - K^{(n)}\| \leq \frac{2C_0\, \pi^{m/2}}{(m - \alpha)\Gamma(m/2)}\, \eta_n^{m-\alpha} \longrightarrow 0, n \to \infty .$$

The region S is finite and the kernel $K^{(n)}(t, \tau)$ for fixed n is bounded (its modulus is not larger than $C_0\, \eta_n^{-\alpha}$) hence the operator $K^{(n)}$ is compact. By theorem 1.4, also the operator K is compact.

4^0. The operators considered in 3^0, are also compact under certain more general conditions. For example, it is sufficient for S to be a sectional Lyapunov surface of dimension m, embedded in Euclidean space of larger dimension. The proof of this assertion will be left for the reader.

We note yet that for a certain change in condition (1.1.3) the operator (1.1.2) is compact in the space $L_p(S)$, $1 < p < \infty$. However, in this space, the operator is compact with a weaker singularity.

We will now dwell on one particular case which is important for the sequel. Consider the operator T, defined by

$$(Tx)(t) = y(t) = \int\limits_{R^2} r^{-2}[\alpha(\tau) - \alpha(t)]\, e^{in\theta}\, x(\tau)\, d\tau , \tag{1.1.10}$$

where t and τ are points in the R^2, $r = \mid \tau - t \mid$, $\theta = arctg \frac{\tau_2 - t_2}{\tau_1 - t_1}$, n is any integer, α is a function bounded in R^2 which satisfies the inequality

$$\mid \alpha(\tau) - \alpha(t) \mid \leq c \mid r \mid^\lambda \, [(1 + \mid \tau \mid^2)(1 + \mid t \mid^2)]^{-\lambda/2} , \tag{1.1.11}$$

where C and λ are positive constants.

We will show that under these conditions operator (1.1.10) is compact in $L_2(R^2)$.

As is known, the stereographic projection (for details, cf. [25, 54])

$$\xi_k = 2t_k/(1+|t|^2),\ k=1,2;\ \xi_3 = (|t|^2-1)/(|t|^2+1),$$

where $|t|^2 = t_1^2 + t_2^2$, maps the Euclidean plane R^2 into sphere $\sum$:

$$\sum_{k=1}^{3}\xi_k^2 = 1,$$

the so-called Riemann sphere. If we denote by

$$x'(\xi) = \left(\frac{|t|^2+1}{2}\right)x(t),\ y'(\xi) = \left(\frac{|t|^2+1}{2}\right)y(t),\qquad(1.1.12)$$

then (1.1.10) becomes

$$y'(\xi) = \int_\Sigma \frac{\alpha(\tau)-\alpha(t)}{r^2}\,e^{in\theta}\,x'(\eta)\,d_\eta\Sigma,\qquad(1.1.13)$$

where $d_\eta\Sigma$ is the element of the surface of the sphere Σ, η is the point on this sphere, into which goes the point $\tau \in R^2$ under the stereographic projection. The operator on the right hand side of (1.1.13) is compact in $L_2(\Sigma)$. It is readily verified that Transformation (1.1.12) isometrically maps $L_2(R^2)$ into $L_2(\Sigma)$. Hence operator (1.1.10) is compact in $L_2(R^2)$.

Note condition (1.1.11) has a simple geometric meaning: If under the stereographic projection points t and τ go over into ξ and η, respectively, and $\rho = |\xi - \eta|$, then

$$|\alpha(\tau)-\alpha(t)| \le C2^{-\lambda}\rho^\lambda.\qquad(1.1.11a)$$

This assertion remains true, if is a compact operator, if R^2 is replaced by a Euclidean space R^m of any dimensionality m, multiplier $e^{in\theta}$ of the spherical function and exponent -2 to $-m$. Under these conditions, the operator T is compact in $L_2(\Sigma)$ for any m.

A derivation of the last result cf. [25, 54].

5^0. We will assume now that $X = Y$ and that X is a, in general, complex space. Consider the equation

$$x - \lambda T x = (I - \lambda T)x = f,\qquad(1.1.14)$$

where T is a compact operator acting from X into X, λ is a numerical, in general, complex parameter, x and f are elements of the space X which is known and given. Finally, I is the identity operator in X.

In his classical work, F. Riesz [35] has shown that the basic theorems of Fredholm remain true for equations of the form (1.1.14). A proof may be found in [35] as well as in the monographs [10, 12]. We will now formulate Fredholm's basic theorems.

Theorem 1.5. With the exclusion of some finite or enumerable manifold of the values of the parameter λ, referred to as the characteristic values of the operator T, or equation (1.1.14), this equation has for a fixed value of λ one and only one solution for any given right hand side f. If the manifold of the characteristic numbers is enumerable, then it has the unique point of accumulation $\lambda = \infty$.

Non-characteristic values λ is called regular. If a λ is regular, the inverse operator $(I - \lambda T)^{-1}$ exists; it is defined for the entire space X and is bounded there.

Theorem 1.6. If λ_0 is a characteristic number of a compact operator T, then it is a characteristic number of the adjoint operator T^*, and conversely.

Theorem 1.7. If λ_0 is a characteristic number of the compact operator T, then the homogeneous adjoint equations

$$x - \lambda_0 T x = 0, \qquad\qquad (1.1.15)$$

$$y - \lambda_0 T^* y = 0 \qquad\qquad (1.1.16)$$

have one and the same number of solutions. This number is finite and positive.

Theorem 1.8. In order for the non-homogeneous equation (1.1.14) with a compact operator T to be soluble, it is necessary and sufficient that the free term f be orthogonal to the solutions of the homogeneous, adjoint equation (1.1.16).

We shall now elucidate the last theorem. Let y be any solution of (1.1.16). Then $y \in X^*$ and, therefore, y is a bounded, linear functional on X. The orthogonality of y and f signifies that the value of the functional y on the element f equals zero. This relationship may be given the form $y(f) = 0$ or $(y, f) = 0$.

Theorems 1.5 - 1.8 yield the so-called Fredholm alternative which plays an important role in the study of the solubility of equations with compact operators. Let T be a compact operator which acts of the Banach space X. If the homogeneous equation (1.1.15) has only the trivial (i.e., identically zero) solution, then the non-homogeneous equation (1.1.14) has a solution which, besides, is unique for any $f \in X$. However, if (1.1.15) has a non-trivial solution, then (1.1.14) either has no solution or has a manifold of solutions.

§2. The concept of symbol. Examples

1^0. A manifold $\mathcal{R}$ is called an associative ring, if for its elements the operations of addition and multiplication with the following properties are defined: A sum and a product of elements of $\mathcal{R}$; $\mathcal{R}$ is an Abelian group with respect to addition, since addition is commutative and associative, and there exists in $\mathcal{R}$ the element 0, referred to as the zero of the ring, such that $\forall a \in \mathcal{R}$, $a + 0 = 0 + a = a$ and for every element $a \in \mathcal{R}$ there exists an "opposite' element $-a$, such that $a + (-a) = -a + a = 0$. Furthermore, multiplication is associative and distributive: if $a, b, c \in \mathcal{R}$, then $(ab)c = a(bc)$, $(a + b)c = ac + bc$, $a(b + c) = ab + ac$.

In the sequel, we will consider only associative rings which we will refer to as simple rings.

Multiplication in a ring could be non-commutative so that, in the general case, $ab \neq ba$. If any two elements a, $b \in \mathcal{R}$ are such that $ab = ba$, then these elements are said to commute. If any two elements of a ring commute, the ring is said to be commutative.

A ring may have the element 1 (unit) such that $\forall a \in \mathcal{R}$, $a \cdot 1 = 1 \cdot a = a$. In that case, it is called a ring with a unit.

Let $\mathcal{R}$ and $\mathcal{T}$ be two rings. Ring homomorphism of $\mathcal{R}$ in $\mathcal{T}$ (we will simply write "homomorphism" below) is the mapping σ of the ring $\mathcal{R}$ onto the ring $\mathcal{T}$ which satisfies the following conditions: 1) To any element of the ring $\mathcal{R}$ there corresponds one and only one element of the ring $\mathcal{T}$; 2) the sum and product of elements of $\mathcal{R}$ become under the transformation σ the sum and product of corresponding elements of $\mathcal{T}$.

2^0. Let there be given some ring $\mathcal{R}$ of linear operators with ordinary actions of operator addition and multiplication, and another ring $\mathcal{T}$ with elements of an arbitrary kind. We will assume that we have succeeded to find some homomorphism of the ring $\mathcal{R}$ on the ring $\mathcal{T}$.

By strength of the definition of ring homomorphism, if to the operators $A, B \in \mathcal{R}$ there correspond elements $a, b \in \mathcal{T}$, then to the operators $A + B$ and AB there correspond the elements $a + b$ and ab of the ring $\mathcal{T}$. In this case, we will say that $\mathcal{T}$ is a ring of symbols for $\mathcal{R}$; if to the $A \in \mathcal{R}$ there corresponds the element $a \in \mathcal{T}$, then a is said to be the symbol of the operator A. We will then write

$$a = Smb\, A. \qquad (1.2.1)$$

A ring of symbols is not unique: If $\mathcal{T}_1$ is some ring on which $\mathcal{T}$ maps homomorphly, then $\mathcal{T}_1$ likewise may be taken as a ring of symbols of the ring $\mathcal{R}$.

Note the two trivial cases which are excluded from the considerations below:

a) $\mathcal{T} = \mathcal{R}$, and the homomorphism σ referred to above is the identity transformation. In this case, the symbol of any operator of $\mathcal{R}$ coincides with the very same operator.

b) $\mathcal{T} = \{0\}$; in this case the symbol of any operator of $\mathcal{R}$ equals zero.

Note yet the fact that an ideal ring of symbols corresponds to an ideal ring of operators.

3^0. Let the ring $\mathcal{R}$ of operators contains the identity operator which we denote by I, and let $Smb\, I = i$. Furthermore, let A be an arbitrary operator of $\mathcal{R}$, and a its symbol. It follows from the identity $A = AI = IA$ that $a = ai = ia$. This means that the symbol of the identical operator is equal to the unit of the ring of symbols. Moreover, let the element a be irreversible in the ring of symbols; we will show that then the operator A is not reversible in $\mathcal{R}$. In fact, we assume that there exists in $\mathcal{R}$ an operator $B = A^{-1}$ and that $Smb\, B = b$. It follows from $AB = BA = I$ that $ab = ba = i$; in spite of the assumption, the symbol a is found to be reversible.

4^0. Let $\mathcal{R}$ be a ring of bounded operators which comprises simple and compact operators. By theorem 1.2, the last form in $\mathcal{R}$ a double-sided ideal. The zero element of the ring of symbols itself represents a double-sided ideal of this ring. Below we will everywhere select the zero element of a ring of symbols from the symbols of any compact operator of $\mathcal{R}$. It will follow from this that a mapping of a ring of symbols on to a ring of operators is not single-valued: If an operator $A \in \mathcal{R}$ and $Smb\, A = a \in \mathcal{T}$, then also $Smb\, (A+T) = a$, where T is any compact operator of $\mathcal{R}$.

5^0. We will now present some examples:

1) Let X be a space $C^{(\infty)}[\alpha, \beta]$, where α and β are real numbers, finite of infinite, and $\mathcal{R}$ is the ring of linear, differential operators with constant coefficients acting in X. As ring $\mathcal{T}$ of symbols, one can select the ring of polynomials with ordinary operations of addition and multiplication; the symbol of a differential operator $\mathcal{R}$ is then its characteristic polynomial.

2) Consider in the same space $C^{(\infty)}[\alpha, \beta]$ the wider ring $\mathcal{R}$ of linear, differential operators with variable coefficients of the form

$$A = \sum_{k=0}^{n} a_k(t)\, \frac{d^k}{dt^k}\,. \tag{1.2.2}$$

Let $a_k \in C^{(\infty)}[\alpha, \beta]$. Then we adopt as ring of symbols the ring of polynomials of a new variable ξ with coefficients which depend on t, if on this ring addition is defined in the ordinary manner, while multiplication obeys the rule

$$\sum_{k=0}^{n} a_k(t)\,\xi^k \,\sum_{k=0}^{m} b_k(t)\,\xi^k \;=\; \sum_{k=0}^{n+m} c_k(t)\,\xi^k\,;$$

$$c_k(t) = \sum_{p=0}^{n}\sum_{q=0}^{n} \binom{p}{q}\, a_p(t)\, b_{k-q}^{(p-q)}(t)\,. \tag{1.2.3}$$

The superscript $(p-q)$ in these formulas denote the derivative of corresponding order; it must be assumed that $b_{k-q} \equiv 0$, if $k - q < 0$ or $k - q > m$. Multiplication according to (1.2.3) is associative and distributive, but not commutative. The symbol of the operator (1.2.2) is the polynomial

$$Smb\,A = \sum_{k=0}^{n} a_k(t)\,\xi^k\,. \tag{1.2.4}$$

3) Let $\mathcal{R}$ be the manifold of operators of the form

$$A = a(t)I + T\,, \tag{1.2.5}$$

where I is the unit operator in some Banach space B the elements of which are functions defined on some measurable manifold $M \subset R^m$ (R^m is the Euclidean space of dimension m), t is the variable point on M, T is an operator, compact in B. We will assume regarding the functions $a(t)$ that the multiplication operator on such a function is bounded in B. Obviously, the manifold of these functions forms a ring. The manifold $\mathcal{R}$ is a ring with respect to the ordinary operations of addition and multiplication of operators. In fact, if $A_k = a_k(t)I + T_k$, $k = 1,2$, be operators taken from $\mathcal{R}$, then $A_1 + A_2 = [a_1(t) + a_2(t)]I + (T_1 + T_2) \in \mathcal{R}$, hence the multiplication operator on $a_1(t) + a_2(t)$ is bounded and the operator $T_1 + T_2$, by theorem 1.1 is compact in B. Furthermore, consider the product

$$A_1 A_2 = a_1(t)\,a_2(t)\,I + [a_1(t)T_2 + a_2(t)T_1 + T_1\,T_2]\,.$$

The first term is bounded, the second in square brackets is compact in B. Thus, $\mathcal{R}$ is a ring. Select from the ring of symbols $\mathcal{T}$ the above mentioned ring of functions $a(t)$. If A is the operator (1.2.5), then let by definition

$$Smb\,A = a(t)\,; \tag{1.2.6}$$

thereby we accept that the symbol of any compact operator of $\mathcal{R}$ equals zero.

6^0. The concept of the symbol was first introduced by S.G. Mikhlin in 1936 in his papers [17, 18] for singular, integral operators on the two-dimensional plane and in [19] for such operators on sufficiently smooth, two-dimensional manifolds. In the same year, G. Giraud [51] published without proof formulas for the symbol of a multi-dimensional, singular operator, generalizing the formulas of [17, 18]. Such a proof was published by the authors in [22] several years after the death of Giraud. They presented another, simpler proof in the monograph [54]. Paper [23] defines the symbol of a singular integral extended over a sufficiently smooth manifold of any dimensionality. It is also noted there that the concept of symbol is uniquely formulated and proves to be of some use also for one-dimensional, singular integral operators.

In the work of J.J. Cohn and L. Nirenberg [13], the concept of the symbol is extended to pseudo-differential operators. The general concept of symbol, presented above, was first published by the authors in [54].

§3. Regularization

1^0. Let X and Y be Banach spaces, A a linear, bounded operator, defined throughout X and acting from X to Y; in the notation of §1, 2^0: $A \in (X \to Y)$. We will say that an operator A admits regularization, if there exists a linear, bounded operator R, defined on the manifold of values of the operator A and acting from Y to X such that

$$R A = I + T, \tag{1.3.1}$$

where I is the unit and T a compact operator in X. In this case, we will refer to R as the left, regularizatory operator of A. The right regularization and right regularizator is defined in an analogous manner.

Some comments on the notation follow. Let A be a linear operator, acting from the Banach space X into the Banach space Y. We will denote by $D(A)$ the region of definition of the operator A and by $Im(A)$ the domain of its values. Thus, if $x \in D(A)$, then $x \in X$ and the expression $A x$ make sense; it is an element of the manifold $Im(A) \subset Y$. Furthermore, throughout part I, we will denote by the symbol T, with or without index, compact operators.

If R is a left (right) regularizator of an operator A, then for any compact operators T_1 and T_2, acting in corresponding spaces, $R + T_1$ is the left (right) regularizator of the operator $A + T_2$.

Equations

$$A x = f \tag{1.3.2}$$

and

$$RAx = Rf, \qquad (1.3.3)$$

where R is the left regularizator, in the general case, are not equivalent; any solution of (1.3.2) satisfies (1.3.3), but the inverse is not true: If x is a solution of (1.3.3), then it satisfies the equation $Ax = f + y_0$, where y_0 satisfies the equation $Ry = 0$. Equations (1.3.2) and (1.3.3) are equivalent, if the equation $Ry = 0$ has the unique solution $y = 0$; in that case, we will speak of the equivalent, left regularization of the operator A. An analogous reasoning holds for right regularization.

If an operator admits left as well as right regularization, then we will speak of double-sided regularization. If then the left and right regularizators are equal to each other, we will refer to a double-sided regularizator of an operator. In the sequel, we will just deal with double-sided regularizations, hence we will simply refer to "regularization" and "regularizators" instead of to "double-sided regularization" and "double-sided regularizators".

As a rule, we will denote by an asterisk adjoint spaces and adjoint operators. We recall that X^* is the space of bounded functionals which act on elements of the Banach space X, while the adjoint operator A^* acts from Y^* to X^* and is defined by

$$(Ax, y) = (x, A^*y). \qquad (1.3.4)$$

Solutions of the homogeneous equation

$$Ax = 0 \qquad (1.3.5)$$

are called zeroes of the operator A; if A is a linear, bounded operator, then the manifold of its zeroes forms in X a subspace, the dimensionality of which we will call the number of zeroes of the operator A.

2^0. We will now consider the problem of the dimensionality of (1.3.2). We will denote by y^* an arbitrary zero of the operator A^*. Let equation (1.3.2) be soluble and let x be any of its solutions. Then

$$(f, y^*) = (Ax, y^*) = (x, A^*y^*) = 0. \qquad (1.3.6)$$

If the orthogonality condition (1.3.6) is not only necessary, but also sufficient for the solubility of (1.3.2), then we will say that the operator A is normally soluble. The problem of normal solubility is resolved on the basis of

Theorem 1.9. If A is the bounded, linear operator $(A \in X \to Y)$, defined on the entire space X, then the closure $\overline{Im\,A}$ of the manifold of its values is a subspace of the space Y which is orthogonal to the subspace of the zeroes of the operator A^*.

We will denote this last subspace, as is usually done, by $N(A^*)$ and let y^* be any element of this subspace.

If $x \in X$, then $f = Ax \in Im\,A$, and, by (1.3.6) $f \perp y^*$; this means that $Im\,A$ is orthogonal to $N(A^*)$. Let such that $\overline{f} \in \overline{Im\,A}$. Then there exists a sequence $f_n \in Im\,A$, that $\overline{f} = \lim_{n\to\infty} f_n$. Elements of the space Y^* are continuous functionals, hence $(\overline{f}, y^*) = \lim_{n\to\infty} (f_n, y^*) = 0$; this proves that $\overline{Im\,A} \perp N(A^*)$.

Corollary 1.1. An operator A which is linear, bounded and defined on an entire space, is normally soluble if and only of the manifold $Im\,A$ of its values is closed.

Theorem 1.10. An operator which admits left regularization has no more than a finite number of zeroes.

Let R be the left regularizator of the operator A. Then the equation

$$R\,A\,x = x + T\,x = 0 \qquad (1.3.7)$$

has only a finite number of zeroes. Equation (1.3.5) each solution of which satisfies equation (1.3.7) has the same property.

Corollary 1.2. If an operator A admits regularization, and R is its regularizator, then

$$\begin{aligned} R\,A = I + T,\ \ A\,R = I + T_1 \\ A^*\,R^* = I^* + T^*,\ \ R^*\,A^* = I^* + T_1^*. \end{aligned} \qquad (1.3.8)$$

These relations prove that each of the operators A, A^*, R, R^* admits left regularization; by theorem 1.10, the subspace of the zeroes of each of these operators has finite dimension.

Theorem 1.11. If an operator A admits left regularization, then it is normally soluble.

Let $f \in Im\,A$, then there is an element $x_0 \in X$ which satisfies the equation $Ax_0 = f$. By theorem 1.10, the space $N(A)$ of the zero operator A has finite dimension. If x_i, $i = 1, 2, ..., n$, is the base of the subspace $N(A)$, then the general solution of the equation $Ax = f$ is

$$x = x_0 + \sum_{k=1}^{n} a_k\,x_k\,,\ \ a_k = const. \qquad (1.3.9)$$

Riesz in his paper [35] has proved that among the elements (1.3.9) is at least one element with minimal norm * ; we will denote it by $\tilde{x}$. We will now show that the exists a constant c such that

$$\forall f \in Im\, A, \ \|\tilde{x}\| \leq c \, \|f\| . \tag{1.3.10}$$

If R is a left regularizator for A, then $RA\tilde{x} = \tilde{x} + T\tilde{x} = Rf$ and the operator T is compact in X. We will show that the ratio $\|\tilde{x}\|/\|f\|$ is bounded. Assume that this is not true. Then there exists a subsequence $\{f^{(\nu)}\} \subset Y$ and corresponding subsequence $\{\tilde{x}^{(\nu)}\} \subset X$, such that $\|f^{(\nu)}\| \to 0$ and $\|x^{(\nu)}\| = 1$. From the bounded subsequence $\{\tilde{x}^{(\nu)}\}$ one may separate a subsequence $\{\tilde{x}^{(\nu_j)}\}$ such that the expression $T\,\tilde{x}^{(\nu_j)}$ tends to some limit which we will denote by x^0. It follows from the equation $\tilde{x}^{(\nu_j)} + T\,\tilde{x}^{(\nu_j)} = R\,f^{(\nu_j)}$ that $\tilde{x}^{(\nu_j)} \to x^0$. At the same time, one has $A\,\tilde{x}^{(\nu_j)} = f^{(\nu_j)} \to 0$. Since the operator A is bounded, one has $A\,\tilde{x}^{(\nu_j)} \to A\,x^{(0)}$ and, consequently, $A\,x^{(0)} = 0$, hence $A(x^{(\nu)} - x^{(0)}) = f^{(\nu)}$. Of all the solutions of the equation $Ax = f^{(\nu)}$, the solution $\tilde{x}^{(\nu)}$ has the smallest norm, hence $\|\tilde{x}^{(\nu)} - Ax^{(0)}\| \geq 1$. However, this contradicts the limit equality: $\tilde{x}^{(\nu_j)} \to x^{(0)}$; thus the boundedness of the ratio $\|\tilde{x}\|/\|f\|$ has been proved.

Next let $f \in \overline{Im\, A}$. There exists a sequence $\{f_k\} \subset Im\, A$ such that $f_k \to k$, $k \to \infty$. We will relate to every element f_k an element $\tilde{x}_k$ with least norm, satisfying the equation $A\,\tilde{x}_k = f_k$. By operating on this equation with the regularizator R, we find that $\tilde{x}_k + T\,\tilde{x}_k = R\,f_k$. We have shown that the ratio $\|\tilde{x}_k\|/\|f_k\|$ is bounded; besides, $\|f_k\| \leq const$. Hence it follows that the norms of the elements $\tilde{x}_k$ are bounded overall, and one may select a sequence $\{\tilde{x}_{kj}\}$ such that $T\,\tilde{x}_{kj}$ tends to some limit. The regularizator R is bounded; therefore $R\,f_{kj} \to R\,f$. It now follows from the equation $A\,\tilde{x}_{kj} = f_{kj}$ noted above that $\tilde{x}_{kj}$ tends to some limit $x^{(0)}$. Going to the limit in equation $A\,\tilde{x}_{kj} = f_{kj}$, we obtain $A\,x_0 = f$. However, then $f \in Im\, A$, and the manifold $Im\, A$ is closed.

Note 1.1. Fredholm's theorem 1.8 establishes the normal solubility of operators of the form (1.1.14). It follows from the just proved theorem 1.11, since the operator on the left hand side of this equation had the obvious regularizator $R = I$.

Note 1.2. Monograph [25] studies the results of this paragraph for the more general case of closed operators.

---

* This statement is readily proved, if X is a Hilbert space. One can then assume that the base $(x_1, \ldots, x_n)$, is ortho-normal; hence $\|x\|^2 = \|x_0\|^2 + 2Re \sum_{k=1}^{n} (x_0, x_k)\bar{a}_k + \sum_{k=1}^{n} |\,a_k\,|^2$, which has a minimum for $a_k = (x_0, x_k)$.

Note 1.3. Let $\mathcal{R}$ be the ring of bounded operators, acting from the Banach space X into the Banach space Y, and $\mathcal{T}$ the ring of symbols for the ring $\mathcal{R}$, where the symbol of any compact operator of $\mathcal{R}$ is a zero of the ring $\mathcal{T}$, and, conversely, an operator from $\mathcal{R}$, the symbol of which equals zero, is compact. If $A \in \mathcal{R}$ and the symbol $a = Smb\,A$ reverses into $\mathcal{T}$, then the operator A admits regularization: Any operator with the symbol a^{-1} is a regularizator.

§4. The index of an operator

1^0. Let A be a bounded operator, $A \in (X \to Y)$, and let either one of the operators A and A^* have a finite number of zeroes. Then the difference between the zeroes of these operators is called the index of the operator A; it is denoted by $Ind\,A$.

Theorem 1.12. If A admits regularization, then for any compact operator $T \in (X \to Y)$

$$Ind\,(A + T) = Ind\,A. \tag{1.4.1}$$

Let R be a regularizator of the operator A. The adjoint equations with compact operators $RAx = 0$ and $A^*R^*y = 0$ have one and the same number ρ of zeroes. Let n, n^*, m, m^* be the numbers of zeroes of the operators A, A^*, R, R^*, respectively. By Corollary 1.2, all these numbers are finite. Hence, by the way, it follows that the index of the operator A is finite. Obviously, $n \leq \rho$, $m^* \leq \rho$. We will next compute the number ρ.

Let x_j, $j = 1, \cdots, n$, and y_k, $k = 1, \cdots, m$ be the bases of the subspaces of the zeroes $N(A)$ and $N(R)$, respectively. If x satisfies the equation $RAx = 0$, then

$$A\,x = \sum_{k=1}^{m} c_k\,y_k\,, \quad c_k = const. \tag{1.4.2}$$

By theorem 1.11, it is necessary and sufficient for solubility of (1.4.2) that

$$\sum_{k=1}^{m} c_k(y_k, y_j^*) = 0\,, \quad j = 1, 2, ..., n^*\,, \tag{1.4.3}$$

where $\{y_j^*\}$ is the base of the subspace $N(A^*)$ of the zeroes of the operator A^*. We will denote by s the rank of the matrix of the elements (y_k, y_j^*); clearly $s \leq m$ and $s \leq n^*$. The general solution of the system (1.4.3) depends on $m - s$ arbitrary constants and equation (1.4.2) has $\rho = n + m - s$ linearly independent solutions.

The operator A^* is a regularizator of the operator R^*. Repeating the preceding reasoning, we find that $\rho^* = m^* + n^* - s^*$. Hence

$$Ind\,A = n - n^* = m^* - m = -Ind\,R.\qquad(1.4.4)$$

We apply reasoning, analogous to the preceding one, to the operator $A + T$ which satisfies the conditions of the theorem: It is bounded and has the same regularizator as R. By (1.4.4)

$$Ind\,(A + T) = -Ind\,R,$$

and, consequently, $Ind\,(A + T) = Ind\,A$.

Theorem 1.13. Let A and B be normally soluble. bounded operators with finite indices, $A \in (X \to Y)$, $B \in (Y \to X)$.

Then

$$Ind\,(BA) = Ind\,(AB) = Ind\,A + Ind\,B.\qquad(1.4.5)$$

We will denote the bases of the subspaces of the zeroes of the operators A, A^*, B, B^*, respectively, by

$$x_1, x_2, ..., x_n;\ x_1^*, x_2^*, ..., x_{n^*}^*;$$

$$y_1, y_2, ..., y_m;\ y_1^*, y_2^*, ..., y_{m^*}^*.$$

Let s be the rank of the matrix of numbers (x_j^*, y_k). Repeating the reasoning of theorem 1.12, we find that the operator BA has $n + m - s$ zeroes, while the adjoint operator A^*B^* has $m^* + n^* - s$ zeroes. Hence follows (1.4.5).

Theorem 1.14. We will retain the notation and conditions of theorem 1.12 and let $C \in (X \to Y)$ be such an operator that $\|R\| \cdot \|C\| < 1$. Then

$$Ind\,(A + C) = Ind\,A.$$

We have $RA = I + T$, hence $R(A + C) = I + RC + T$. The, $\|RC\| \le \|R\| \cdot \|C\| < 1$; by a well known theorem of Banach the inverse operator $(I + RC)^{-1}$ exists, is defined on the entire space of X and is bounded. Now (T' being a compact operator)

$$R(A + C) = (I + RC)[I + (I + RC)^{-1}T] = (I + RC)(I + T').\qquad(1.4.6)$$

Hence it is seen that $A + C$ has the left regularizator $(I + RC)^{-1}R$ and, consequently, the number of zeroes of the operator $A + C$ is finite (theorem 1.10). Moreover, $AR = I + T_1$, hence $(A + C)R = I + CR + T_1 = (I + CR)(I +$

T_2); proceeding to adjoint operators, we find $R^*(A^* + C^*)(I + R^*C^*)^{-1} = I + T_2^*$. Thus, the operator $(A^* + C^*)(I + R^*C^*)^{-1}$ has the left regularizator R^*, and by the same theorem 1.10, the number of zeroes of this operator is finite. It is readily verified that

$$dim\, N[(A^* + C^*)(I + R^*C^*)^{-1}] = dim\, N(A^* + C^*).$$

Once the number of zeroes of the operators $A + C$ and $A^* + C^*$ are finite, then $Ind\,(A + C)$ is finite. By theorem 1.13, it follows from (1.4.6) that

$$Ind\,R(A + C) = Ind\,R + Ind\,(A + C) = Ind\,(I + RC) + Ind\,(I + T').$$

The last part of this equality equals zero: $Ind(I + T') = 0$, by theorems 1.5 and 1.8, while $Ind(I + RC) = dim\,N(1 + RC) - dim\,N(1 + C^*R^*) = 0$, because the operator $I + RC$ is invertible. Now $Ind(A + C) = -Ind\,R$ and, by (1.4.4), $Ind(A + C) = Ind\,A$.

Theorem 1.15. For a bounded operator to admit equivalent, left regularization, it is necessary and sufficient that it be normally soluble and have a finite non-negative index.

Let the operator A admits left equivalent regularization and R be a left equivalent regularizator. By theorem 1.10, the operator A is normally soluble, while by theorem 1.2 the number of its zeroes is finite. We will denote by ω_k, $k = 1, 2, ..., n$ the base of the subspace of the zeroes of the equations with with compact operators $(RA)^*\omega = A^*R^*\omega = 0$. The equation $RAx = Rf$ and with it the equation $Ax = f$, equivalent to it, are soluble if and only of

$$(Rf, \omega_k) = (f, R^*\omega_k) = 0\,, \quad k = 1, 2, ..., n\,. \tag{1.4.7}$$

The elements $\psi_k = R^*\omega_k$ satisfy the equation $A^* = 0$. This equation does not have solutions which are linearly independent of the ψ_k. In fact, let $\tilde{\psi}$ be such a solution. For the equation $Ax = f$ to be soluble it is necessary that $(f, \tilde{\psi}) = 0$, while this contradicts the sufficiency of condition (1.4.7). Thus, the operator A^* has no more than n zeroes. At the same tome, the operator A exactly n zeroes in as much as has the operator RA. Hence it follows that $Ind\,A \geq 0$.

Next, we will prove the sufficiency of the theorem. Let $\{\varphi_j\}, j = 1, 2, ..., n$ and $\{\psi_j\}$, $j = 1, 2, ..., n^*$ be the bases of the subspaces of the zeroes of the operators A and A^*, respectively, where $n^* \leq n$. As usually, we assume that A acts from the Banach space X into the Banach space Y; we will denote by $N(A)$ the subspace of the zeroes of the operator A. We construct linear functionals $\alpha_j \in X^*$, $j = 1, 2, ..., n$, orthogonal to the elements φ_j, so that $(\alpha_j, \varphi_k) = \delta_{ik}$. The method of construction of such functionals is given, for example, in [10]. For any element $x \in X$, we set $x'' = \sum_{k=1}^{n} (\alpha_k, x)\varphi_k$ and denote $x' = x - x''$. The join of the elements x' forms the subspace $X' \subset X$, orthogonal to the functionals α_k. Obviously, X is the direct sum of the subspaces X' and X''.

Let n^* be a number such that $n^* > 0$. We can construct the linear functionals $\omega_k \in Y$, $k = 1, 2, \ldots, n^*$, orthogonal to the elements Ψ_j, $j = 1, 2, \ldots, n^*$. We will denote subspace of Y with the base $\{\omega_k\}$ by Y'' and Y' will be subspace of Y orthogonal to all the elements Ψ_j. It is easy to see that Y is the direct sum of the subspaces X' and X''.

Let A' be the contraction of the operator A on the subspace X', so that $D(A') = X'$ and $A'x = Ax$, $x \in X'$. The operator A is normally soluble, hence $Im\, A = Y'$ and, consequently, $Im\, A' = Im\, A = Y'$. Finally, A' has only the trivial zero $x = 0$: if $A'x = Ax = 0$, then $x \in X'' = N(A)$; however, X' and X'' intersect only in the zero, hence $x = 0$. It follows from this that there exists the inverse operator $(A')^{-1}$, defined on the entire space Y' and closed operator, because it is inverse operator to the bounded operator A'. However, then the operator $(A')^{-1}$ is bounded.

We will extend the operator $(A')^{-1}$ to some operator R, defined on the entire space Y; for this purpose, let

$$Rw_j = \varphi_j\,, \; j = 1, 2, ..., n^*\,.$$

We will show that R is a left equivalent regularizator for A. To start with, R does not have zeroes beside the trivial one $x = 0$. In fact, let $Ry = 0$. Set $Y = Y' + Y''$, where $y' \in Y'$, $y'' \in Y''$. Then $0 = Ry = Ry' + Ry'' = (A')^{-1}y' + Ry''$, where $(A')^{-1}y' \in X'$, $Ry'' \in N(A)$. However, X' and $X'' = N(A)$ intersect only in the zero element, hence $(A')^{-1}y' = 0$ and $Ry'' = 0$. Then $y' = y'' = 0$, and , finally, $y = 0$.

It is now clear that the equations $Ax = f$ and $RAx = Rf$ are equivalent. If $RAx = R$, then $R(Ax - f) = 0$ and $Ax = f$.

It remains to prove that R is a left regularizator for A, i.e., that the operator $RA - I$ is compact in X. Let $x = x' + x''$, where $x' \in X'$ and $x'' \in X'' = N(A)$. Then $Ax = Ax' = A'x' \in Y'$. However, on the subspace Y', one has $R = (A')^{-1}$, hence

$$RAx = x' = x - x'' = x - \sum_{j=1}^{n^*} (\alpha_j, x)\, \varphi\,;$$

the subtrahend is finite dimensional, and therefore a compact operator.

It has been assumed above that $n^* > 0$. For $n^* = 0$, the proof is simplified; then $Y' = Y$ and $R = (A')^{-1}$.

In conclusion, we note that the requirement of boundedness of the operator A in theorem 1.15. is unnecessary.

§5. Noether operators and Noether's theorems

1^0. A linear operator A which acts out of a Banach space X and is bounded to it is said to be Noetherian if it is normally soluble and has a finite index.

Theorems 1.10, 1.11 and the method of proof of theorem 1.12 yield

Theorem 1.16. If a bounded operator admits left regularization, then it is Noetherian.

2^0. Let the linear operator $A \in (X \to Y)$ have the following properties: 1) It is normally soluble; 2) it has a finite index. Under these conditions, we will say that Noether's theorems apply to it. Clearly, Noether's theorems are true for certain operators and, in particular, for operators which admit left regularization.

Fredholm's theorems 1.5 - 1.8 are particular cases of Noether's theorems.

F.Noether proved "Noether's theorems" in 1921 for a very important class of one-dimensional. singular, integral equations (cf. chapter 2).

CHAPTER 2
ONE-DIMENSIONAL SINGULAR INTEGRAL EQUATIONS

The theory of one-dimensional, singular integral equations, as an independent, scientific discipline, began in the Twentieth Century, almost contemporary with Fredholm's theory, in the work of D. Hilbert and H. Poincaré. The efforts of F. Noether (1921) and T. Carleman (1922) were of fundamental importance for this theory. Its further developments occurred in the work of F.D. Gakhov and his students, the Tbilissi Mathematical School headed by N.I. Muskhelishvili and I.N. Vekua, and the Kishinev Mathematical School. Of well known significance were also the efforts of the authors of this work on the theory and applications of these equations. In more recent years, important studies were due to S. Prössdorf and his students; in particular, their work on the approximate solution of these equations should be mentioned.

The above–mentioned results are contained in the works: [30, 54, 4, 43, 34, 6, 20].

§1. The Cauchy operator on Hoelder functions

1^0. Let L be a rectifiable curve in the complex plane and t some point on L. We will specify a number $\varepsilon > 0$ and denote by L_ε that part of L which lies outside the circle with radius ε with centre at t. Moreover, let τ be a variable point on L and let the function $\varphi(\tau)$ be summable on L_ε for any $\varepsilon > 0$. If there exists the limit

$$\lim_{\varepsilon \to 0} \int_{L_\varepsilon} \varphi(\tau)\, d\tau \,,$$

then it is called the principal value of the integral (a term introduced by O. Cauchy), and represented by the symbol

$$\int_L \varphi(\tau)\, d\tau \tag{2.1.1}$$

If the integral (2.1.1) exists even if it be improper, then it coincides with its principal value. In the contemporary literature, as a rule, the principal value of an integral is called a singular integral (some authors use the term special).

2^0. We will now present an important example of a singular integral. Let Γ be a closed Lyapunov contour in the complex plane, bounding some finite region Ω^+. This region may also be multiply connected so that Γ may comprise a finite number of Lyapunov curves $\Gamma_1, \Gamma_2, \dots \Gamma_n$, regarding which we will assume that they do not have points of intersection nor have pairwise common points.

We will recall here the definition of a Lyapunov curve. A closed, rectifiable curve L without self intersection is said to be Lyapunovian is it has the following properties.

Let s be the length of arc of L from some point of reference to the point $t \in L$, and l the length of the entire curve. This curve may be given the parametric form

$$t = f(s), \tag{2.1.2}$$

where $f(s)$ a periodic, complex function with period l; it is assumed that its first derivative satisfies a Hoelder condition

$$| f'(\sigma) - f'(s) | \leq C \, | \sigma - s |^\beta \; ; C, \beta = const; \; 0 < \beta \leq 1; \tag{2.1.3}$$

and its modulus is positively bounded above and below:

$$m \leq | f'(s) | \leq M; \; m, M = const, \; 0 < m \leq M < \infty. \tag{2.1.4}$$

Let $u(\tau)$ be a function, given on Γ and satisfying on this contour a Hoelder condition

$$\forall t, \tau \in \Gamma, \; | u(\tau) - u(t) | \leq c \, | \tau - t |^\alpha; \; c, \alpha = const; \; 0 < \alpha \leq 1. \tag{2.1.5}$$

We will consider the, in general, unbounded integral

$$\int_\Gamma \frac{u(\tau)}{\tau - t} \, d\tau, \tag{2.1.6}$$

and prove that it exists as a singular integral. For the sake of definiteness, we will assume that the contour Γ is traveled in the positive direction so that the region Ω^+ is located on the left hand side.

We will construct a circle with radius ε with centre at the point t and by Γ_ε that part of Γ which lies outside this circle. The integrand in (2.1.6) is continuous on Γ_ε and the integral

$$\int_{\Gamma_\varepsilon} \frac{u(\tau)}{\tau - t} \, d\tau, \tag{2.1.7}$$

exists. We will show that, as $\varepsilon \to 0$, the integral tends to some limit.

3^0. Integral (2.1.7) can be rewritten in the form

$$\int_{\Gamma_\varepsilon} \frac{u(\tau)}{\tau - t}\, d\tau = \int_{\Gamma_\varepsilon} \frac{u(\tau) - u(t)}{\tau - t}\, d\tau + u(t) \int_{\Gamma_\varepsilon} \frac{d\tau}{\tau - t}. \tag{2.1.8}$$

The integrand in the first integral on the right hand side has the estimate $0(|\tau - t|^{\alpha-1})$; the integral converges absolutely and uniformly to the integral

$$\int_{\Gamma} \frac{u(\tau) - u(t)}{\tau - t}\, d\tau,$$

which represents a continuous function of t on the contour Γ. The second integral in (2.1.8) may be evaluated. Let the contour Γ consist of n simple closed curves. For the sake of definiteness, we will consider the case $n = 2$ (Fig. 1). For example, let $t \in \Gamma_1$, then

$$\int_{\Gamma_2} \frac{d\tau}{\tau - t} = 0, \quad \int_{\Gamma_\varepsilon} \frac{d\tau}{\tau - t} = \int_{\Gamma_{1\varepsilon}} \frac{d\tau}{\tau - t} = i\vartheta\,;$$

where $\Gamma_{1\varepsilon}$ is that part of Γ_1 which lies outside the circle $|\tau - t| < \varepsilon$; the value of the angle ϑ is shown in Fig. 1. As $\varepsilon \to 0$, this angle tends to π, and it follows from (2.1.8) that the singular integral (2.1.6) exists and is equal to

$$\int_{\Gamma} \frac{u(\tau)}{\tau - t}\, d\tau = \int_{\Gamma} \frac{u(\tau) - u(t)}{\tau - t}\, d\tau + \pi i\, u(t). \tag{2.1.9}$$

Next, let $t \in \Gamma_2$ (Fig. 2). We construct again the circle $|\tau - t| < \varepsilon$ and denote by $\Gamma_{2\varepsilon}$ that part of Γ_2 that lies outside this circle. In this case

$$\int_{\Gamma_1} \frac{d\tau}{\tau - t} = 2\pi i, \quad \int_{\Gamma_{2\varepsilon}} \frac{d\tau}{\tau - t} = i\,\vartheta_1\,;$$

as $\varepsilon \to 0$, this time $\vartheta_1 \to -\pi$, and we arrive again at (2.1.9). The same result is obtained for $n > 2$.

In the sequel, we will employ the notation

$$t \in \Gamma;\quad \frac{1}{\pi i} \int_{\Gamma} \frac{u(\tau)}{\tau - t}\, d\tau = (S\, u)(t). \tag{2.1.10}$$

The operator S is called a Cauchy operator, and the expression $d\tau/(\tau - t)$ a Cauchy kernel.

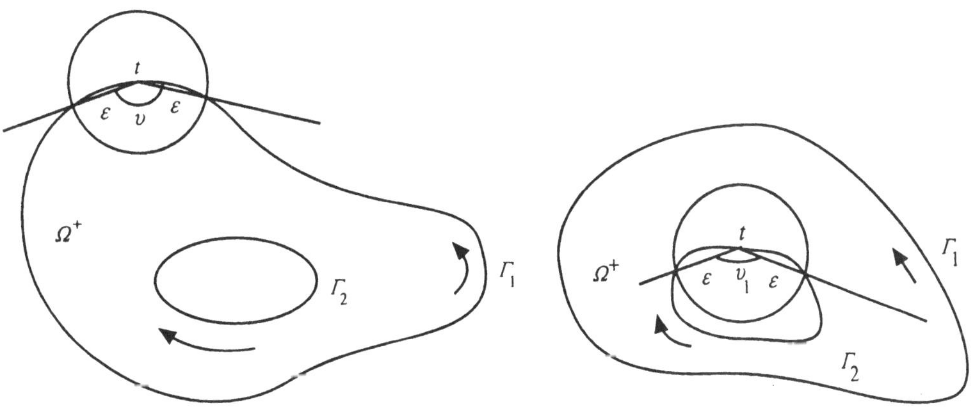

Fig. 1   Fig. 2

4^U. The integral on the right hand side of (2.1.9) converges uniformly for $t \in \Gamma$, hence the singular integral (2.1.9) not only exists, but represents itself a function of t which is continuous on Γ. True is also the stronger assertion of

Theorem 2.1. If Γ is a Lyapunov curve and the density $u(\tau)$ satisfies a Hoelder condition with index $\alpha < 1$, then the singular integral (2.1.6) satisfies a Hoelder condition with the same index α.

A proof of this important theorem is presented, together with the necessary bibliographic references in the monograph [30].

In the theory of one-dimensional, singular integral equation the theorem of Sokhotskii-Plemelj on the limiting values of a Cauchy type integral plays an important role.

Theorem 2.2. Let Γ be a Lyapunov contour, bounding a finite region Ω^+ and let Ω^- be the complement of the complex plane to $\Omega^+ \cup \Gamma$. Furthermore, let $u(\tau)$ be a function which is given on Γ and satisfies on this contour a Hoelder condition, and

$$U(z) = \frac{1}{2\pi i} \int_\Gamma \frac{u(\tau)}{\tau - z}\, d\tau$$

be a Cauchy type integral with the density $u(\tau)$. Then there exist the limiting values

$$U^+(t) = \lim_{z\in\Omega^+,\, z\to t} U(z), \quad U^-(t) = \lim_{z\in\Omega^-,\, z\to t} U(z),$$

defined by

$$U^\pm(t) = \pm\frac{1}{2}\,u(t) + \frac{1}{2}\,(S\,u)(t)\,. \qquad (2.1.11)$$

A proof of theorem 2.2 can be found in monographs [15, 30].

We will note yet two important cases.

1) If the density $u(\tau)$ is continued analytically into Ω^+ and the continued function $u(z)$ is continuous in $\bar{\Omega}^+ = \Omega^+ \cup \Gamma$, then $U(z) = u(z)$, $z \in \Omega^+$ and one finds from (2.1.11), under the assumption that $u(\tau)$ satisfies a Hoelder condition, that

$$(S\,u)(t) = u(t)\,. \qquad (2.1.12)$$

2) Let the function $u(\tau)$, satisfying, as before, on Γ a Hoelder condition, be analytically continued to each of the regions, comprising the open manifold Ω^-, and let these analytic continuations be continuous in the closure $\bar{\Omega}^- = \Omega^- \cup \Gamma$. We will assume in addition that $u(\infty) = 0$. Then $U(z) = -u(z)$, $z \in \Omega^-$, and $U(z) = 0$, $z \in \Omega^+$. It follows from (2.1.11) that in this case

$$(S\,u)(t) = -u(t)\,. \qquad (2.1.13)$$

5^0. **Theorem 2.3.** If Γ is a closed Lyapunov curve and the density $u(\tau)$ satisfies on Γ a Hoelder condition, then

$$(S^2\,u)(t) = u(t)\,. \qquad (2.1.14)$$

As before, we will denote by $U(z)$ the Cauchy type integral with density $u(\tau)$. By (2.1.11), $(Su)(t) = 2U^+(t) - u(t)$; by theorem 2.1, $(Su)(t)$ satisfies on Γ a Hoelder condition, and the expression $(S^2u)(t) = (S(Su))(t)$ makes sense and represents itself a function which likewise satisfies on Γ a Hoelder condition. Thus, by (2.1.11)

$$(S^2u)(t) = (S(2U^+u))(t) = 2U^+(t) - (S\,u)(t) = u(t)\,,$$

which bears the name of Poincaré - Bertrand formula; it may be given the form (as usually, I being the identity operator)

$$S^2 = I\,. \qquad (2.1.15)$$

§2. Cauchy operator in L_2. The general singular operator

1^0. We introduce now the manifold $L_2(\Gamma)$ of functions, measurable and quadratically summable on Γ, i.e., such that, if $U \in L_2(\Gamma)$, then $\int_\Gamma |u(\tau)|^2 \, d\sigma < \infty$, $d\sigma = |d\tau|$.

If we introduce in $L_2(\Gamma)$ the scalar product and norm by means of the formulas

$$(u, \nu) = \int_\Gamma u(\tau)\overline{\nu(\tau)} \, d\tau, \; \|u\|^2 = \int_\Gamma |u(\tau)|^2 \, d\sigma, \qquad (2.2.1)$$

then $L_2(\Gamma)$ becomes a Hilbert space. We recall the well known fact that a manifold of functions satisfying on Γ a Hoelder condition with any fixed index is dense in $L_2(\Gamma)$.

2^0. The basic result of this section is stated by

Theorem 2.4. The Cauchy operator (2.1.1) is bounded in $L_2(\Gamma)$.

Proof. It involves several steps:

a) Let γ be the unit circle with centre at the origin of coordinates. If $\tau \in \gamma$, then $\tau = e^{i\sigma}$, $-\pi \le \sigma \le \pi$. Moreover, if the real function $u(\tau)$ satisfies on Γ a Hoelder condition, then, as follows easily from the Sokhotskii-Plemelj formulas, the Schwartz integral

$$\Phi(z) = \frac{1}{2\pi i} \int_\gamma u(\tau) \frac{\tau + z}{\tau - z} \, d\tau$$

represents inside γ a holomorphic function the real part of which tends to $u(t)$ as $z \to t \in \gamma$.

We now let $\Phi(z) = u(z) + i\,v(z)$; $u(z)$ and $v(z)$ are two harmonical functions of x and y $(z = x + iy)$ in the circle $|z| < 1$. Let $z \to t = e^{is}$, $-\pi \le s \le \pi$, when the formulas of Sokhotskii-Plemelj readily reduce to the new formula

$$v(t) = -\frac{1}{2\pi} \int_{-\pi}^{\pi} u(\tau) \, ctg \, \frac{\sigma - s}{2} \, d\sigma; \; d\sigma = |d\tau|; \qquad (2.2.2)$$

the integral on the right hand side exists as a singular integral.

The kernel $ctg\,(\sigma - s)/2$ is referred to as the Hilbert kernel.

It is readily verified that

$$\int_{-\pi}^{s-\varepsilon} ctg \, \frac{\sigma - s}{2} \, d\sigma + \int_{s+\varepsilon}^{\pi} ctg \, \frac{\sigma - s}{2} \, d\sigma = 0,$$

hence (2.2.2) may be rewritten in the form

$$
v(t) = -\frac{1}{2\pi} \lim_{\varepsilon \to 0} \left\{ \int\limits_{-\pi}^{s-\varepsilon} [u(\tau) - u(t)] \, ctg \, \frac{\sigma - s}{2} \, d\sigma \right.
$$

$$
\left. + \int\limits_{s+\varepsilon}^{\pi} [u(\tau) - u(t)] \, ctg \, \frac{\sigma - s}{2} \, d\sigma \right\}. \tag{2.2.3}
$$

It is now seen that, if the function $u(t)$ satisfies Hoelder condition, then the right hand side of (2.2.3) tends to the corresponding singular integral uniformly in s.

We note yet that (2.2.3) may be rewritten

$$
\int\limits_{-\pi}^{\pi} u(\tau) \, ctg \, \frac{\sigma - s}{2} \, d\sigma = \lim_{\varepsilon \to 0} \int\limits_{-\pi}^{\pi} K_\varepsilon(\sigma - s) \, u(\tau) \, d\tau, \tag{2.2.4}
$$

where

$$
K_\varepsilon(\sigma - s) = \begin{cases} ctg\,(\sigma - s)/2, & |\,\sigma - s\,| \geq \varepsilon \\ 0, & |\,\sigma - s\,| < \varepsilon. \end{cases}
$$

b) Correctness of the formula of change if order of integration's: If $f(t)$ and $g(t)$ satisfy a Hoelder condition, then

$$
\int\limits_{-\pi}^{\pi} f(t) \left[\int\limits_{-\pi}^{\pi} g(\tau) \, ctg \, \frac{\sigma - s}{2} \, d\sigma \right] ds
$$

$$
= \int\limits_{-\pi}^{\pi} g(t) \left[\int\limits_{-\pi}^{\pi} f(\tau) \, ctg \, \frac{\sigma - s}{2} \, ds \right] d\sigma. \tag{2.2.5}
$$

In fact, the integral on the left hand side equals

$$
\int\limits_{-\pi}^{\pi} f(t) \left[\lim_{\varepsilon \to 0} \int\limits_{-\pi}^{\pi} g(\tau) \, K_\varepsilon(\sigma - s) \, d\sigma \right] ds.
$$

Performing the limit uniformly, the last expression becomes

$$
\lim_{\varepsilon \to 0} \int\limits_{-\pi}^{\pi} \int\limits_{-\pi}^{\pi} f(t) \, g(\tau) \, K_\varepsilon(\sigma - s) \, ds \, d\sigma
$$

$$
= \lim_{\varepsilon \to 0} \int\limits_{-\pi}^{\pi} g(\tau) \left[\int\limits_{-\pi}^{\pi} f(t) \, K_\varepsilon(\sigma - s) \, ds \right] d\sigma. \tag{2.2.6}
$$

The inner integral on the right hand side tends uniformly to its limit, hence the right hand side and the left hand side of (2.2.5) are equal between them.

b) Let $n \geq 0$ be an integer and $u(t) = \cos ns$. The function which is harmonic in the circle $\mid z \mid < 1$ is $\rho^n \cos ns$, $\rho = \mid z \mid$; its harmonic adjoint function which vanishes at the centre of that circle is $\rho^n \sin ns$ by (2.2.2)

$$\frac{1}{2\pi} \int_{-pi}^{\pi} \cos n\sigma \cdot ctg \frac{\sigma - s}{2} \, d\sigma = -\sin ns \, .$$

In the same way, we find

$$\frac{1}{2\pi} \int_{-pi}^{\pi} \sin n\sigma \cdot ctg \frac{\sigma - s}{2} \, d\sigma = \cos ns \, .$$

The last two formulas yield

$$\frac{1}{2\pi} \int_{-pi}^{\pi} e^{in\sigma} \, ctg \frac{\sigma - s}{2} \, d\sigma = \left\{ \begin{array}{l} 0 \, , \, n = 0 \, ; \\ ie^{ins} \, , \, n > 0 \, ; \\ -ie^{ins} \, , \, n < 0 \, . \end{array} \right. \qquad (2.2.7)$$

which is true for all n.

c) We return now to the integral (2.2.2) and expand the functions $u(t)$ and $v(t)$ in Fourier series:

$$u(t) = \sum_{n=-\infty}^{+\infty} a_n \, e^{ins} \, , \quad v(t) = \sum_{n=-\infty}^{+\infty} b_n \, e^{ins} \, ,$$

$$a_n = \frac{1}{2\pi} \int_{-\pi}^{\pi} u(t) \, e^{-ins} \, ds \, , \quad b_n = \frac{1}{2\pi} \int_{-\pi}^{\pi} v(t) \, e^{-ins} \, ds \, .$$

Moreover,

$$b_n = -\frac{1}{4\pi^2} \int_{-\pi}^{\pi} e^{-ins} \left[\int_{-\pi}^{\pi} u(t) \, ctg \frac{\sigma - s}{2} \, d\sigma \right] ds \, .$$

By the result of b), the order of integration may be reversed:

$$b_n = -\frac{1}{4\pi^2} \int_{-\pi}^{\pi} u(\sigma) \left[\int_{-\pi}^{\pi} e^{-ins} \, ctg \frac{\sigma - s}{2} \, ds \right] d\sigma \, ;$$

and, by (2.2.7)

$$b_n = \begin{cases} 0\,, & n = 0\,; \\ ia_n\,, & n > 0\,; \\ -ia_n\,, & n < 0\,. \end{cases}$$

Thus

$$\|u\|_{L_2}^2 = 2\pi \sum_{n=-\infty}^{+\infty} |\,a_n\,|^2\,; \quad \|v\|_{L_2}^2 = 2\pi \left[\sum_{n=1}^{\infty} |\,a_n\,|^2 + \sum_{n=-1}^{-\infty} |\,a_n\,|^2 \right]. \quad (2.2.8)$$

We will denote an integral operator with a Hilbert kernel by the letter P, so that $v = -Pu$. It follows now from (2.2.8) that $\|v\|_{L_2} \le \|u\|_{L_2}$; this means that the operator P is bounded on $L_2(\gamma)$ and $\|P\|_{L_2} \le 1$. Thus, if $a_0 = 0$, then $\|v\|_{L_2} = \|u\|_{L_2}$, and, finally, $\|P\|_{L_2} = 1$. In the sequel. we will omit the subscript L_2 in the notation for the norm so that the symbol $\|\cdot\|$ will always denote a norm in L_2.

It is readily proved that on the circle γ the Cauchy and Hilbert kernels are linked by the relation

$$\frac{1}{\pi i}\,\frac{d\tau}{\tau - t} = \left[-\frac{1}{2\pi}\,ctg\,\frac{\sigma - s}{2} + \frac{1}{2\pi} \right] d\sigma\,.$$

We will now expand $u(\tau)$ in a Fourier series and use the preceding results to obtain

$$(S_\gamma u)(t) = i \sum_{n=1}^{+\infty} a_n\,e^{ins} + a_0 - i \sum_{n=-1}^{-\infty} a_n\,e^{ins}\,;$$

where S_γ is a Cauchy operator on the contour γ, hence it follows that

$$\|S_\gamma u\| = \|u\|\,, \quad \|S_\gamma\| = 1\,.$$

Thus, theorem 2.4 has been proved for the case when the contour of integration is a unit circle.

d) We will now proceed to the general case, when Γ is a set of Lyapunov curves which in pairs do not have common points. If t and τ belong to different curves Γ_j, then the Cauchy kernel is bounded and the corresponding operator is bounded in $L_2(\Gamma)$. Therefore it is sufficient to consider the case, when Γ consists of a single closed Lyapunov curve Γ_j. We will select a unit length to make the length of the curve Γ_j equal to 2π; as above, we denote by s and σ, respectively, the lengths of arc of the curve Γ_j from some starting point to the points t and τ. We can give the curve Γ_j the parametric representation $t = f(s)$, where f satisfies the conditions (2.1.3) and (2.1.4); obviously, one has $\tau = f(\sigma)$.

e) We will prove now the indentity

$$\frac{1}{\pi i(\tau - t)} \frac{d\tau}{d\sigma} = -\frac{1}{2\pi} \, ctg \, \frac{\sigma - s}{2} + P(s, \sigma), \qquad (2.2.9)$$

where $P(s, \sigma)$ is a kernel with a weak singularity (it is bounded, if the index β in the inequality (2.1.3) equals unity). Obviously, one has $d\tau/d\sigma = f'(\sigma)$.

We consider now the difference

$$\frac{f'(\sigma)}{\tau - t} - \frac{i}{2} \, ctg \, \frac{\sigma - s}{2} = \frac{2(e^{i\sigma} - e^{is}) f'(\sigma) - i(\tau - t)(e^{i\sigma} - e^{is})}{2(\tau - t)(e^{i\sigma} - e^{is})}. \qquad (2.2.10)$$

We will fix a sufficiently small number $\varepsilon > 0$. If $|\tau - t| \geq \varepsilon$, then the fraction (2.2.10) is bounded. In order to obtain an estimate for (2.2.10) for $|\tau - t| < \varepsilon$, we see that (2.1.4) yields $|\tau - t| \leq M\sqrt{2} \, |\sigma - s|$ so that for sufficiently small ε

$$|e^{i\sigma} - e^{is}| = 2\left| \sin \frac{\sigma - s}{2} \right| \geq 2\left| \sin \frac{|\tau - t|}{M\sqrt{2}} \right| \geq \frac{4}{\pi M\sqrt{2}} \, |\tau - t|; \quad (2.2.11)$$

hence follows an estimate for the denominator in (2.2.10), true for any position of the points $t, \tau \in \Gamma_j$:

$$|2(\tau - t)(e^{i\sigma} - e^{is})| \geq C_1 \, |\tau - t|^2, \; C_1 = const. \qquad (2.2.12)$$

Next, we estimate the numerator in (2.2.10):

$$|2(e^{i\sigma} - e^{is})f'(\sigma) - i(\tau - t)(e^{i\sigma} - e^{is})|$$
$$\geq 2\,|(e^{i\sigma} - e^{is})f'(\sigma) - i(\tau - t)e^{is}| + |(\tau - t)(e^{i\sigma} - e^{is})|.$$

The second term on the right hand side yields a bounded value for (2.2.10). We must still study the first term. It does not exceed the quantity

$$2\,|(e^{i\sigma} - e^{is})f'(\sigma) - f'(s)| + |2(e^{i\sigma} - e^{is})f'(s) - i(\tau - t)e^{is}|. \qquad (2.2.13)$$

We see from Inequalities (22.1.3) and (2.2.11) that the first term in (2.2.13) yields in (2.2.11) a term with a weak singularity which for $\beta = 1$ is bounded. The subtrahend in it may be given the form

$$i(\tau - t)e^{is} = i\,e^{is} \int_s^\sigma f'(\xi)\,d(\xi) = i\,e^{is}(\sigma - s)f'(s) + O(|\,(\sigma - s)\,|^{1+\alpha});$$

where we have used inequality (2.1.3). The minuend in the same term can be transformed as follows:

$$(e^{i\sigma} - e^{is})f'(s) = i\,f'(s)\,e^{is} \int_0^{\sigma-s} e^{i\xi}\,d\xi = if'(s)\,e^{is}(\sigma - s) + O((\sigma - s)^2),$$

hence

$$\mid 2(e^{i\sigma} - e^{is})f'(s) - i(\tau - t)e^{is}\mid = O(\mid \sigma - s \mid^{1+\beta}),$$

which yields in (2.2.10) the term with the weak singularity, if $\beta < 1$, and the bounded term if $\beta = 1$. Collecting our results, we see that the function $P(s,\sigma)$, entering into (2.2.9), is bounded for $\beta = 1$ and that it has a weak singularity for $\beta < 1$.

f) It follows from the identity (2.2.9) that

$$(S_j u)(t) := \frac{1}{\pi i}\int_{\Gamma_j} \frac{u(\tau)}{\tau - t}\,d\tau = \frac{1}{\pi i}\int_{\Gamma_j} \frac{u(\tau)\,f'(\sigma)}{\tau - t}\,d\sigma = (pu)(s) + (Pu)(s),$$

where P is an operator with a Hilbert kernel and p is an operator with a weak singularity. Both of them are bounded in $L_2(-\pi,\pi)$, where $\|P\|_{L_2(-\pi,\pi)} = 1$.

1^0. We will set now $Q = \|p\|_{L_2(-\pi,\pi)}$, when

$$\int_{-\pi}^{\pi} \mid (S_j u)(t)\mid^2 ds \le (1 + Q)^2 \int_{-\pi}^{\pi} \pi \mid u(t)\mid^2 ds\,;$$

the last term here may be given the form

$$\int_{\Gamma_j} \frac{\mid (S_j u)(t)\mid^2}{\mid f'(t)\mid}\mid dt\mid \le (1 + Q)^2 \int_{\Gamma_j} \frac{\mid u(\tau)\mid^2}{\mid f'(t)\mid}\mid dt\mid.$$

By (2.1.4), we find

$$\frac{1}{M}\|S_j u\|_{L_2(\Gamma_j)}^2 \le \frac{(1 + Q)^2}{m}\|u\|_{L_2(\Gamma_j)},$$

or $\|S_j\|_{L_2(\Gamma_j)} \le (1 + Q)\sqrt{M/m}$. This completes the proof of theorem 2.4.

2^0. **Corollary 2.1.** If $u, v \in L_2(\Gamma)$, then one has the formula for inversion of the order of integration

$$\int_\Gamma u(t)\left[\int_\Gamma \frac{v(\tau)}{\tau - t}\, d\tau\right] dt = \int_\Gamma v(\tau)\left[\int_\Gamma \frac{u(t)}{\tau - t}\, dt\right] d\tau. \qquad (2.2.14)$$

We will still retain the definitions of the number ε and the contour Γ_ε introduced above and introduce the new notation

$$v_\varepsilon(\tau) = \begin{cases} v(\tau), \tau \in \Gamma_\varepsilon \\ 0, \tau \in \Gamma/\Gamma_\varepsilon \end{cases}; \; u_\varepsilon(\tau) = \begin{cases} u(\tau), \tau \in \Gamma_\varepsilon \\ 0, \tau \in \Gamma/\Gamma_\varepsilon \end{cases};$$

$$V(t) = \int_\Gamma \frac{v(\tau)}{\tau - t}\, d\tau, \; V_\varepsilon(t) = \int_\Gamma \frac{v_\varepsilon(\tau)}{\tau - t}\, d\tau = \pi\, i(Sv_\varepsilon)(t).$$

Obviously, $v_\varepsilon(\tau) \to v(\tau)$ in the norm of $L_2(\Gamma)$. By theorem 2.4, in the same metric, $V_\varepsilon(\tau) \to V(\tau)$. As is known, the scalar product is continuous in L_2, hence

$$\lim_{\varepsilon \to 0} \int_\Gamma u_\varepsilon(t)\left[\int_\Gamma \frac{v_\varepsilon(\tau)}{\tau - t}\, d\tau\right] dt = \lim_{\varepsilon \to 0}(u_\varepsilon, \bar{V}_\varepsilon) = (u, \bar{V}) = \int_\Gamma u(t) \int_\Gamma \frac{v(\tau)}{\tau - t}\, d\tau\, dt.$$

In an analogous manner, we may show that

$$\lim_{\varepsilon \to 0} \int_\Gamma v_\varepsilon(\tau)\left[\int_\Gamma \frac{u_\varepsilon(t)}{\tau - t}\, dt\right] d\tau = \int_\Gamma v(\tau)\left[\int_\Gamma \frac{u(t)}{\tau - t}\, dt\right] d\tau.$$

The functions $u_\varepsilon(t)/(\tau - t)$ and $v_\varepsilon(\tau)/(\tau - t)$ are summable on Γ_ε; by Fubini's theorem

$$\int_\Gamma u_\varepsilon(t)\left[\int_\Gamma \frac{v_\varepsilon(\tau)}{\tau - t}\, d\tau\right] dt = \int_\Gamma v_\varepsilon(t)\left[\int_\Gamma \frac{u_\varepsilon(t)}{\tau - t}\, dt\right] d\tau.$$

Setting $\varepsilon \to 0$ and using the two preceding limiting relations, we arrive at (2.2.14).

3^0. In the theory of singular integral equations, an important role is played by

Theorem 2.5. Let the contour Γ satisfy the conditions of §1. If $c(t)$ is continuous on Γ, then the operator T, defined by

$$(T\,u)(t) = (S\,c\,u)(t) - c(t)(S\,u)(t) = \frac{1}{\pi i} \int_\Gamma \frac{c(\tau) - c(t)}{(\tau - t)}\, u(\tau)\, d\tau \qquad (2.2.15)$$

is compact in $L_2(\Gamma)$.

This theorem is obvious of $c(t)$ satisfies a Hoelder condition - in that case the kernel of the integral in (2.2.15) has a weak singularity. In the general case, the continuous function $c(t)$ can be represented as the limit of $\{c_n(t)\}$ a sequence of Hoelder functions, uniformly convergent on Γ; for example, one can subject the function $c(t)$ to averaging in the sense of S.L. Sobolev (cf. [39]). We introduce the notation

$$(T_n\,u)(t) = (S\,c_n\,u)(t) - c_n(t)(S\,u)(t) = \frac{1}{\pi i} \int_\Gamma \frac{c_n(\tau) - c_n(t)}{(\tau - t)}\, u(\tau)\, d\tau\,.$$

The operator T_n is compact in $L_2(\Gamma)$. Estimating the norm of the difference

$$\|Tu - T_n u\| \le \|S[(c - c_n)u] + (c - c_n)\,S\,u\| \le 2\|S\| \cdot \max_{t \in \Gamma} c(t) - c_n(t) \cdot \|u\|,$$

we find

$$\|T - T_n\| \le 2\|S\| \cdot \max_{t \in \Gamma} |\, c(t) - c_n(t)\, | \to 0, n \to \infty.$$

By theorem 1.4, the operator T is compact in $L_2(\Gamma)$.

4^0. Above, we have established (theorem 2.3) formula (2.2.14), viz: $(S^2\,u)(t) = u(t)$, under the assumption that $u(t)$ is a Hoelder function. This formula remains true if $u \in L_2(\Gamma)$. In fact, we can construct a sequence $\{U_n\}$ of Hoelder functions which converges to u in the norm of $L_2(\Gamma)$. Then $(S^2 u_n)(t) = u_n(t)$. The operator S, and with it also the operator S^2, is bounded, hence it is also continuous in $L_2(\Gamma)$; going to the limit in the last equality, we verify that (2.2.14) remains true for functions of the class $L_2(\Gamma)$.

§3. Symbol and regularization

1^0. We will consider the class $\mathcal{R}$ of operators of the form

$$(Au)(t) = a(t)\,u(t) + b(t)\,(Su)(t) + (Tu)(t), \qquad (2.3.1)$$

acting in the space $L_2(\Gamma)$ where the functions $a(t)$ and $b(t)$ are continuous on Γ, S is the Cauchy operator (2.1.10), T is an operator, compact in $L_2(\Gamma)$, and Γ is a contour which satisfies the conditions of §1. It is readily seen that the operators of this class form a ring. In fact, let

$$(A_k)(t) = a_k(t)\,u(t) + b_k(t)\,(Su)(t) + (T_k u)(t)\,, \quad k = 1, 2,$$

be two such operators. Obviously, the sum $A_1 + A_2$ — belongs to the same class $\mathcal{R}$; we will study the product $A_1 A_2$. By theorems 2.3 and 2.5, we easily find that this product has the form (2.3.1) in which $a(t) = a_1(t)\,a_2(t) + b_1(t)\,b_2(t)$, $b(t) = a_1(t)\,b_2(t) + a_2(t)\,b_1(t)$ and T is some operator compact in $L_2(\Gamma)$. Note that multiplication in $\mathcal{R}$ is commutative apart from a compact term.

Theorem 2.6. If the operator (2.3.1) is compact in $L_2(\Gamma)$, then $a(t) = b(t) \equiv 0$.

Together with the operator (2.3.1) also the simpler operator $a(t)I + b(t)S$ is compact. Multiplying it from the left by the bounded operator $a(t)I - b(t)S$, and using theorems 2.3 and 2.5, we find that also the operator $(a^2(t) - b^2(t))I$ is compact. As we will now show, it follows from this that $a^2(t) - b^2(t) = 0$. In fact, otherwise there would exist a constant $\eta_0 > 0$ and an arc $\Gamma_0 \subset \Gamma$ such that $|\,a^2(t) - b^2(t)\,| \geq \eta_0$, $t \in \Gamma_0$, when, by theorem 1.2, the identity operator is compact in the infinitely dimensional space $L_2(\Gamma_0)$, – which, as we know, is impossible. Thus, $a(t) = \pm b(t)$, and we arrive at the fact that one of the operators $a(t)(I \pm S)$ is compact. For example, let this be the operator $a(t)(I + S)$. We will show that then $a(t) \equiv 0$. If this were not the case, there would exist an arc $\Gamma_1 \subset \Gamma$ and a number $\eta_1 > 0$ such that $|\,a(t)\,| \geq \eta_1$, $t \in \Gamma_1$, and the operator $I + S$ is compact in $L_2(\Gamma)$. Consider now the manifold of functions $u(t)$, continuous on Γ and admitting analytic continuation into the space D^+. On this manifold, the operator S coincides with I (cf. (2.1.12)), and this manifold is infinitely dimensional, hence it follows that the identity operator is compact in $L_2(\Gamma)$ on an infinite manifold; as noted already, this is impossible $a(t) \equiv 0$, and therefore $b(t) \equiv 0$. In analogous manner, one can study the assumption of the compactness of the operator $a(t)(I - S)$.

2^0. For the ring of the operators $\mathcal{R}$ of 1^0, one readily constructs a corresponding ring of symbols. Let us denote by θ the independent variable which assumes only the two different values $+1$ and -1 so that $\theta^2 = 1$, and consider the manifold of functions of the form $a(t) + b(t)\theta$, where the functions $a(t)$ and $b(t)$ are continuous in t on Γ. This manifold forms a ring with respect to the normal operations of addition and multiplication; denote it by $\mathcal{T}$. As symbol of the singular operator (2.3.1) we take the function of the continuous variables t and θ

$$Smb\, A = a(t) + b(t)\,\theta\,. \tag{2.3.2}$$

Such a definition is correct: the sum and product of operators of the form (2.3.1) obviously correspond to the sum and product of their symbols (2.3.2). Note that, according to our definition, the symbol of any operator in $L_2(\Gamma)$, is identically zero, and that the symbol of the Cauchy operator equals θ. It is also obvious that, according to the given operator from $\mathcal{R}$, its symbol is defined in a unique manner, and that, according to the given symbol from $\mathcal{T}$, the corresponding operator from $\mathcal{R}$, is determined exactly apart from an arbitrary term compact in $L_2(\Gamma)$.

3^0. In 3^0 of §2 of chapter 1, we have explained the special role of reversible symbols: If the symbol of an operator is not reversible in $\mathcal{T}$, then the operator A is not reversible in R. From other side if the symbol of operator A is reversible then the operator A admits left regularization and, according to theorem 1.16, such an operator is Noetherian.

We will assume now that the symbol $a(t) + b(t)\theta$ is reversible. Hence there exist functions $a(t)$ and $b(t)$, continuous in Γ, which satisfy the identity $[a_1(t) + b_1(t)\theta][a(t) + b(t)\theta] = 1$ hence

$$a_1(t) + b_1(t)\theta = \frac{1}{a(t) + b(t)\theta} = \frac{a(t)}{a^2(t) - b^2(t)} - \frac{b(t)}{a^2(t) - b^2(t)}\theta,$$

and, consequently,

$$a_1(t) = \frac{a(t)}{a^2(t) - b^2(t)}, \quad b_1(t) = -\frac{b(t)}{a^2(t) - b^2(t)}. \tag{2.3.3}$$

For the functions (2.3.3) to be continuous, it is necessary and sufficient that

$$a^2(t) + b^2(t) \neq 0, \forall t \in \Gamma. \tag{2.3.4}$$

We will say that the symbol (2.3.2) does not degenerate, if condition (2.3.4) is fulfilled; however, if this condition is violated, be it only at one point, then we will say that the symbol (2.3.2) degenerates.

From the results of §§2, 3 of chapter 1 and from the results proved just now follows

Theorem 2.7. If the symbol of the singular operator (2.3.1) does not degenerate, then this operator admits left regularization and, consequently, it is a Noether operator. The left regulariztor is

$$R = a_1(t) + b_1(t)S + T_1, \tag{2.3.5}$$

where $a_1(t)$ and $b_1(t)$ are defined by (2.3.3) and T_1 is an arbitrary operator compact in $L_2(\Gamma)$.

Note 2.1. If the symbol of the operator (2.3.1) degenerates, then, as is readily seen, this operator does not admit left regularization in the ring $\mathcal{R}$.

Note 2.2. In the case considered here, the ring of symbols if commutative. Hence it is easily concluded that, if the symbol does not degenerate, then the corresponding operator admits left as well as right regularization, and the regulariztors are equal to each other.

§4. Introduction to boundary value problems

1^0. Let A be the operator (2.3.1). We will now consider the singular integral equation

$$(Au)(t) = f(t),\ f \in L_2(\Gamma);\qquad (2.4.1)$$

we will assume that the coefficients $a(t)$ and $b(t)$ satisfy on Γ a Hoelder condition, and that the symbol of the operator A does not degenerate. Then equation (2.4.1) may be reduced to a certain boundary value problem of the theory of analytic functions. This problem has an elementary solution, if $T = 0$, which leads to the complete solution of (2.4.1). If $T \neq 0$, the same method reduces (2.4.1) to an equivalent equation with a compact operator.

The reduction of a singular integral equation to a boundary value problem of analytic functions was first employed by T. Carleman [48] in 1922 for the case when the contour of integration Γ is a segment of the real axis; F.D. Gakhov (cf. [4]) applied Carleman's method to the case of a sufficiently smooth closed contour, bounding by a finite, simply-connected region; the general case of a multiply-connected region has been considered by B.V. Khvedelidze [43].

We will write $f(t) - (Tu)(t) = f_1(t)$ and assume that the function $f_1(t)$ is known. If equation (2.4.1) has its solution in the class $L_2(\Gamma)$, then $f_1 \in L_2(\Gamma)$ and equation (2.4.1) assumes the simpler form

$$a(t)\,u(t) + b(t)\,(Su)(t) = f_1(t)\,.\qquad (2.4.2)$$

2^0. We will introduce the notation

$$m = \tfrac{1}{2\pi i} \int\limits_{\Gamma} d\,ln\,\tfrac{a(\tau)-b(\tau)}{a(\tau)+b(\tau)} = \tfrac{1}{2\pi}\left[arg\,\tfrac{a(\tau)-b(\tau)}{a(\tau)+b(\tau)}\right]_{\Gamma},$$

$$m_k = \tfrac{1}{2\pi i} \int\limits_{\Gamma_k} d\,ln\,\tfrac{a(\tau)-b(\tau)}{a(\tau)+b(\tau)} = \tfrac{1}{2\pi}\left[arg\,\tfrac{a(\tau)-b(\tau)}{a(\tau)+b(\tau)}\right]_{\Gamma_k},\qquad (2.4.3)$$

$k = 0, 1, 2, ..., n$, so that

$$m = \sum_{k=0}^{n} m_k\,.$$

Here Γ_0 is the curve which bounds Ω^+ outside; by the symbol $[...]_\Gamma$, where Γ is a closed contour, we will denote the increment of the function in square brackets for a circuit of Γ in a direction selected on this contour. Inside each of the curves Γ_k, $k = 1, 2, ..., n$, we select some point c_k and set

$$\Pi(z) = \prod_{k=1}^{n} (z - c_k)^{m_k} , \qquad (2.4.4)$$

where z is an arbitrary point of the complex plane. We will place the origin of coordinates inside the region Ω^+.

We will prove that the function

$$v(t) = ln \frac{a(t) - b(t)}{a(t) + b(t)} - m \, ln \, t + ln \, \Pi(t) \qquad (2.4.5)$$

is continuous on Γ. It will be sufficient to verify that $[v(t)]_{\Gamma_k} = 0$, $k = 0, 1, 2, ..., n$.

We note that Γ_0 is traveled in a counter-clockwise direction, the k, $1 \leq k \leq n$ in a clockwise direction. If $k = 0$, then

$$[ln \, \Pi(t)]_{\Gamma_0} = \sum_{j=1}^{n} m_j [ln(t - c_j)]_{\Gamma_0} = 2\pi i \sum_{j=1}^{n} m_j ,$$

$$[m \, ln \, t]_{\Gamma_0} = 2\pi i m , \quad \left[ln \frac{a(t) - b(t)}{a(t) + b(t)} \right]_{\Gamma_0} = 2\pi i m_0$$

and

$$[v(t)]_{\Gamma_0} = 2\pi i \left(m_0 - m + \sum_{j=1}^{n} m_j \right) = 0 .$$

Moreover, if $1 \leq k \leq n$, then

$$[ln \, \Pi(t)]_{\Gamma_k} = 2\pi i m_k , \quad [m \, ln \, t]_{\Gamma_k} = 0 , \quad \left[ln \frac{a(t) - b(t)}{a(t) + b(t)} \right]_{\Gamma_k} = 2\pi i m_k$$

and $[v(t)]_{\Gamma_k} = 0$.

3^0. We will introduce the Cauchy type integral

$$\Phi(z) = \frac{1}{2\pi i} \int_{\Gamma} \frac{u(\tau)}{\tau - z} \, d\tau , \ \Psi(z) = \frac{1}{2\pi i} \int_{\Gamma} \frac{v(\tau)}{\tau - z} \, d\tau , \qquad (2.4.6)$$

and write

$$F(z) = \Phi(z)\, e^{-\Psi(z)}\,. \tag{2.4.7}$$

Using the formula of Sokhotskii-Plemelj (2.1.11), we readily reduce (2.4.1) to

$$\Pi(t)\, F^+(t) - t^m\, F^-(t) = \frac{\Pi(t)\, e^{-\Psi^+}\, f_1(t)}{a(t) + b(t)} := f_2(t)\,. \tag{2.4.8}$$

The next step depends on the sign of the number m in (2.4.3)

I) Let $m = 0$. Then we introduce the function $F_1(z)$ which we set equal to $\Pi(z)\, F(z)$ in Ω^+ and $F(z)$ in Ω^-. equation (2.4.8) then reduces to

$$F_1^+(t) - F_1^-(t) = f_2(t)\,. \tag{2.4.9}$$

Noting that $F^-(\infty) = 0$, we find that (2.4.9) is satisfied by the Cauchy type integral

$$F_1(z) = \frac{1}{2\pi i} \int_\Gamma \frac{f_2(\tau)}{\tau - z}\, d\tau\,. \tag{2.4.10}$$

The solution of (2.4.10) is unique. In fact, if problem (2.4.9) had yet another solution $D(z)$, then, letting $F_1(z) - \Omega(z) = \varphi(z)$, we have $\forall t \in \Gamma$, $\varphi^+(t) - \varphi^-(t) = 0$, hence it follows that $\varphi^-(z)$ is the analytic continuation of the function $\varphi^+(z)$ in Ω^-; the function $\varphi(z)$ turns out to be holomorphic in the entire complex plane and, consequently, $\varphi(z) = const$. However, $\varphi(\infty) = 0$, hence $\varphi(z) \equiv 0$.

Since it is not difficult to derive the formula for $u(t)$, this will be left to the reader.

2) Let $m > 0$. We will seek a solution of (2.4.8) for which holds the estimate $F(z) = O(z^{-m-1})$ at infinity. This time, we denote by $F_1(z)$ the function which equals $\Pi(z)F(z)$ in Ω^+ and $z^m F(z)$ in Ω^-. As before, the function $F_1(z)$ solves problem (2.4.9), and its solution is given by (2.4.10); starting from this formula, one may obtain one of the solutions of problem (2.4.8) and, consequently, one of the solutions of problem (2.4.1).

However, the given case, the solution of (2.4.8) is not unique. In order to verify this fact, we will consider the corresponding homogeneous problem

$$\Pi(t)\, F_0^+(t) - t^m F_0^-(t) = 0\,,\ F_0^-(\infty) = 0\,. \tag{2.4.11}$$

equation (2.4.11) shows that the function $\Pi(z)\, F_0^+$ is holomorphic in Ω^+ analytically continues throughout the entire complex plane, may have, at infinity, a pole of order $m - 1$; hence

$$F_0^-(z) = \sum_{k=0}^{m-1} a_k\, z^{k-m}\,;\ \ F_0^+(z) = \frac{1}{\Pi(z)} \sum_{k=0}^{m-1} a_k\, z^k\,;$$

where a_k is an arbitrary constant.

3) The remains the study of the case $m < 0$. In this case, we will set $F_1(z)$ equal to $\Pi(z)\, F^+(z)$ in Ω^+, and to $z^m F^-(z)$ in Ω^-; the function $F_1(z)$ is holomorphic in Ω^+ as well as in Ω^-, where it has at infinity a zero of order no less than $-m+1$. We will once again proceed to problem (2.4.9) which has the unique solution (2.4.10). It leads us to the solution of the initial problem if and only if the Cauchy type integral

$$\frac{1}{2\pi i} \int\limits_\Gamma \frac{f_2(\tau)}{\tau - z}\, dt$$

has at infinity a zero of order no less than $-m + 1$; expanding this integral in a Laurent series about the point at infinity, we realize that then, in turn, the identity

$$\int\limits_\Gamma t^k\, f_2(t)\, dt\,,\ \ k = 0, 1, 2, \ldots, -(m+1)\,, \tag{2.4.12}$$

is necessary and sufficient.

§5. Evaluation of the index

1^0. The results of sections $1^0 - 3^0$ of §4 allow us to evaluate the index of the singular integral operator (2.3.1), if its symbol does not degenerate. By theorem 2.7, such an operator admits regularization; it follows then from theorems 1.12 and 1.4 that one may assume for the evaluation of the index of this operator that $T = 0$ and that the coefficients $a(t)$ and $b(t)$ satisfy a Hoelder condition. This fact leads to the determination of the number of zeroes of the two adjoint operators A_0 and A_0^*, where $(A_0 u)(t) = a(t)\, u(t) + b(t)\,(Su)(t)$, and the coefficients $a(t)$ and $b(t)$ are Hoelderian. It is seen form the results of 3^0 of §4 that the number of zeroes of the operator A_0 equals zero for $m \leq 0$ and m for $m > 0$.

2^0. We will now derive the formula for the operator

$$(A_0^*)(t) = \overline{a(t)}\, v(t) + (S^*(\bar{b}v))(t)\,.$$

We assume that the equation of the contour Γ can be written in the form $t = t(s)$, where s is a real parameter. Also let $\tau = t(\sigma)$. By definition of the adjoint operator, $\forall u, v \in L_2(\Gamma)$, $(Su, v) = (u, S^*v)$, we have

$$(Su, v) = \frac{1}{\pi i} \int_\Gamma \overline{v(t)} \left[\int_\Gamma \frac{v(\tau)}{\tau - t} \, d\tau \right] ds \, .$$

We invert now the order of integration (cf. (2.2.14)):

$$(Su, v) = \frac{1}{\pi i} \int_\Gamma u(\tau) \left[\int_\Gamma \frac{\overline{v(t)}}{\tau - t} \, ds \right] d\tau = \frac{1}{\pi i} \int_\Gamma u(t) \left[\int_\Gamma \frac{\overline{v(\tau)}}{t - \tau} \, d\sigma \right] dt \, .$$

Since $\tau = t(\sigma)$, we may write $d\sigma = d\tau / t'(\sigma)$ and, moreover,

$$(Su, v) = -\frac{1}{\pi i} \int_\Gamma u(t) \left[\int_\Gamma \frac{\overline{v(\tau)/t'(\sigma)}}{\tau - t} \, d\tau \right] t'(s) \, ds \, =$$

$$= -\frac{1}{\pi i} \int_\Gamma u(t) \left[\int_\Gamma \frac{\overline{v(\tau)t'(s)}}{(\tau - t)\, t'(\sigma)} \, d\tau \right] ds =$$

$$= \int_\Gamma u(t) \left[\overline{\frac{1}{\pi i} \int_\Gamma \frac{v(\tau)}{\bar\tau - \bar t} \, \frac{\overline{t'(s)}}{t'(\sigma)} \, d\bar\tau} \right] ds \, .$$

hence follows the formula for S^*:

$$(S^*v) = \frac{1}{\pi i} \int_\Gamma \frac{v(\tau)}{\bar\tau - \bar t} \, \frac{\overline{t'(s)}}{\overline{t'(\sigma)}} \, d\bar\tau \, . \tag{2.5.1}$$

The equation $A_0^* v = 0$ can now be given the form

$$\overline{a(t)} \, v(t) + \frac{1}{\pi i} \int_\Gamma \frac{b(\tau) v(\tau)}{\bar\tau - \bar t} \, \frac{\overline{t'(s)}}{\overline{t'(\sigma)}} \, d\bar\tau = 0 \, . \tag{2.5.2}$$

This equation may be replaced by a somewhat simpler, equivalent equation. For this purpose, we set $v(t)/\overline{t'(s)} = \overline{w(t)}$ and replace all quantities in (2.5.2) by their conjugate complex values to arrive at the equation

$$a(t) \, w(t) - (Sbw)(t) = 0 \, . \tag{2.5.3}$$

Our next objective is the determination of the number of zeroes of this equation.

3^0. We will consider the auxiliary singular equation

$$a(t)\,w_1(t) - b(t)\,(Sw_1)(t) = 0\,.\tag{2.5.4}$$

The operator on the left hand side of this equation differs from A_0 only by the sign of $b(t)$; the number of zeroes for the new operator, determinable from the first of formulas (2.4.3), equals $-m$, hence follows that the number of zeroes of equation(2.5.4) is zero, if $m \geq 0$, and $-m$, if $m < 0$.

We will now show that one can establish between the solution of equations (2.5.3) and (2.5.4) a mutually single-valued correspondence which yields the relations

$$b(t)\,w(t) = w_1(t),\quad a(t)\,w(t) = (Sw_1)(t)\,.\tag{2.5.5}$$

In fact, let $w(t)$ be the solution of (2.5.3). The first of relations (2.5.5) determines the function $w_1(t)$, the second relation (2.5.5) follows from the first and the equality (2.5.3):

$$a(t)\,w(t) = S(bw)(t) = (Sw_1)(t)\,.$$

Multiplying now the first term in (2.5.5) by $a(t)$, the second by $b(t)$, and subtracting, we find that $w_1(t)$ satisfies (2.5.4).

Conversely, let w_1 solve equation (2.5.4). By supposition, the symbol of the operator A does not degenerate, hence $a(t)$ and $b(t)$ vanish simultaneously anywhere on Γ. We now find from the first relation (2.5.5) the value of $w(t)$ at all those points at which $b(t) \neq 0$, and from the second at those points where $a(t) \neq 0$. Thus, if simultaneously $a(t) \neq 0$ and $b(t) \neq 0$, then one obtains for $w(t)$ two values; however, these values coincide, hence by the strength of (2.5.4) one finds $w_1(t)/b(t) = (Sw_1)(t)/a(t)$. Finally, the function $w(t)$ will be defined in a single-valued manner on the entire contour Γ; for this purpose, equation (2.5.3) must be fulfilled. In fact, if $b(t) \neq 0$, then, replacing in (2.5.4) w_1 and Sw_1 by relations (2.5.4) and (2.5.5), we obtain $b(t)[a(t)\,w(t) - (Sbw)(t)] = 0$. However, if $b(t) = 0$, then the second relation (2.5.5) yields $w(t) = (Sw_1)(t)/a(t)$, and the left hand side of (2.5.3) assumes the form $(Sw_1)(t) - (Sw_1)(t) = 0$.

4^0. **Theorem 2.8.** The index of the operator (2.3.1) with a non-degenerating symbol is given by

$$Ind\,A = m = \frac{1}{2\pi}\left[arg\,\frac{a(t) - b(t)}{a(t) + b(t)}\right]_\Gamma\,.\tag{2.5.6}$$

It follows from the result proved in 3^0 that the number of zeroes of the operator A_0^* equals zero, if $m \geq 0$, it equals $-m$ if $m < 0$. Using the number of zeroes of the operator A_0 stated in 1^0, we find that $Ind\,A_0 = m$. However, since $Ind\,A = Ind\,A_0$, formula (2.5.6) is proved.

§6. Systems of singular equations

1^0. We will consider systems of the form

$$\sum_{k=1}^{n} \left\{ a_{jk}(t)\, u_k(t) + b_{jk}(t)\,(Su_k)(t) + (T_{jk}u_k)(t) \right\}$$
$$= f_j(t)\,,\; j = 1, 2, ..., n\,, \tag{2.6.1}$$

where S is the Cauchy operator, described in §1, $a_{jk}(t)$ and $b_{jk}(t)$ are functions, continuous on Γ, T_{jk} are operators, compact in $L_2(\Gamma)$, the given functions f_j and the unknowns u_k are elements of the same space, and n is finite.

To start with we will simplify the notation of system (2.6.1). We will introduce the Hilbert space $L_2^{(n)}(\Gamma)$, the elements of which are vector functions $w(t) = (w_1(t), w_2(t), ..., w_n(t))$ with components $w_k \in L_2(\Gamma)$; the scalar product and norm in $L_2^{(n)}$ are given by

$$(v, w)_{L_2^{(n)}(\Gamma)} = \int_{\Gamma} \sum_{k=1}^{n} v_k(t)\,\overline{w_k(t)}\, ds\,;\; \|w\|^2_{L_2^{(n)}(\Gamma)} = (w, w)_{L_2^{(n)}(\Gamma)}\,. \tag{2.6.2}$$

Note that the vector functions $f(t) = (f_1(t), f_2(t), ..., f_n(t))$ and $u(t) = (u_1(t), u_2(t), ..., u_n(t))$ are elements of the space $L_2^{(n)}(\Gamma)$.

We will introduce the square matrices $a(t)$, $b(t)$, T of order n with the respective elements $a_{jk}(t)$, $b_{jk}(t)$, T_{jk}. System (2.6.1) now assumes the form

$$(Au)(t) = a(t)\,u(t) + b(t)\,(Su)(t) + (Tu)(t) = f(t)\,. \tag{2.6.3}$$

Obviously, the operator A is bounded, the operator T is compact in $L_2^{(n)}(\Gamma)$ and the manifold of the operators of the type (2.6.3) form a ring to be denoted by $\mathcal{R}$.

For the ring of operators $\mathcal{R}$, we may construct the ring of symbols $\mathcal{T}$, if we select as symbol of the operator (2.6.3) the matrix function

$$Smb\, A = a(t) + b(t)\,\theta\,, \tag{2.6.4}$$

where $t \in \Gamma$ and $\Theta = \pm 1$; we will call the matrix (2.6.4) the symbolic matrix of the operator A.

2^0. Using the fact that $\Theta^2 = 1$, we may to reduce the determinant of the symbolic matrix (the symbolic matrix of the operator (2.6.3)) to the form $det\, Smb\, A = \alpha(t) + \beta(t)\,\Theta$, where $\alpha(t)$ and $\beta(t)$ arc certain functions, continuous on Γ. The symbol (2.6.4) has an inverse element in the ring $\mathcal{T}$ if and only if the functions $\alpha(t) \pm \beta(t)$ do not vanish anywhere on Γ; under this condition, we will say that the symbol of the operator (2.6.3) does not degenerate. A non-degenerating symbol has an inverse element in the ring $\mathcal{T}$ hence a singular matrix operator with a non-degenerating symbol admits regularization: Any singular operator the symbolic matrix of which equals $(Smb\,A)^{-1} = [a(t) + b(t)\,\theta]^{-1}$ is a regulariztor of the singular operator A. It follows from theorem 1.16 that a matrix singular operator with non-degenerating symbol is a Noetherian operator.

3^0. The method of reduction to a boundary value problem, studied in §4, has been extended to systems of singular, one-dimensional integral equations in the work of N.I. Muskhelishvili and N.P. Vekua, exposed in detail in the monograph [30], where also the formula for the index of a system of non-degenerating symbols may be found:

$$Ind\,A = \frac{1}{2\pi} \left[arg\, \frac{\alpha(t) - \beta(t)}{\alpha(t) + \beta(t)} \right]_\Gamma . \tag{2.6.5}$$

§7. Equations on open contours and with discontinuous coefficients

1^0. We shall start with a simple example. Let Γ_1 be a smooth open arc without self-intersections on the complex plane $\mathcal{C}$, a_1 and b_1 constants,

$$(A_1\varphi)(t) \equiv a_1\,\varphi(t) + \frac{b_1}{\pi i} \int\limits_{\Gamma_1} \frac{\varphi(\tau)\,d\tau}{\tau - t} = f(t),\ t \in \Gamma_1 . \tag{2.7.1}$$

We will assume that $a_1^2 - b_1^2 \neq 0$; we will denote the ends of the arc Γ_1 by α and β. We will introduce as new unknown function the Cauchy integral with density $\varphi(t)$:

$$\Phi(z) = \frac{1}{2\pi i} \int\limits_{\Gamma_1} \frac{\varphi(\tau)\,d(\tau)}{\tau - z},\ z \notin \Gamma_1 .$$

The formulas of Sokhotskii-Plemelj remain true in the case of open contours of integration; we must then select the directions i and e as shown in Fig. 3. The arc Γ_1 is traveled in the direction from α to β. We note yet that in this case the formulas of Sokhotskii-Plemelj are only true for internal points of the arc Γ_1; at the ends α and β these formulas loose their meaning.

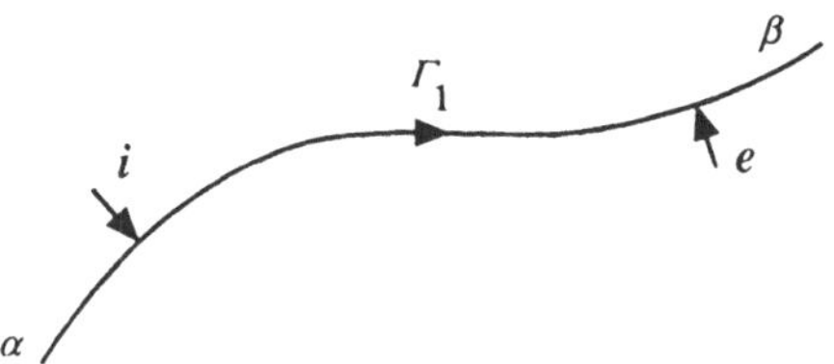

Fig. 3

We will write equation (2.7.1) in the form

$$(a_1 + b_1)\,\Phi_i(t) - (a_1 - b_1)\,\Phi_e(t),\ t \in \Gamma_1\,.$$

In the present case, the solution of the corresponding homogeneous problem $(a_1 + b_1)\,\omega_i(t) - (a_1 - b_1)\,\omega_e(t) = 0$ is readily constructed. For example, such a solution is the function

$$\omega(z) = \left(\frac{z - \alpha}{z - \beta}\right)^m,\ m = \frac{1}{2\pi i}\,ln\,\frac{a_1 + b_1}{a_1 - b_1}\,. \tag{2.7.2}$$

Since $ln\,z$ is multi-valued, and its different values differ only by an integral multiple of $2\pi i$, then (2.7.2) defines m exactly apart from an arbitrary integer term. We will select this term so that $0 \le Re\,m < 1$. For this purpose, it is sufficient to select the value of $arg[(a_1 + b_1)/(a_1 - b_1)]$ which lies between the limits $0 \le arg[(a_1 + b_1)/(a_1 - b_1)] < 2\pi$. For such a choice of m both functions $\omega(z)$ and $\omega^{-1}(z)$ are summable on Γ_1.

Letting now $\Phi(z) = \Psi(z)\,\omega(z)$, we find that $\Psi(z)$ satisfies the equation

$$\Psi_i(t) - \Psi_e(t) = \frac{f(t)}{a_1 + b_1}\left(\frac{t - \beta}{t - \alpha}\right)^m.$$

It follows directly from the formulas of Sokhotskii-Plemelj that one can, for example, set

$$\Psi(z) = \frac{1}{2\pi(a_1 + b_1)}\int_{\Gamma_1}\left(\frac{\tau - \beta}{\tau - \alpha}\right)^m f(\tau)\,\frac{d\tau}{\tau - t},$$

and, consequently,

$$\Phi(z) = \frac{1}{2\pi(a_1 + b_1)} \left(\frac{z-\alpha}{z-\beta}\right)^m \int\limits_{\Gamma_1} \left(\frac{\tau-\beta}{\tau-\alpha}\right)^m f(\tau)\,\frac{d\tau}{\tau-z}\,.$$

Thus one of the solutions of equation (2.7.1) has been obtained. Using the formula $\varphi(t) = \Phi_i(t) - \Phi_e(t)$, we find

$$\begin{aligned}
\varphi(t) = {}& \frac{a_1}{a_1^2 - b_1^2}\, f(t) \\
& - \frac{b_1}{\pi\, i(a_1 + b_1)} \left(\frac{t-\alpha}{t-\beta}\right)^m \int\limits_{\Gamma_1} \left(\frac{\tau-\beta}{\tau-\alpha}\right)^m f(\tau)\,\frac{d\tau}{\tau-t}\,.
\end{aligned} \tag{2.7.3}$$

In order to find all solutions of (2.7.1), it is sufficient to add to the solution (2.7.3) the general solution of the homogeneous problem

$$a_1\,\varphi_0(t) + \frac{b_1}{\pi\, i} \int\limits_{\Gamma_1} \frac{\varphi_0(\tau)}{\tau - t}\, d\tau = 0\,. \tag{2.7.4}$$

By the same method, we set

$$\Phi_0(z) = \frac{1}{2\pi\, i} \int\limits_{\Gamma_1} \frac{\varphi_0(\tau)}{\tau - z}\, d\tau\,, \quad \Phi_0(z) = \omega(z)\,\Psi_0(z)\,. \tag{2.7.5}$$

We note that $\Phi_0(\infty) = \Psi_0(\infty) = 0$. As before, we arrive at the equation

$$\Psi_{0i}(t) - \Psi_{0e}(t) = 0\,,$$

which signifies that the analytic function $\Psi_0(z)$ is continuous on a passage through the line Γ_1. Hence only the points α and β can be special points of this function. We will seek the solution $\varphi_0(t)$ of equation (2.7.4) in the class $L_p(\Gamma_1)$ with some value of $p > 1$. Then $\Phi_{0i}(t) \in L_p(\Gamma_1)$ and it follows from (2.7.5) that β is a regular point of $\Psi_0(t)$, while α may either be a regular point or a first order pole of the same function. In the last case, one has $p\,Re(1 - m) < 1$. Hence

$$\Psi_0(z) = \frac{c'}{z-\alpha}\,, \quad c' = const\,; \quad \Phi_0(z) = \frac{c'}{(z-\alpha)^{1-m}(z-\beta)^m}\,,$$

and the solution of the homogeneous equation (2.7.4) is

$$\varphi_0(t) = \frac{c}{(t-\alpha)^{1-m}(t-\beta)^m} \, , \quad c = c'(1 - e^{2\pi i m}) .$$

The case $m = 0$ (and, consequently, $b_1 = 0$) may be excluded from our consideration; then $0 < Re\, m < 1$ and $\varphi_0(t) \in L_p(\Gamma_1)$ for $1 < p < min\{(Re\, m)^{-1},\ (1 - Re\, m)^{-1}\}$. This means that the solution of (2.7.4) exists in the class $L_p(\Gamma_1)$ for sufficiently small $p - 1 > 0$.

The general solution of (2.7.1) can be presented in the form

$$\varphi_0(t) = \frac{c}{(t-\alpha)^{1-m}(t-\beta)^m} + \frac{a_1}{a_1^2 - b_1^2} \, f(t) - \frac{b_1}{a_1^2 + b_1^2} \left(\frac{t-\alpha}{t-\beta}\right)^m \int_{\Gamma_1} \left(\frac{\tau-\beta}{\tau-\alpha}\right)^m f(\tau) \frac{d\tau}{\tau-t} , \tag{2.7.6}$$

where c is an arbitrary constant. One can obtain for $\varphi(t)$ a formula into which α and β enter in a more symmetric manner:

$$\varphi_0(t) = \frac{c}{(t-\alpha)^{1-m}(t-\beta)^m} + \frac{a_1}{a_1^2 - b_1^2} \, f(t) - \frac{b_1}{(a_1^2 + b_1^2)\, \pi\, i\, (t-\alpha)^{1-m}(t-\beta)^m} \times$$
$$\times \int_{\Gamma_1} \frac{(\tau-\alpha)^{1-m}(\tau-\beta)^m}{\tau-t} \, f(\tau)\, d\tau . \tag{2.7.7}$$

In particular, for the equation

$$\frac{1}{\pi\, i} \int_{\Gamma_1} \frac{\varphi(\tau)}{t-\tau} \, d\tau = f(t), \ t \in \Gamma_1 ,$$

we have $a_1 = 0$, $b_1 = 1$, $m = 1/2$; in correspondence with (2.7.6) and (2.7.7), we obtain two equivalent formulas for the solution

$$\varphi(t) = \frac{c}{\sqrt{t-\alpha}\,\sqrt{t-\beta}} + \frac{1}{\pi\, i} \sqrt{\frac{t-\alpha}{t-\beta}} \int_{\Gamma_1} \sqrt{\frac{\tau-\alpha}{\tau-\beta}} \, \frac{f(\tau)}{\tau-t} \, d\tau ,$$

$$\varphi(t) = \frac{c}{\sqrt{t-\alpha}\,\sqrt{t-\beta}} + \frac{1}{\pi\, i} \frac{1}{\sqrt{t-\alpha}\,\sqrt{t-\beta}} \int_{\Gamma_1} \frac{\sqrt{\tau-\alpha}\,\sqrt{\tau-\beta}}{\tau-t} \, f(\tau)\, d\tau .$$

2^0. We will now consider equation (2.7.1) for the case when $a_1 = a_1(t)$ and $b_1 = b_1(t)$ are continuous functions, and also the

$$A\varphi \equiv a(t)\,\varphi(t) + \frac{b(t)}{\pi i} \int_\Gamma \frac{\varphi(\tau)}{\tau - t}\, d\tau = f(t)\,, \ t \in \Gamma\,, \qquad (2.7.8)$$

where Γ is a sufficiently smooth closed curve, surrounding the origin, and $\Gamma_1 \subset \Gamma$.

Equation (2.7.1) reduces to equation (2.7.8) with the coefficients

$$a(t) = \begin{cases} a_1(t)\,, t \in \Gamma_1 \\ 1\,, t \in \Gamma/\Gamma_1 \end{cases}\,, \ \ b(t) = \begin{cases} b_1(t)\,, t \in \Gamma_1 \\ 0\,, t \in \Gamma/\Gamma_1 \end{cases}\,.$$

In fact, for

$$\chi(t) = \begin{cases} 1\,, t \in \Gamma_1 \\ 0\,, t \in \Gamma/\Gamma_1 \end{cases}$$

we have

$$(1 - \chi)\,A\,(\chi\varphi) = 0\,, \ (1 - \chi)\,A\,(1 - \chi)\varphi = (1 - \chi)\varphi\,,$$

hence

$$A\,\varphi = (\chi + 1 - \chi)\,A(\chi\varphi + (1 - \chi)\varphi) = \chi A(\chi\varphi) + (1 - \chi)\varphi + \chi A(1 - \chi)\varphi\,.$$

We write now

$$A_{12}\varphi \equiv \chi(t)\frac{b_1(t)}{\pi i} \int_\Gamma \frac{[1 - \chi(\tau)]\varphi(\tau)\, d\tau}{\tau - t}\,,$$

when the operator A, acting on the space $L_p(\Gamma) = L_p(\Gamma_1) \times L_p(\Gamma/\Gamma_1)$, may be presented in the form of the block matrix

$$A\varphi = \begin{pmatrix} A_1 & A_{12} \\ 0 & I \end{pmatrix} \begin{pmatrix} \chi\varphi \\ (1 - \chi)\varphi \end{pmatrix} = \begin{pmatrix} A_1\varphi + A_{12}\varphi \\ (1 - \chi)\varphi \end{pmatrix}\,,$$

acting in the space $L_p(\Gamma_1) \times L_p(\Gamma/\Gamma_1)$. Obviously,

$$A = \begin{pmatrix} A_1 & A_{12} \\ 0 & I \end{pmatrix} = \begin{pmatrix} I & A_{12} \\ 0 & I \end{pmatrix} \begin{pmatrix} A_1 & 0 \\ 0 & I \end{pmatrix}\,, \qquad (2.7.9)$$

and, since we may always invert the first factor on the right hand side of (2.7.9), the operator A has a left (right) inverse if and only if A_1 has a left (right) inverse. This is the standard method (cf. [34]) of reduction of equation (2.7.1) to an equation on a closed contour with an operator which has, generally speaking, discontinuous coefficients.

3^0. Prior to proceeding to the exposition of the general theory of singular operators of the form (2.7.8) with piecewise continuous coefficients, we will recall some concepts of algebra.

A ring A (cf. §2, chapter 1) is called an algebra, if for any element $x \in A$ and any number α the element $\alpha x \in A$ is defined, where

1. $\alpha(\beta x) = (\alpha\beta)x$,
2. $I \cdot x = x$,
3. $(\alpha + \beta)x = \alpha x + \beta x$,
4. $\alpha(x + y) = \alpha x + \alpha y$.

Everywhere in this section, the number field of the algebra considered is the field of complex numbers. A ring for which $C \subset A$, obviously, is an algebra. In the algebra of complex valued functions, the field C may be identified with the union constant function. In the algebra of operators, the field C is identified with the manifold of operators of the form αI, $\alpha \in C$, where I is the unit operator.

By definition, the algebra A is a linear space. If in A the norm $\|x\|$ of an element $x \in A$ is defined under the conditions

1. $\|I\| = 1$,
2. $\|x \cdot y\| \leq \|x\| \cdot \|y\|$, then A is called a normalized algebra. If, in addition, a normalized algebra A is complete (i.e., if it is a normalized space), then it is referred to as a Banach algebra.

The mapping $F : X \to Y$ is called a homomorphism of an algebra X into an algebra Y, if it satisfies the conditions

1. $F(x + y) = Fx + Fy$,
2. $F(\alpha x) = \alpha F x$,
3. $F(xy) = Fx \cdot Fy$.

Two algebras X and Y are said to be isomorphic, if there exists a mutually single-valued mapping F which is a homomorphism of the algebra X into the algebra Y.

Let $a(t)$ and $b(t)$ be piecewise continuous matrix functions of order $n \times n$, defined on Γ and continuous on Γ/K, where K is the manifold of the points of discontinuity.

The manifold of operators of the form (2.7.8) forms a normalized algebra, contained in the algebra of all linear, continuous operators, acting in $L_p^{(n)}(\Gamma) = L_P(\Gamma) \times ... \times L_p(\Gamma)$. The completion of this algebra by an operator norm will be denoted by $\phi_p^n(\Gamma, K)$. We will then say that the Banach algebra

$\phi_p^n(\Gamma, K)$ is a closure of an algebra generated by operators of the form (2.7.8).

In correspondence with the local principle of I.B. Simonenko (cf. [36, 37]), consideration is given to algebras of canonical representatives Λ_p^n and $\Omega_p^n \subset \Lambda_p^n$ on the real straight line R^1. An algebra Ω_p^n consists of operators of the form (2.7.8) with $\Gamma = R^1$ and constant matrix functions $a(t) = const$, $b(t) = const$. The algebra Λ_p^n represents the closure of the algebras, generated by the operators

$$(S_{R^1} f)(t) = \int\limits_{R^1} f(\tau) \frac{d\tau}{\tau - t}$$

and by operators of multiplication on matrix functions which are piecewise constant on the rays R_+^1, $R_-^1 = R^1 / \bar{R}_+^1$.

In order to establish a link between the algebras $\phi_p^n(\Gamma, X)$ and Λ_p^n, we will introduce the required definitions.

Operators A_1 and B_1 are said to be locally equivalent at the point 0, if for any $\varepsilon > 0$ one can find a neighbourhood $V_1 \ni 0$ such that

$$\|(A_1 - B_1)P_{V_1}\| < \varepsilon, \quad \|P_{V_1}(A_1 - B_1)\| < \varepsilon,$$

where P_{V_1} is the operator of multiplication the characteristic function of the manifold V_1. This concept will be generalized to the case of operators acting in different spaces.

Let $t_0 \in \Gamma$. We will denote by $\varphi = \varphi_{t_0}$ the mapping of some arc $U \subset \Gamma$, containing t_0 as internal point, which maps isometrically (i.e., with conservation of distance) this arc on to an interval V of the straight line R^1, linking the point t_0 to the point 0 and conserving orientation. The straight line is assumed to be orientated so that motion from smaller to larger numbers corresponds to the positive direction of passage.

We will denote by Q_φ the operator which every vector function $f \in L_p^{(n)}(R^1)$ applies in correspondence to the vector function $Q_\varphi f$ according to the rule: $(Q_\varphi f)(x) = f(\varphi(x))$ for all x for which the mapping φ is defined and $Q_\varphi f(x) = 0$ for the remaining $x \in \Gamma$. Thus, $Q_\varphi : L_p^n(R^1) \to L_p^n(\Gamma)$ is a linear, continuous operator.

By definition, an operator A at a point t_0 is φ_{t_0} quasi–equivalent to an operator B at the point 0, if the operators $A_1 = Q_{\varphi^{-1}} A Q_\varphi$ and $B_1 = P_v B P_v$ are locally equivalent at the point 0.

If $A \in \Phi_p^n(\Gamma, K)$, $t_0 \in \Gamma$, then an operator B is called the local representative of the operator A at the point t_0 if the operator A at the point t_0 is φ_{t_0} quasi–equivalent t_0 the operator B at the point 0.

It may be shown (cf. [37]) that at any point t_0 of a contour Γ there exists a unique local representative of an operator $A \in \Phi_p^n(\Gamma, K)$ which we will denote by $A_t \in \Lambda_p^n$. A mapping $\alpha : \Gamma \to \Lambda_p^n$ yielded by the equality: $\alpha(t_0) = A_{t_0}$ is thereby defined. It has the properties

1) $\alpha(t_0) \in \Omega_p^n$ for any $t_0 \in \Gamma/K$;

2) α is continuos at the points of the manifold Γ/K, while at points of the manifold K it is either continuous or has first order discontinuities;

3) $\lim\limits_{t \to t \mp 0} \alpha(t) = \Psi_\mp(\alpha(t_0))$, where $\Psi_-(\Psi_+)$ is a continuous homomorphism which compares at the point $x < 0$ $(x > 0)$ with an operator A from Λ_p^n its local representative A_x from Ω_p^n. Obviously, if $A \in \Omega_p^n$, then $\Psi_\mp(A) = A$.

We will denote by $\Psi_p^n(\Gamma, K)$ the Banach algebra of operator-valued functions α with values in Λ_p^n, defined on Γ and satisfying conditions 1) - 3) with the norm

$$\|\alpha\| = sup\,\|\alpha(t)\| \, .$$

We will define the mapping $\sigma = \sigma_{p,\Gamma,K}^n$ of an algebra $\Phi_p^n(\Gamma, K)$ on to $\Psi_p^n(\Gamma, K)$ by the formula

$$(\sigma(A))(t) = A_t \, , \ t \in \Gamma \, .$$

This mapping is an (operatorial) symbol, as is demonstrated by

Theorem 2.9. 1) The mapping $\sigma = \sigma_{p,\Gamma,K}$ is homomorphic, acting on $\Psi_p^n(\Gamma, K)$. 2) $Ker\,\sigma$ consists of all operators, compact in $L_p(\Gamma)$. 3) An operator A from $\Phi_p^n(\Gamma, K)$ is Noetherian if and only if $(\sigma A)(t)$ is reversible for each t of Γ. 4) A regularizator B of a Noetherian operator A from $\Phi_p^n(\Gamma, K)$ likewise belongs to $\Phi_p^n(\Gamma, K)$, where $\sigma B = (\sigma A)^{-1}$.

Note 2.3. The assertion formalized above permits to reduce the question regarding the Noether property of an operator $A \in \Phi_p^n(\Gamma, K)$ to the question regarding the invertibility of the "model" (canonical) operator $\sigma(A)(t)$ at "characteristic" points $t \in \Gamma$.

Example 2.1. Let Γ be an oriented Lyapunov contour on the complex plane. An operator of the form (2.7.8) with matrix functions a and b, which are continuous on Γ, is locally equivalent at a point t_0 to the operator

$$a(t_0)\,\varphi(t) + \frac{b(t_0)}{\pi i} \int\limits_\Gamma \frac{\varphi(\tau)\,d\tau}{\tau - t} \, , \ t \in \Gamma \, .$$

The local representative A_{t_0} is contained in Ω_p^n and equals

$$A_{t_0} = a(t_0) + b(t_0) S_{R^1} \, .$$

We will write down in more detail the structure of the algebra Λ_p^n of all possible (by the strength of 1) of theorem 2.9) canonical representatives of operators of the form (2.7.8) with piecewise continuous coefficients.

We will denote by $\bar{R}^1$ the extended real straight line $\bar{R}^1 = R^1 \cup \{\pm\infty\}$, by $C(\bar{R}^1)$ the continuous functions with the norm

$$\|f\|_{C(R^1)} = \sup_{x \in R^1} |f(x)|,$$

and by $V(R^1)$ the space of the functions of bounded variation with norm

$$\|f\|_{C(R^1)} = \sup_{x \in R^1} |f(x)| + V_{-\infty}^{+\infty}(f).$$

Moreover, let T_φ be the operator of multiplication on the bounded measurable function φ on R^1, and the space M_p^p $(1 < p < \infty)$ consist of those φ for which the operator

$$F^{-1} T_\varphi F : L_2(R^1) \cap L_p(R^1) \to L_p(R^1)$$

(F is the Fourier Transform on R_1) is continued to a continuous operator in $L_p(R^1)$. The norm $\varphi \in M_p^p$, by definition, equals the norm of this continued operator. As is known (cf. [8])

$$V(R^1) \in \cap M_p^p.$$

We will denote the closure of the linear manifold $C(R^1) \cap V(R^1)$ in the norm M_p^p by $C_p(\bar{R}^1)$. Finally, let $\mathcal{Z}_p^n$ be the algebra of the matrix functions of order $2n$, continuous on $\bar{\mathcal{R}}^1$, with elements from $C_p(\bar{R}^1)$ the values of which at the points $\pm\infty$ are block-diagonal matrices with two blocks of dimensionality n. The norm in the algebra $\mathcal{Z}_p^n$ is determined by

$$\|a\|_{\mathcal{Z}_p^n} = \sum_{j=1}^{2n} \sum_{i=1}^{2n} \|a_{ij}\|_{C_p(\bar{R}^1)}.$$

Theorem 2.10. The algebra of Λ_p^n is the topologically isomorphous to the algebra $\mathcal{Z}_p^n$. There exists the unique topological isomorphism $\mu : \Lambda_p^n \to \mathcal{Z}_p^n$, which satisfies the conditions

1) If A is the operator of multiplication on the matrix function

$$a(t) = \begin{cases} a_- & t < 0 \\ a_+ & t > 0 \end{cases},$$

where $a_\pm$ is a constant matrix of order n, then

$$\mu A = \begin{pmatrix} a_- & 0 \\ 0 & a_+ \end{pmatrix} ;$$

2)

$$(\mu S_{R^1})(x) = \begin{pmatrix} [th(\pi y)]I_n & -i[ch(\pi y)]^{-1}I_n \\ i[ch(\pi y)]^{-1}I_n & -[th(\pi y)]I_n \end{pmatrix},$$

where $y = x - (1/p - 1/2)i$, I_n is the unit matrix of order n, and the functions $th(\cdot)$ and $ch(\cdot)$ are extended on $\bar{R}^1$.

A certain representation on the algebra Λ_p^n and symbol μ yields

Theorem 2.11. If A is an operator of the form

$$(A_\alpha f)(x) - \frac{\mid x \mid^\alpha}{\pi i} \int\limits_{R^1} \frac{f(t)\, dt}{\mid t \mid^\alpha (t - x)},$$

where $-1/p < \alpha < 1 - 1/p$, then $A_\alpha \in \Lambda_p^n$ and $(\mu\, A_\alpha)(x) = (\mu\, S_{R^1})(x - \alpha\, i)$, $x \in \bar{R}^1$.

Besides, Λ_p^n is the closure of an algebra, continued by the operator A_α and the operators of multiplication on matrix functions which are piecewise constant. constant on the rays R_+^1, $R^1/\bar{R}_+^1$.

By theorem 2.9, the Noetherian property of the operator (2.7.8) with piecewise continuous coefficients is linked to the invertibility of its local representatives in the algebra Λ_p^n. One has

Theorem 2.12. An operator $A \in \Lambda_p^n$ is invertible if and only if the symbol $(\mu A)(x)$ is invertible at each point $x \in \bar{R}^1$.

§8. Wiener-Hopf integral equations

1^0. We shall consider the equation with difference kernel on the semi-axis (Wiener-Hopf equation)

$$(Wg)(t) \equiv g(t) - \int\limits_0^\infty K(t - s)\, g(s)\, ds = f(t), \ t > 0, \qquad (2.8.1)$$

where $K \in L_1(R^1)$. We will set $g(t) = g_+(t) - g_-(t)$ for $t \in R^1$, where $g_+(t) = g(t)$ for $t < 0$ and $g_+(t) = 0$ for $t < 0$, when (2.8.1) becomes

$$g_+(t) - \int\limits_{-\infty}^\infty K(t - s)\, g_+(s)\, ds = g_-(t) + f_+(t), \ t \in \bar{R}^1. \qquad (2.8.2)$$

If F is the Fourier transform of functions from $L_2(R^1)$, i.e.,

$$(FK)(t) = \int_{-\infty}^{\infty} e^{it\lambda} K(\lambda)\, d\lambda\,,\ t \in R^1\,,$$

and $g, f_+ \in L_2(R^1)$, then (2.8.2) is equivalent to

$$G_+(t) - (FK)(t) \cdot G_+(t) = G_-(t) + F_+(t)\,, \tag{2.8.3}$$

where $G_\pm(t) = (Fg_\pm)(t)$, $F_+(t) = (Ff_+)(t)$. The transition from (2.8.2) to (2.8.3) employed the convolution formula of Fourier transforms. We will introduce as new unknown function the Cauchy integral with density $(Fg)(t)$:

$$G(z) = \frac{1}{2\pi i} \int_{-\infty}^{\infty} \frac{(Fg)(t)}{t-z}\, dt\,,\ z \in R^1\,.$$

By strength of the identity

$$G(z) = \frac{1}{2\pi i} \int_{-\infty}^{\infty} \frac{(Fg_+)(t)}{t-z}\, dt - \frac{1}{2\pi i} \int_{-\infty}^{\infty} \frac{(Fg_-)(t)}{t-z}\, dt =$$

$$= \frac{1}{2\pi i} \int_{-\infty}^{\infty} g_+(\tau) \int_{-\infty}^{\infty} \frac{e^{2\pi i\tau t}}{t-z}\, dt\, d\tau - \frac{1}{2\pi i} \int_{-\infty}^{\infty} g_-(\tau) \int_{-\infty}^{\infty} \frac{e^{2\pi i\tau t}}{t-z}\, dt\, d\tau$$

we find

$$G(z) = \begin{cases} G_+(z)\,, & Im\,z > 0\,, \\ G_-(z)\,, & Im\,z < 0\,. \end{cases}$$

Consequently, $G_+(x) = G_i(x)$, $G_-(x) = G_e(x)$, $x \in R^1$ and the determination of $G(z)$ from (2.8.3) is a problem of linear conjugation, analogous to what has already been considered in §4. Equation (2.8.1) has been studied extansively in [14, 7, 5, 50, 34, 3]; many examples have been treated in [5, 31]. We will formulate here certain classical results (cf. [7, 52]).

Let $\alpha(W)$ denote the dimensionality of the kernel * of the operator W, $\beta(W)$ that of its co-kernel *, so that $ind\,W = \alpha - \beta$.

---

* Kernel and cokernel of the operator W are called the subspaces $N(W)$ and $N(W^*)$, respectively.

Theorem 2.13. If $k \in L_1(R^1)$ and $f \in L_p(R^1_+)$, $1 \le p \le \infty$ are given functions, then W is a Noether operator (normally resolvable) with finite numbers α and β if and only if

$$1 - (Fk) \ne 0, \ t \in \dot{R}^1 = R^1 \cup \{\infty\}, \tag{2.8.4}$$

where F denotes a Fourier transform.

For the fulfillment of condition (2.8.4), one has the formulas

$$\alpha(W) = max\{-\kappa, 0\}, \ \beta(W) = max\{0, \kappa\}$$

where $\kappa = \frac{1}{2\pi}\left[arg(1 - (Fk)(\xi))\right]_{+\infty}^{-\infty}$.

Besides, any solution of the homogeneous equation, corresponding to (2.8.1), belongs to $L_1(R^1_+) \cap C_0(R^1_+)$, $C_0(R^1_+) = \{\varphi \in C(R^1_+) : \lim_{t \to \infty} \varphi(t) = 0\}$. For $-\kappa > 0$, there exists the base of the space of homogeneous solution $Ker\ W$ which forms a d-chain, i.e., the base $\varphi_1, ..., \varphi_{|\kappa|-1}$, where $\varphi_1, ..., \varphi_{|\varphi|-1}$ are absolutely continuous and

$$\varphi_{j+1} = \frac{d}{dt}\varphi_j, \ j = 1, ..., |\kappa| - 1; \ \varphi_j(+0) - 0, \ j = 1, ..., |\kappa| - 2;$$

where $\varphi_{|\kappa|-1}(+0) \ne 0$.

The basic step in the proof of this theorem is the representation (factorization) of the function $1 - F(k)$ in the form

$$1 - (Fk)(t) = a_-(t)\left(\frac{t-i}{t+i}\right)^\kappa a_+(t), \tag{2.8.5}$$

where the functions $a_\pm(t)$ are non-zero on $\dot{R}^1$ and analytically continued into the upper and lower half-plane of the complex plane $z = t + is$, respectively; here $|a_\pm(z)|$ are bounded above and below by positive constants (cf. corollary 2.5 of §9).

2^0. The question of factorization of functions will be considered further in this and the following sections, but we will introduce here the concept of the generalization of the Wiener-Hopf operator (cf. [50, 52]).

Let P be an orthogonal projector in (separable) Hilbert space, A a linear operator with domain of definition $D(A)$ and domain of values $Im(A)$ then

$$T_p(A) = PAP \tag{2.8.6}$$

is called the generalized Wiener–Hopf operator.

Example 2.2.

$$(Ag)(t) = g(t) - \int\limits_{-\infty}^{\infty} K(t - s)\, g(s)\, ds \,,$$

is acting in $L_2(R^1)$, and projector

$$(Pg)(t) = g_+(t) = \begin{cases} g(t)\,, t > 0\,, \\ 0\,, t < 0\,. \end{cases} \tag{2.8.7}$$

Example 2.3. For a measurable, bounded function $\Phi(t)$ (t), we define the operator

$$(Ag)(t) = (F^{-1}\, \Phi F g)(t)\,, \quad t \in R^n\,.$$

If $ess\ inf \mid \Phi(t) \mid > 0$ for $t \in R^n$, then A maps $L_2(R^n)$ mutually single-valued on to itself.

Lemma 2.1. Let $\varphi \in L_p(R^1)$, $1 \le p \le 2$, $c \in R^1$ and

$$g(x) = e^{-icx}(S_{R^1}\, e^{icy}\, \varphi(y))(x) = \frac{1}{\pi i} \int\limits_{R^1} \frac{e^{ic(y-x)}\varphi(y)\, dy}{y - x}\,,$$

then the Fourier transform of the function g equals $sgn(c - \lambda)F\varphi$.

Proof. The manifold of functions with compact support is a dense manifold in $L_p(R^1)$, hence one may assume that the φ is contained in the segment $[-a, a]$, $a \in R^1$. By the Sokhotskii-Plemelj formulas, we find

$$g(x) = \frac{1}{2\pi i} \lim_{\varepsilon \to 0} \int\limits_{R^1} \left[\frac{e^{ic(y-x)}}{y - x - i\varepsilon} + \frac{e^{ic(y-x)}}{y - x + i\varepsilon} \right] \varphi(y)\, dy\,.$$

Furthermore,

$$(Fg)(\lambda) = \frac{1}{2\pi i} \lim_{N \to \infty} \lim_{\varepsilon \to 0} \int\limits_{-N}^{N} e^{i\lambda x} \int\limits_{-a}^{a} \left[\frac{e^{ic(y-x)}}{y - x - i\varepsilon} + \frac{e^{ic(y-x)}}{y - x + i\varepsilon} \right] \varphi(y)\, dy\, dx =$$

$$= -\frac{1}{2\pi i} \lim_{N \to \infty} \lim_{\varepsilon \to 0} \int\limits_{-a}^{a} e^{i\lambda y}\, \varphi(y) \int\limits_{-N}^{N} \left[\frac{e^{i(\lambda-c)(x-y)}}{x - y + i\varepsilon} + \frac{e^{i(\lambda-c)(x-y)}}{x - y - i\varepsilon} \right] dy\, dx =$$

$$= -\frac{1}{2\pi i} \int\limits_{-a}^{a} e^{i\lambda y}\, \varphi(y) \lim_{N\to\infty} \lim_{\epsilon\to 0} \int\limits_{-N-y}^{N-y} \left[\frac{e^{i(\lambda-c)x}}{x+i\varepsilon} + \frac{e^{i(\lambda-c)x}}{x-i\varepsilon} \right] dy\, dx\,.$$

Using again the Sokhotskii-Plemelj formulas, we obtain

$$(Fg)(\lambda) = -\frac{1}{\pi i} \int\limits_{-a}^{a} e^{i\lambda y}\, \varphi(y) \lim_{N\to\infty} \int\limits_{-N-y}^{N-y} \frac{e^{i(\lambda-c)x}}{x}\, dx\, dy =$$

$$= -\frac{2}{\pi} \int\limits_{-a}^{a} e^{i\lambda y}\, \varphi(y)\, dy \lim_{N\to\infty} \lim_{\epsilon\to 0} \int\limits_{-N}^{N} \frac{sin(\lambda-c)x}{x}\, dx = -sgn(\lambda-c)(F\varphi)(\lambda)\,.$$

Corollary 2.2. Let $P_{R^1} = 1/2(I + S_{R^1})$, $Q_{R^1} = 1/2(I - S_{R^1})$, then

$$P_{R^1} = F^{-1}\left(\frac{1 - sgn\,\lambda}{2}\right) F\,, \quad Q_{R^1} = F^{-1}\left(\frac{1 + sgn\,\lambda}{2}\right) F\,.$$

Example 2.4. The singular integral equation

$$a(t)\varphi(t) + b(t)(S\varphi)(t) = f(t)$$

can be represented, for $a^2 - b^2 \neq 0$, $t \in \dot{R}^1$, in the form

$$(P_{R^1}\Phi P_{R^1}\varphi)(t) = (P_{R^1}(a-b)^{-1}f)(t)\,, \quad \Phi = (a-b)^{-1}(a+b)\,,$$

or, after the introduction of the orthogonal projector $P = F^{-1}P_{R^1}F$, in the form

$$T_P(A)\Psi = P(F^{-1}(a-b)^{-1}f)\,, \quad \Psi = P(F^{-1}\varphi)\,.$$

By corollary 2.1, the operator P has the form (2.8.7).

Some examples of operators of the type $T_P(A)$ have been described in the survey paper [52], where also applications to the diffraction theory of electro-magnetic and elastic waves can be found.

3^0. We will now study a method of solution of the equation (cf. [52])

$$T_P(A)\varphi \equiv PAP\varphi = f \tag{2.8.8}$$

based on factorization of the operator A. We will assume to start with that A is a positive definite operator in the separable Hilbert space H:

$$(Au, u) \geq \delta\|u\|^2\,, \quad u \in H\,, \quad \delta > 0\,.$$

We will denote by $Q = I - P$ the orthogonal projector subsidiary to P and let $f = Pv$. Equation (2.8.8) is equivalent to

$$(Q + AP)u = v \qquad (2.8.9)$$

a more correct, true next assertion.

Corollary 2.3. Let H be a Banach space, A and P linear operators, $P^2 = P$. Then the operator $T_P(A) = PAP$ in the space HB and the operators $T_1 = AP + Q$, $T_2 = AP + Q$ $(Q = I - P)$ in the space H are simultaneously invertible or not, where

$$\dim Ker\, T_P(A) = \dim Ker\, T_i, \ \dim Ker\, T_P(A)^* = \dim Ker\, T_i^*, \ i = 1, 2.$$

Proof. One readily verifies the identities

$$T_1 = AP + Q = (PAP + Q)(I - QAP),$$

$$T_2 = PA + Q = (I + PAQ)(PAP + Q).$$

Furthermore, since $PQ = QP = 0$, then the operators $D_1 = (I + QAP)$ and $D_2 = (I + PAQ)$ are invertible and $D_1^{-1} = (I - QAP)$, $D_2^{-1} = (I - PAQ)$. Consequently,

$$\dim Ker\, T_1 = \dim Ker(PAP + Q), \ \dim Ker\, T_2 = \dim Ker(PAP + Q).$$

If $u = Pv \in Ker\, T_p(A)$, then $u \in Ker(PAP + q) = Ker\, T_1$.

Conversely, if $u \in Ker\, T_1$, then $PAP = 0$ and $Qu = 0$. Hence $u \in PH \cap Ker\, T_p(A)$. The remaining assertions of the corollary are proved in an analogous manner.

Since the operator A in (2.8.8) is positive definite, then the operators $A^{1/2}$ and $A^{-1/2}$ are defined. Let the operator A admit factorization, i.e.,

$$A = A_- A_+ ,$$

where A_- and A_+ are linear continuous operators, invertible in H and $Im(A_+ P) = Im(P)$, $Im(A_- Q) = Im(Q)$. Then equation (2.8.9) can be rewritten

$$(Q + AP)u = A_-(A_-^{-1}Q + A_+ P)u = v \qquad (2.8.10)$$

or, after obvious transformations,

$$A_-^{-1}Qu + A_+ Pu = QA_-^{-1}v + PA_-^{-1}v .$$

Using the orthogonality of the terms $A_-^{-1}Qu$ and A_+Pu, we find the unique solution of (2.8.10)

$$u = A_-QA_-^{-1}v + A_+^{-1}PA_-^{-1}v.$$

The solution of the initial equation (2.8.8) can now be written in the form

$$\varphi = Pu = PA_+^{-1}PA_-^{-1}f = A_+^{-1}PA_-^{-1}f. \qquad (2.8.11)$$

We will now prove that the operator A admits factorization. Let $\{p_j\}$, $\{q_j\}$, $\{p_j^0\}$, $\{q_j^0\}$ be orthonormalized systems in the subspaces PH, QH, $A^{1/2}PH$, $A^{-1/2}QH$, respectively. We define the operators

$$A_+ = UA^{1/2}, \quad A_- = (UA^{-1/2})^{-1},$$

where U is an operator, acting according to the formula

$$Uv = \sum_j (v, p_j^0)p_j + \sum_j (v, q_j^0)q_j.$$

We will verify that $Im(A_+P) = Im(P)$. In fact, for $v = Pu$, we have

$$A_+Pu = UA^{1/2}Pu = \sum_j (A^{1/2}v, p_j^0)p_j + \sum_j (A^{1/2}v, q_j^0)q_j.$$

The terms of the second sum vanish, since

$$(A^{1/2}v, q_j^0) = (v, A^{1/2}q_j^0) = (Pu, A^{1/2}q_j^0) = 0.$$

Thus, we have proved

Theorem 2.14. Let A be a complete operator in separable Hilbert space, H and P orthogonal projectors, then the Wiener-Hopf equation (2.8.8) has the unique solution (2.8.11) for all $f \in Im(P)$.

Proof. The condition of positiveness of A is not necessary for the invertibility of $T_{\dot{p}}(A)$. In order to formulate a criterion of invertibility for $T_{\dot{p}}(A)$, we will employ the concept of the strong elliptic operator. A linear, continuous operator A, acting in Hilbert space H is said to be strongly elliptic, if there is satisfied the condition

$$Re(Au, u) \geq \delta\|u\|^2, \delta > 0, u \in H. \qquad (2.8.12)$$

The reader can find in [50] the next statements.

Theorem 2.15. Let A be a linear continuous operator, invertible in H. The operator $T_{\tilde{p}}(A)$ will be invertible on $Im(P)$ if and only if there exists an operator B, invertible on H, such that $Im(BP) = Im(P)$ and the operator AB is strongly elliptic on H.

Theorem 2.16. Let A be a linear, continuous operator, invertible on H. The operator $T_{\tilde{p}}(A)$ will be invertible on $Im(P)$ if and only if there exist operators $A_{\pm}$, invertible on H, such that $Im(A_+P) = Im(P)$, $Im(A_-Q) = Im(Q)$ with $Q = I - P$ and $A = A_-A_+$ (i.e. in order for the operator A to admit factorization.)

Corollary 2.4. Under the conditions of theorem 2.16,

$$[T_{\tilde{p}}(A)]^{-1} = A_+^{-1} P A_-^{-1} \,.$$

We will consider as an example the operator $A = F^{-1}\Phi F$ in the space $L_2(R^n)$. The condition of strong ellipticity of an operator is equivalent to the following condition: for almost all $\xi \in R^n$, $Re\,\Phi(\xi) \geq \delta > 0$. Hence this condition is sufficient for the invertibility of $T_{\tilde{p}}(A)$ on $Im(P)$, where P is an arbitrary orthogonal projector in $L_2(R^n)$.

A generalization of the results studied here to the case of unbounded operators is given in [55].

§9. Factorization of functions and matrix functions

1^0. The method of factorization, employed in the preceding sections for the solution of equation of the convolution type, is one of the basic analytic methods of solution of the problems of the theory of elasticity (cf. [31, 38, 42]). We will now study several transitional aspects of this method. Another approach and its link to non-linear equations of the form

$$u(x) - \int_0^\infty u(t)\,u(x+t)\,dt = f(x)$$

has been exposed in the survey [1].

Let Γ be a smooth, closed curve on the complex plane, dividing it into two regions: the inner region Ω^+, containing the origin of coordinates, and the outer region Ω^-, containing the point at infinity. We will use the notation $G^\pm = \Omega^\pm \cup \Gamma$.

A factorization of a continuous function $a(z)$, $z \in \Gamma$, is called its representation in the form

$$a(z) = a_-(z)z^\kappa a_+(z), \; z \in \Gamma, \tag{2.9.1}$$

where κ is some integrand and the functions $a_\pm(z)$ admit continuations, analytic in $\Omega^\pm$ and continuous in $G^\pm$, where

$$a_+(z) \neq 0, \; z \in G^+ \, ; \; a_-(z) \neq 0, \; z \in G^- \, .$$

It follows directly from (2.9.1) that

$$ind\, a = \frac{1}{2\pi}\big[arg\, a\big]_\Gamma = \frac{1}{2\pi}\big[arg\, z^\kappa\big]_\Gamma = \kappa \, ,$$

in order for the factorization to be unique. In fact, let

$$a_-^1(z)z^\kappa\, a_+^1(z) = a_-^2(z)z^\kappa\, a_+^2(z) \, ,$$

when

$$a_-^1(z)[a_-^1(z)]^{-1} = a_+^2(z)[a_+^1(z)]^{-1} \, , \; z \in \Gamma \, . \qquad (2.9.2)$$

From the analyticity of $a_-^{1,2}(z)$ in parts of the expanded plane Ω^- follows the existence of constants $c_1, c_2 \neq 0$ such that

$$a_-^1(z) \to c_1 \, , \; a_-^2(z) \to c_2 \, , \; |z| \to \infty \, .$$

Relations (2.9.2) permit determination of a function which is analytic and bounded in the extended complex plane. Then, by Liouville's theorem, one has for some constant c

$$a_+^2(z) = ca_+^1(z) \, , \; a_-^1(z) = ca_-^2(z) \, , \; z \in G^\pm \, .$$

The condition of normalization: $a_-(\infty) = 1$ permits single- valued determination of the constant c.

The class of continuous functions $C(\Gamma)$ on the closed, smooth curve Γ forms a Banach algebra. This means that the complete linear space $C(\Gamma)$ is an algebra with respect to the operation of multiplication which is an agreement with a uniform norm, i.e.,

$$\|a \cdot b\|_{C(\Gamma)} \leq \|a\|_{C(\Gamma)} \cdot \|b\|_{C(\Gamma)} \, .$$

It is known that not every continuous function admits factorization. However, for functions from some subalgebras of the algebra $C(\Gamma)$ factorizations are possible. As an example of such a subalgebra serves $H_\mu(\Gamma)$ - the union of all complex functions, defined on Γ and satisfying a Hoelder condition with index $\mu \in (0,1)$.

Theorem 2.17. Every function $a \in H_\mu(\Gamma)$, $0 < \mu < 1$ which does not vanish anywhere, admits the factorization (2.9.1), where

$$a_\pm(t) \in H_\mu(\Gamma).$$

Proof. For $z \in \Omega^\pm$, we set

$$a_\pm(z) = exp\{\pm b(z)\}, \quad b(z) = \frac{1}{2\pi i} \int\limits_\Gamma \frac{ln\,[t^\kappa a(t)]}{t - z}\, dt.$$

Since $0 \in \Omega^+$, then

$$|\,ln[t_1^{-\kappa} a(t_1)] - ln[t_2^{-\kappa} a(t_2)]\,| \le c\,|\,t_1^{-\kappa} a(t_1) - t_2^{-\kappa} a(t_2)\,| \le$$

$$\le c\{|\,t_1^{-\kappa} - t_2^{-\kappa}\,||\,a(t_1)\,| + |\,t_2^{-\kappa}\,||\,a(t_1) - a(t_2)\,|\} \le c_1\,|\,t_1 - t_2\,|^\mu,$$

where the constant c_1 depends on a, κ and Γ, i.e.,

$$S_\Gamma(ln\,[t^{-\kappa} a(t)]) \in H_\mu(\Gamma),$$

and it follows from the Sokhotskii-Plemelj formulas that $b_\pm(t) \in H_\mu(\Gamma)$ and

$$b_+(t) = \frac{1}{2} ln\,[t^{-\kappa} a(t)] + \frac{1}{2} S_\Gamma\left(ln\,[\tau^{-\kappa} a(\tau)]\right)(t).$$

$$b_-(t) = -\frac{1}{2} ln\,[t^{-\kappa} a(t)] + \frac{1}{2} S_\Gamma\left(ln\,[\tau^{-\kappa} a(\tau)]\right)(t).$$

The fulfillment of (2.9.1) is now readily verified.

2^0. We will now formulate a general criterion (cf. [34]) for the factorization of functions. For this purpose, certain special concepts will be introduced.

Let $R(\Gamma)$ be the union of all rational functions which do not have poles on Γ, and $\mathcal{U}(\Gamma)$ the Banach algebra of the functions from $C(\Gamma)$. Let $R(\Gamma) \subset \mathcal{U}(\Gamma)$ and $R(\Gamma)$ be dense in $\mathcal{U}$ according to the norm $\|\cdot\|_{\mathcal{U}(\Gamma)}$, then $\mathcal{U}(\Gamma)$ is called an R-algebra.

We will denote by $\mathcal{U}(\Gamma)$ the closure of the rational functions which do not have poles inside (outside) Γ according to the norm $\|\cdot\|_{\mathcal{U}(\Gamma)}$; by $\mathcal{U}^\circ$ the closure of the rational functions which decrease at and have poles outside infinity Γ according to the norm $\|\cdot\|_{\mathcal{U}(\Gamma)}$. Obviously, $\mathcal{U}^\circ(\Gamma) \subset \mathcal{U}(\Gamma)$.

Theorem 2.18. Let $\mathcal{U}(\Gamma)$ be some R-algebra. In order that every function $a(z) \in \mathcal{U}(\Gamma)$ shall not vanish anywhere on Γ shall admit the factorization

(2.9.1) with $a^{\pm}(z) \in\in \mathcal{U}^{\pm}(\Gamma)$, it is necessary and sufficient that $\mathcal{U}(\Gamma)$ should disintegrate the algebra, i.e.,

$$\mathcal{U}(\Gamma) = \mathcal{U}^{+}(\Gamma) + \mathcal{U}^{\circ -}(\Gamma).$$

Note 2.4. The Banach algebra $H_{\mu}(\Gamma)$ is a disintegrating algebra, not being an R-algebra (manifold $R(\Gamma)$ not dense in $H_{\mu}(\Gamma)$, $0 < \mu < 1$). However, by theorem 2.17, every function from $H_{\mu}(\Gamma)$ which does not vanish on Γ admits factorization.

In the case $\Gamma = R^{1}$, factorization of functions $f \in C(R^{1})$, having finite limits $f(-\infty) = f(\infty)$, can be defined in the following manner. A function $f(\lambda)$, $\lambda \in R^{1}$, admits factorization, if $f(\lambda(z))$ as a function of z, $\mid z \mid = 1$, is factorisable, where

$$\lambda = -i\,\frac{z+i}{z-i}\,, \quad z = i\,\frac{\lambda - i}{\lambda + i}$$

is the conformal mapping which transforms the unit circle on to the real axis R^{1}, augmented by the point at infinity.

Thus, the role of the regions $\Omega^{\pm}$ play the upper $(Im\,\lambda > 0)$ and lower $(Im\,\lambda < 0)$ half-planes, respectively, and by a factorization $f(\lambda)$, $\lambda \in R^{1}$ is understood a representation in the form

$$a(\lambda) = a_{-}(\lambda)\Big(\frac{\lambda - i}{\lambda + i}\Big)^{\kappa} a_{+}(\lambda)\,, \quad \lambda \in R^{1}\,, \tag{2.9.3}$$

where κ is an integer and the function $a_{+}(\lambda)$ (respectively $a_{-}(\lambda)$) admits continuation, analytic in the open half-plane $Im\,\lambda > 0$ (respectively, $Im\,\lambda < 0$) and continuous in the closed half-plane $Im\,\lambda \geq 0$ (respectively, $Im\,\lambda \leq 0$), where

$$a_{+}(\lambda) \neq 0,\ (Im\,\lambda \geq 0),\ a_{-}(\lambda) \neq 0,\ (Im\,\lambda \leq 0).$$

Analogously, one calls an R-algebra in the case $\Gamma = R^{1}$ a Banach algebra $\mathcal{U}(R^{1}) \subset C(R^{1})$, containing all rational functions with poles off the real axis which form a dense manifold in $\mathcal{U}(R)$. Obviously, the linear shell of functions

$$\Big(\frac{\lambda - i}{\lambda + i}\Big)^{k}\,, \quad k = 0, \pm 1, \pm 2, \ldots, \tag{2.9.4}$$

is dense in the R-algebra. The spaces $\mathcal{U}^{+}(R^{1})$, $\mathcal{U}^{-}(R^{1})$, $\mathcal{U}^{\circ}(R^{1})$ are defined as the closures in the norm $\mathcal{U}(R^{1})$ of the linear shell of functions of the form (2.9.4) for $k = 0, 1, 2, \ldots$, $k = 0, -1, -2, \ldots$ and $k = -1, -2, \ldots$, respectively. Theorem 2.18 yields immediately.

Theorem 2.19. Let $\mathcal{U}(R^1) \subset C(R^1)$ be some R-algebra. In order that every function $a(\lambda) \in \mathcal{U}(R^1)$ which does not vanish on the real axis $(-\infty \leq a \leq +\infty)$ admit factorization in the form (2.9.3) with $a_{\pm}(\lambda)$ and $a_{\pm}^{-1}(\lambda)$ from $\mathcal{U}^{\pm}(R^1)$, it is necessary and sufficient that $\mathcal{U}(R^1)$ be a dissociate algebra, i.e.,

$$\mathcal{U}(R^1) = \mathcal{U}^+(R^1) + \mathcal{U}^{\circ-}(R^1).$$

We will consider an example of a dissociate R-algebra , linked to the Wiener-Hopf equation of §8. We will denote by $W(R^1)$ the class of functions of the form $a(x) = c + (Fk)(x)$, where $c \in C$ is a constant, $k \in L_1(R^1)$, F a Fourier transform.

Theorem 2.20. The Wiener class $W(R^1)$ forms a dissociate Banach algebra with the norm

$$\|a\|_{W(R^1)} = |\,c\,| + \|k\|_{L_1(R^1)}.$$

Proof. Let $a_i(x) = c_i + (Fk_i)(x)$, $i = 1, 2$. Then $a_1 a_2 = c_1 c_2 + F(c_1 k_2 + c_2 k_1) + (Fk_1)(Fk_2)$. It is known that $F(k_1 * k_2) = Fk_1 \cdot Fk_2$, where

$$\int\limits_{-\infty}^{\infty} |\,k_1 * k_2(x)\,|\ dx = \int\limits_{-\infty}^{\infty} \int\limits_{-\infty}^{\infty} |\,k_1(y)k_2(y-x)\,dy\ dx \leq \|k_1\|_{L_1}\|k_2\|_{L_1},$$

hence $a_1 \cdot a_2 \in W(R^1)$. Furthermore

$$\|a_1 a_2\|_{W(R^1)} = |\,c_1 c_2\,| + \|c_1 k_2 + c_2 k_1 + k_1 * k_2\|_{W(R^1)} \leq \|a_1\|_{W(R^1)} \cdot \|a_2\|_{W(R^1)}.$$

We will verify that a function of the form (2.9.4) belongs to $W(R^1)$. We will set

$$f_1(t) = \begin{cases} 2e^{-t}, & t > 0 \\ 0, & t \leq 0 \end{cases}, \quad f_2(t) = \begin{cases} 0, & t \geq 0 \\ 2e^{t}, & t < 0 \end{cases},$$

when

$$\frac{\lambda - i}{\lambda + i} = 1 - Ff_1(\lambda), \quad \frac{\lambda + i}{\lambda - 1} = 1 - Ff_2(\lambda).$$

Note that $(f_1 * f_2)(\lambda) = 2\lambda f_1(\lambda)$, hence the linear shell of functions of the form (2.9.4) comprises all functions of the form

$$c - (Fr)(t), \quad r(\lambda) = f_1(\lambda)\,p_1(\lambda) + f_2(\lambda)\,p_2(\lambda), \tag{2.9.5}$$

where p_1, p_2 are arbitrary polynomials of non-negative degree λ. The manifold of functions $r(\lambda)$ of the form (2.9.5) is dense in $L_1(R^1)$, hence it has been proved that $W(R^1)$ is an R-algebra. We will show that $W(R^1)$ is a dissociate Banach algebra. Functions of the form $F(f_1 p_1)(\lambda)$ and $F(f_2 p_2)(\lambda)$, where p_1, p_2 are arbitrary polynomials, admit analytic continuation into the upper half-plane $Im\,\lambda \geq 0$ and lower half-plane $Im\,\lambda < 0$ and are dense in $W^{\pm}(R^1)$, respectively. The subalgebra $W^{\circ -}(R^1)$ comprises all functions of the form $(Fk)(x)$, where $k(x) \in L_1(R^1)$ and $k(x) = 0$ for $x \geq 0$.

Let $\{a_n(t)\}_{n=1}^{\infty}$ be a sequence of functions of the form (2.9.5)

$$a_n(t) = c_n - F(f_1 p_n + f_2 q_n)(t)$$

which converges in the metric $W(R^1)$ to some function $a(t) \in W(R^1)$, $a(t) = c - Fk$. This means that $c_n \to c$ as a sequence of complex numbers and $f_1 p_n + f_2 q_n \to k$ in $L_1(R^1)$ for $n \to \infty$. It is then obvious that $f_1 p_n \to k_+$ and $f_2 q_n \to k_-$ for $n \to \infty$, where

$$k_+(t) = \begin{cases} k(t)\,, t > 0 \\ \quad 0\,, t \leq 0 \end{cases}, \quad k_- = k - k_+ \,.$$

We have $a = (c - Fk_+) + F(-k_-)$, $c - Fk_+ \in W^+(R^1)$ and $Fk_- \in W^{\circ -}(R^1)$, which completes the proof of theorem 2.20.

Corollary 2.5. Let $k \in L_1(R^1)$ and $a(\lambda) = (1 - Fk)(\lambda) \neq 0$ for $\lambda \in [-\infty, +\infty]$; then $a(\lambda)$ admits the factorization

$$a(\lambda) = a_-(\lambda)\Big(\frac{\lambda - i}{\lambda + i}\Big)^{\kappa} a_+(\lambda)\,,$$

where $a_{\pm}(\lambda) \in W^{\pm}(R^1)$, $a_{\pm}^{-1}(\lambda) \in W^{\pm}(R^1)$.

3^0. The conditions imposed on the function $a(\lambda)$, admitting factorization, have been formulated in theorem 2.17 and corollary 2.5. They reduce to a restriction on the smoothness of the function $a(\lambda) \neq 0$, and non-degeneracy $a(\lambda) \neq 0$, $\lambda \in \Gamma$.

The first of these conditions ($a \in H_{\mu}(\Gamma)$ or $a \in W(R^1)$) does not allow immediate application of factorization to the function $a \in \Pi C(R^1)$, i.e., to functions which have finite limiting values at every point of the real axis, including the point at infinity. For example, the function $sgn(\lambda) \in \Pi C(R^1)$, with discontinuities at the points $\lambda = 0$, $\lambda = \pm\infty$, does not satisfy the conditions of corollary 2.5. The corresponding, generalized Wiener-Hopf operator (cf. §8)

$$T_p(S_R^1)\varphi(t) \equiv \frac{1}{2\pi} \int\limits_{0}^{\infty} \frac{\varphi(s)\,ds}{s - t}\,, \quad S_{R^1}\varphi(t) = -(F^{-1} sgn(s)\,F\varphi)(t)\,,$$

is not invertible in the space $L_2(R^1_+)$. In fact, with the notation $a_+ = a_+(t) = (1 - sgn(t))/2$, $a_- = a_-(t) = (1 - sgn(t))/2$, by corollary 2.2, the invertibility $T_p(S_{R^1})$ is equivalent to the invertibility $S_{R^1} a_+ + a_-$ in the space $L_2(R^1)$. By theorem 2.10, the symbol $S_{R^1} a_+ + a_- \in \Lambda^1_2$ is equal to

$$\mu(S_{R^1} a_+ + a_-) = \mu(S_{R^1}) \cdot \mu(a_+) + \mu(a_-) = \begin{pmatrix} 1 - i[ch(\pi x)]^{-1} \\ 0 - th(\pi x) \end{pmatrix}.$$

At the point $x = 0$, the symbol does not invert, hence, by theorem 2.12, the operator $S_{R^1} a_+ + a_-$ does not invert in $L_2(R^1)$.

For the functions $a \in \Pi C(R^1)$ the concept of generalized p-factorization, $1 < p < \infty$, has been introduced (for details cf. [36, 8]). We will only note a property of generalized p-factorization: Any function $a \in \Pi C(R^1)$ for which ess $inf \mid a(\lambda) \mid > 0$, $\lambda \in R^1$ admits generalized p-factorization

$$a(\lambda) = a_-(\lambda) \Big(\frac{\lambda - i}{\lambda + i} \Big)^\kappa a_+(\lambda), \ \lambda \in R^1,$$

where the functions $a_\pm^{-1}(\lambda)$ satisfy some supplementary conditions, in particular, $a_\pm^{\pm 1}(\lambda) \in L_\infty(R^1)$. Generalized p-factorization is unique; if $a \in W(R^1)$, then $a_\pm \in W(R^1)$, $a_\pm^{-1} \in W(R^1)$ and

$$a_\pm = exp\Big[\frac{1}{2}(I \pm S_{R^1})b \Big], \ b(\lambda) = ln\Big[a(\lambda) \Big(\frac{\lambda - i}{\lambda + i} \Big)^\kappa \Big].$$

A corresponding generalization of theorem 2.17 to the case of functions $a \in \Pi C(R^1)$ is given in [8].

The condition of non-degeneracy of the functions $a(\lambda)$, $\lambda \in \Gamma$ likewise may be weakened. We will consider, as an example, the problem of linear formation of the adjoint on the real axis

$$\Big[16\, sin^4 \frac{w}{2} - v^2 w^2 \Big] \Phi_e(w) + \Big[16\, sin^4 \frac{w}{2} - v^2 w^2 + 4\alpha^2 \Big] \Phi_i(w) = 0, \quad (2.9.6)$$

where $\Phi_e(w)$ and $\Phi_i(w)$ are the boundary values of functions which are analytic in the lower $(Im\, w < 0)$ and upper $(Im\, w > 0)$ half-planes. Let w_k, $\mid k \mid \le k_\alpha$ be the real roots of the equation

$$16\, sin^4 \frac{w}{2} - v^2 w^2 + 4\alpha^2 = 0,$$

numbered in the order of increasing modulus:

$$w_1 \le w_2 \le ... \le w_{k_\alpha}, \ w_{-k} = w_k, \ \mid k \mid \le k_\alpha,$$

and w_k^0, $\mid k \mid \leq k_0$, the roots of the same equation for $a = 0$. We will denote by

$$P_0(w) = \prod_{|k| \leq k_0} (w - w_k^0), \ P_\alpha(w) = \prod_{|k| \leq k_\alpha} (w - w_k)$$

the polynomials of degree $2k_0$ and $2k_\alpha$, respectively. We rewrite equation (2.9.6) in the form

$$\Phi_i(w) = -\Phi_e(w) = \frac{P_0(w)(1 + w^2)^{k_\alpha - k_0}}{P_\alpha(w)} \ a(w), \ w \in R^1 , \qquad (2.9.7)$$

where

$$a(w) = \frac{P_\alpha(w)\left[16 \sin^4 \frac{w}{2} - v^2 w^2\right](1 + w^2)^{k_\alpha - k_0}}{P_0(w)\left[\sin^4 \frac{w}{2} - v^2 w^2 + 4\alpha^2\right]} .$$

Since $a(w) \neq 0$ for $w \in R^1$ and $a(\pm\infty) = 1$, then $a(w)$ admits factorization. Obviously

$$\kappa = \frac{1}{2\pi} \left[arg\ a(w)\right]_{R^1} = 0$$

and the relations

$$a_\pm(w) = exp\{\pm b(w)\}, \ b(w) = \frac{1}{2\pi i} \int\limits_{-\infty}^{\infty} \frac{ln(a(t))}{t - w} \ dt ,$$

are satisfied. We will represent (2.9.7) in the form

$$\frac{\Phi_i(w)(w + i)^{k_0}}{(w + i)^{k_\alpha} \ exp\ b_i(w)} = -\left[\frac{\Phi_e(w)\ P_0(w)}{(w - i)^{2k_0}}\right] \frac{(w - i)^{k_0 + k_\alpha}}{P_\alpha(w)\ exp\ b_e(w)} . \qquad (2.9.8)$$

We will now seek the solution (2.9.6) for the condition that the functions $\Phi_i(w)$ are continuous on the real axis and $\Phi_i(\pm\infty) = 0$. The left hand side of (2.9.8) is analytically continued into the upper half-plane ($Im\ w > 0$). The right hand side of (2.9.8) is analytically continued into the lower half-plane. It follows from the continuity of $\Phi_i(w)$ and $b_i(w)$ on R^1 that $\Phi_e(w)$ has zeroes at the points w_k, $\mid k \mid \leq k_\alpha$, and it may have poles at the points w_k^0, $\mid k \mid \leq k_0$.

Consequently, equation (2.9.8) permits determination of the analytic function in the complex plane which on the real axis permits the estimate $0\left(\mid w \mid^{k_0 - k_\alpha}\right)$. If this function has at infinity a pole and $k_0 > k_\alpha$, then the solution of the problem of the linear adjoint formation (2.9.6) assumes the form

$$\Phi_i(w) = \frac{e^{b_i(w)} P_{k_0-k_\alpha-k_0-1}(w)}{(w+i)^{k_0+k_\alpha}}\,,\quad \Phi_e(w) = \frac{e^{b_e(w)} P_\alpha(w) P_{k_0-k_\alpha-k_0-1}(w)}{P_0(w)(w-i)^{k_0+k_\alpha}}\,,$$

where $P_{k_0-k_\alpha k_0-1}$ is a polynomial of degree not higher than $k_0 - k_\alpha - 1$.

Many results for equations with degenerate symbols are presented in the monographs [5, 34, 46].

4^0. One encounters in applications frequently the case when one has to construct factorization of a matrix function $a(\lambda)$ of dimension $n \times n$. If $C_{n\times n}(\Gamma)$ denotes the Banach algebra of all matrices of order n the elements of which are continuous functions on Γ, and $a(\lambda) \in C_{n\times n}(\Gamma)$, then one calls right factorization $a(\lambda)$ the representation in the form

$$a(\lambda) = a(\lambda)D(\lambda)a_+(\lambda)\,,\ \lambda \in \Gamma\,, \tag{2.9.9}$$

with the diagonal matrix $D(\lambda) = \left\{\lambda^{\kappa_i}\delta_{ij}\right\}_{j,k=1}^{n}$ where $\kappa_1 \geq \kappa_2 \geq ... \geq \kappa_n$ are certain integers and $a_\pm(\lambda) \in C_{n\times n}(\Gamma)$ are square, matrices of order n which admit continuation, analytic in the regions $D_\pm$ and continuous in $G_\pm$, where

$$det\, a_-(z) \neq 0\ (z \in G_+)\,,\ det\, a_-(z) \neq 0\ (z \in G_-)\,.$$

The factorization of the matrix $a(\lambda)$, which is obtained by changing the position of the factors in (2.9.9), is referred to as left factorization.

Theorem 2.21. If the matrix $a(\lambda) \in C_{n\times n}(\Gamma)$ admits right (or left) factorization, then the numbers $\kappa_j = \kappa_j(a)\ (j = 1,...,n)$ are defined uniquely by the matrix $a(\lambda)$.

Proof. Let the matrix $a(\lambda)$ side by side with the factorization (2.9.9) admit another factorization

$$a(\lambda) = \tilde{a}_-(\lambda)\,\tilde{D}(\lambda)\,\tilde{a}_+(\lambda)\,,$$

where $\tilde{D}(\lambda) = \left\{\lambda^{\kappa_i}\delta_{ij}\right\}_{j,k=1}^{n}$. Then one has the equality

$$b_-(\lambda)\,D(\lambda) = \tilde{D}(\lambda)\,b_+(\lambda)\,,$$

in which $b_-(\lambda) = \tilde{a}_-^{-1}(\lambda)\,a_-(\lambda)$, $b_+(\lambda) = \tilde{a}_+(\lambda)\,a_+^{-1}(\lambda)$. Thus, we obtain for the elements $b_{jk}^\pm$ of the matrices $b_\pm(\lambda)$

$$b_{jk}^-(\lambda)\lambda^{\kappa_k} = \lambda^{\tilde{\kappa}_j}\,b_{jk}^+(\lambda)\,,\ j,k = 1,2,...,n\,.$$

It follows from Liouville's theorem that for $\tilde{\kappa}_j > \kappa_k$ the conditions

$$b_{jk}(\lambda) = b_{jk}(\lambda) = 0.$$

are fulfilled. Now let $\kappa_r < \tilde{\kappa}_r$ for some $r(1 \le r \le n)$, then $\kappa_k < \tilde{\kappa}_j$ for $j = 1,...,r$ and $k = r,...,n$. By what has been proved, $b_j k^+ = 0$ for these values of the indices j and k. Expanding $det\, B_+(\lambda)$ by the elements of the first column, we find that $det\, B_+(\lambda) = 0$. This contradiction proves the theorem.

In correspondence with the types of factorization, the integers $\kappa_j\,(j = 1,...,n)$ are called the left and right partial indices of the matrix $a(\lambda)$. It follows from the representation (2.9.9) that the sum

$$\kappa = \sum_{j=1}^{n} \kappa_j = ind(det\, a(\lambda))$$

does not depend on the type of factorization. The number κ is referred to as the (summary) index of the matrix $a(\lambda)$. The left and right partial indices, generally speaking, do not coincide.

Methods for the construction of factorization of matrix functions which arise in the solution of problems of the theory of elasticity are presented in [45, 29].

CHAPTER 3
TWO–DIMENSIONAL SINGULAR INTEGRAL EQUATIONS

The results of the theory of two–dimensional, singular integral equations extend in an obvious manner to a variety of such equations of any dimensionality, located in Euclidean space of any number of dimensions. We will limit ourselves here to the earlier mentioned simpler cases, having in mind the requirements of relevant applications.

§1. Brief survey of results

1^0. First significant work on multi–dimensional, singular integral equations was published by F.Tricomi in 1926 and beyond. He considered double singular integrals, extended over the entire plane. An important step in Tricomi's work was the introduction of the concept of the characteristics of the singular integral; he studied only singular integrals with "constant" characteristic (cf. below); in terms of this concept, the problem of the existence of a singular integral is simply resolved, but the role of this concept under the stated circumstances is not exhausted. For operators, containing integrals with "constant" characteristic, Tricomi proved that their output likewise may be represented in the form of a singular integral and he derived a formula for its characteristic. With the aid of this formula, Tricomi reduced the problem of the solution of a two–dimensional singular equation to the same problem for some one-dimensional singular equation.

2^0. Along another trend was constructed important work on multi-dimensional singular integrals, published by G.Giraud in 1934. He considered singular integrals, extended over a Lyapunov manifold without edge, lying in a Euclidean space of any diemnsionality. For such integrals, Giraud proved a theorem which is analogous to theorem 2.1 for one-dimensional integrals. In the case when the diemensionality is two and the kernel of the singular operator has a sufficiently simple structure, he succeeded in constructing under several additional conditions a regularizator.

3^0. The work of S.G.Mikhlin, published in 1936, considered two–dimensional integrals, extended over the two-dimensional plane or a two–dimensional Lyapunov manifold without edge, with an, in general, "variable" characteristic. He derived for singular operators, containing such integrals, the concept of symbol and obtained formulas for its evaluation. On the basis of the concept of symbol, he succeeded in the construction of a comparatively simple theory of singular, two dimensional integral equations; his important results comprised the following: 1) he stated conditions under which a singular operator is bounded in L_2; 2) under these conditions, if the symbol does not degenrate, the singular operator is Noetherian.

Generalized results of S.G.Mikhlin, G.Giraud published in the same year 1936 without proof of the formula for the symbol of a singular integral in Euclidean space of any dimensionality; a proof was published by Mikhlin in 1955, several years after Giraud's death. In work, published at the end of the Fifties, Mikhlin, using Giraud's formula for the symbol, extended his theory of singular equations to multi-dimensional space.

4^0. Several important results are contained in the work of A.Kalderon and A.Zigmund, published in the Fifties; we note here the condition, imposed on the characteristics, under which a multi-dimensional singular operator is bounded in L_p, $1 < p < \infty$.

E.Stein (1957) and B.A.Plamenevskii (1968) established conditions of boundedness of singular operators in several weighted spaces.

5^0. The problem of the index of a singular scalar or matrix operator with a non-degenerating symbol plays an important role. S.G.Mikhlin has shown in 1938 that the index of a two-dimensional scalar singular operator is zero. He extended his theorem to operators in spaces of any dimensionality in 1956; a number of cases were then established when the index of a matrix singular operator vanished. The question relating to the index of a matrix, scalar, two-dimensional was studied by A.I.Volpert in 1962. A general rule for the evaluation of the index of operators of a very wide class, of which the class of singular integral operators with non-degenerating symbol is a particular case, was given in the significant work of M.F.Atya and I.M.Zinger in 1963. For singular integral operators, the very complicated structure of Atya and Zinger was simplified in the work of P.T.Sili in 1963 and B.V.Fedocov in 1970. However, it must be noted that the structure of these two authors all the same was insuffciently constructive.

The results which have been referred to above are described in sufficient detail in the monograph [54], where a more or less complete bibliography of the problem is presented

§2. Definition and basic properties of two–dimensional integrals

1^0. Let Γ be a Lyapunov surface in three-dimensional Euclidean space, x and y points of this surface and r the distance between these points in the sense of the internal metric of the surface Γ. We will consider the integral (in general, unbounded)

$$\int_{\Gamma} K(x,y)\,u(y)\,d_y\Gamma\,. \tag{3.2.1}$$

where the point $x \in \Gamma$ is fixed, so that integration occurs with respect to y,

$d_y\Gamma$ is the area element of the surface Γ, $u(y)$ and $K(x,y)$ are on Γ measurable functions, $u(y)$ being summable on Γ and the product $\Gamma^2 K(x,y)$ bounded on Γ. We will call the principal value of the integral (3.2.1) the limit, if it exists,

$$\lim_{\varepsilon \to 0} \int\limits_{\Gamma/(r<\varepsilon)} K(x,y)\,u(y)\,d_y\Gamma\,. \tag{3.2.2}$$

As in the case of one dimension, the pricipal value of an integral is frequently referred to as a singular integral.

2^0. We will consider in more detail the following simpler case. Let Γ be a closes surface in the two-dimensional Euclidean plane; this domain may coincide with the entir plane.

We will introduce polar coordinates with centre at the point x; the polar coordinates of the point y will be denoted by (r,θ). We will assume that the function $u(y)$ satisfies a Hoelder condition

$$\mid u(y) - u(x) \mid\, \le\, Cr^\alpha\,;\ C,\,\alpha = const$$

and that the function $K(x,y)$ (the kernel of the singular integral) has the form

$$K(x,y) = r^{-2} f(x,\theta)\,,$$

where f is a bounded, measurable function of θ for any fixed value of x. The integral (3.2.1) assumes the form

$$\int\limits_{\Gamma} \frac{f(x,\theta)}{r^2}\,u(y)\,dy\,, \tag{3.2.3}$$

where dy is the area element at the point y. The function $f(x,\theta)$ is called the characteristic of the singular integral (3.2.3), the function $u(y)$ its density, and the point x its pole. We will say that a characteristic is constant, if it does not depend on the pole; otherwise, we will say that it is variable.

We will now clarify the conditions of the exisence of the integral (3.2.3). By definition,

$$\int\limits_{\Gamma} \frac{f(x,\theta)}{r^2}\,u(y)\,dy = \lim_{\varepsilon \to 0} \int\limits_{\Gamma/(r<\varepsilon)} u(y)\,\frac{f(x,\theta)}{r^2}\,dy =$$

$$= \lim_{\varepsilon \to 0} \int\limits_{\Gamma/(r<\varepsilon)} [u(y) - u(x)]\,\frac{f(x,\theta)}{r^2}\,dy + \lim_{\varepsilon \to 0} u(x) \int\limits_{\Gamma/(r<\varepsilon)} \frac{f(x,\theta)}{r^2}\,dy\,.$$

Let a_x be a number such that the circle $r < a_x$ with its boundary lies entirle within Γ. The second integral on the right hand side may be presented in the form

$$\int\limits_{\Gamma/(r<a_x)} \frac{f(x,\theta)}{r^2}\,dy + \ln\frac{a_x}{\varepsilon}\int\limits_{-\pi}^{\pi} f(x,\theta)\,d\theta\;;$$

hence it is seen that it is necessary for the existence of the singular integral (3.2.3) that

$$\int\limits_{-\pi}^{\pi} f(x,\theta)\,d\theta = 0\,. \tag{3.2.4}$$

Whenever hereafter considering integrals of the form (3.2.3), we will assume that condition (3.2.4) is fulfilled for any $x \in \Gamma$. Condition (3.2.4) was obtained by F.Tricomi under the assumption that the characteristic is constant.

3^0. Let Γ' be an internal subregion of Γ and the smallest distance between points of the boundaries $\partial\Gamma'$ and $\partial\Gamma$ not less than some fixed, positive number δ. If $x \in \Gamma'$, then one can write

$$\int\limits_{\Gamma} \frac{f(x,\theta)}{r^2}\,u(y)\,dy = \int\limits_{\Gamma\cap(r>\delta)} \frac{f(x,\theta)}{r^2}\,u(y)\,dy+$$

$$+ \lim_{\varepsilon\to 0}\int\limits_{\Gamma\cap(r<\delta)} [u(y) - u(x)]\frac{f(x,\theta)}{r^2}\,dy + \lim_{\varepsilon\to 0} u(x)\frac{\delta}{\varepsilon}\int\limits_{-\pi}^{\pi} f(x,\theta)\,\theta\,d\theta =$$

$$= \int\limits_{\Gamma\cap(r>\delta)} \frac{f(x,\theta)}{r^2}\,u(y)\,dy + \int\limits_{r<\delta} \frac{f(x,\theta)}{r^2}\,[u(y) - u(x)]\,dy \tag{3.2.5}$$

and obtain a representation of the singular integral in the form of a sum of two integrals over summable functions. In particular, if $\Gamma = R^2$ and Γ' any finite region, containing the given point x, then one may set $\delta = 1$, and (3.2.5) simplifies somewhat to

$$\int\limits_{R^2} \frac{f(x,\theta)}{r^2}\,u(y)\,dy = \int\limits_{r>1} \frac{f(x,\theta)}{r^2}\,u(y)\,dy + \int\limits_{r<1} \frac{f(x,\theta)}{r^2}\,[u(y) - u(x)]\,dy\,.$$

$$\tag{3.2.6}$$

§3. Singular integrals over Hoelderian functions

1^0. Let the density of the singular integral (3.2.3) at infinity (if the region Γ is infinite) have the estimate $O(x^{-k})$, $k = const > 0$, and in any finite, internal subregion $\Gamma' \subset \Gamma$ the density satisfies Hoelder condition with fixed index α, $0 < \alpha < 1$.

Theorem 3.1. Let the characteristic $f(x,\theta)$ be bounded, and the singular kernel $K(x,y) = r^{-2} f(x,\theta)$ be continuously differentiable for $x \neq y$ and have the estimate

$$\mid grad_x K(x,y) \mid \le C r^{-3} . \tag{3.3.1}$$

Then the integral (3.2.3) satisfies in Γ' a Hoelder condition with index α.

Proof. Denote integral (3.2.3) by $v(x)$, the first and second terms in (3.2.5) by $w_0(x)$ and $w(x)$, respectively. The function $w_0(x)$ has in Γ' a continuous, first derivative with respect to the Cartesian coordinate of the point x and, consequently, satisfies a Hoelder condition with any index smaller than unity.

We will consider the function

$$w(x) = \int\limits_{|x-y|<\delta} K(x,y)[u(y) - u(x)]\, dy .$$

. Let $h \in R^2$ and $\mid h \mid < \delta/3$. Obviously

$$w(x + h) - w(x) = \int\limits_{|x-y|<\delta-|h|} K(x + h, y)[u(y) - u(x + h)]\, dy -$$

$$- \int\limits_{|x-y|<\delta-|h|} K(x, y)[u(y) - u(x)]\, dy +$$

$$+ \int\limits_{(|x-y+h|<\delta)\cap(|x-y|>\delta-|h|)} K(x + h, y)[u(y) - u(x + h)]\, dy -$$

$$- \int\limits_{\delta-|h|<|x-y|<\delta} K(x, y)[u(y) - u(x)]\, dy . \tag{3.3.2}$$

In the last two integrals, the integrands are bounded and the region of integration has the estimate $O(h)$, hence the sum of these integrals has the same estimate. We will now study the first two integrals in (3.3.2). We shall subdivide each of them into two integrals: One along the circle $\mid x - y \mid < 2 \mid h \mid$, the other along the circular ring $2 \mid h \mid < \mid x - y \mid < \delta - \mid h \mid$. We find

$$\mid K(x,y)[u(y) - u(x)] \mid < C r^{-2+\alpha} ,$$

$$\mid K(x+h,y)[u(y) - u(x+h)] \mid < C\, r_1^{-2+\alpha} , \quad r_1 = \mid x + h - y \mid ,$$

hence

$$\left| \int\limits_{r<2|h|} K(x,y)[u(y) - u(x)]\, dy \right| \le C \int\limits_{r<2|h|} r^{\alpha-2}\, dy = C_1 \mid h \mid^{\alpha} ,$$

$$\left| \int\limits_{r<2|h|} K(x+h,y)[u(y) - u(x+h)]\, dy \right| < \qquad (3.3.3)$$

$$< C \int\limits_{r<2|h|} r_1^{\alpha-2}\, dy < C \int\limits_{r_1<.3|h|} r_1^{\alpha-2}\, dy = C_1 \mid h \mid^{\alpha} .$$

We will join the integrals over the ring $2 \mid h \mid < \mid x - y \mid < \delta - \mid h \mid$ and transform the integrand:

$$K(x+h,y)[u(y) - u(x+h)] - K(x,y)[u(y) - u(x)] =$$

$$= [K(x+h,y) - K(x,y)][u(y) - u(x+h)] - K(x,y)[u(x+h) - u(x)] .$$

During integration over the ring with centre at x, the integral then vanishes from above as a result of (3.2.4). In order to estimate the integral from below, we note that , by Taylor's formula,

$$K(x+h,y) - K(x,y) = (grad_x\, K(\xi,y), h) ,$$

where ξ is some point of the segment $(x, x + h)$, and, by (3.3.1), $\mid K(x+h,y) - K(x,y) \mid \le C \mid h \mid\mid \xi - y \mid^{-3}$. Moreover, $\xi = x + \nu h$, $0 < \nu < 1$ and $\mid \xi - y \mid > r - \mid h \mid$; in the region of integration $r > 2 \mid h \mid$, hence $\mid \xi - y \mid > r/2$. The function $u(x)$ satisfies a Hoelder condition and therefore

$$\mid K(x+h,y) - K(x,y) \mid \cdot \mid u(y) - u(x+h) \mid \le$$

$$\le C \mid h \mid\mid \xi - y \mid^{-3}\mid x + h - y \mid^{\alpha} \le 8C \mid h \mid r^{-3} (r + \mid h \mid)^{\alpha} , \qquad (3.3.4)$$

where $\alpha < 1$ and $(r+ \mid h \mid)^\alpha \leq r^\alpha + \mid h \mid^\alpha$; this expression has the estimate $8C \mid h \mid (r^\alpha + \mid h \mid^\alpha)^{-3}$. The integral of this expression over the ring does not exceed in modulus

$$16\pi C \mid h \mid \int\limits_{2|h|}^{\delta-|h|} r^{\alpha-2}\, dr + 16\pi C \mid h \mid^{1+\alpha} \int\limits_{2|h|}^{\delta-|h|} r^{-2}\, dr \leq C_1 \mid h \mid^\alpha,$$

which proves theorem 3.1.

We note that without difficulties, this theorem may be extended to Lyapunov manifolds. For this purpose, let Γ be a manifold without edge, then the singular integral satisifes a Hoelder condition with index α in the entire manifold.

Theorem 3.1. does not extend to the case $\alpha = 1$, when one can only assert that the singular integral satisfies a Hoelder condition with any index less than unity.

2^0. Let $\Gamma = R^2$. We will introduce into the consideration the class of funtions $A_{\alpha,k}$, defined in R^2 and satisfying the inequalities

$$\mid u(y) - u(x) \mid \leq A(x^2 + 1)^{-k/2}\, r^\alpha\,, \ r \leq 1\,, \tag{3.3.5}$$

$$\mid u(x) \mid \leq B(x^2 + 1)^{-k/2}\,. \tag{3.3.6}$$

where A, B, α, k are positive constants, $0 < \alpha < 1$, $0 < k < 2$, $x^2 = \mid x \mid^2$. We will now denote by $A'_{\alpha,k}$ another class of functions, defined in R^2 and satisfying the inequality (3.2.5) as well as the inequality (for suffiently large $\mid x \mid$)

$$\mid u(x) \mid \leq B(x^2 + 1)^{-k/2}\, ln(x^2 + 1)\,, \tag{3.3.7}$$

with bounds on the quantities A, B, α, k the same as on those for the class $A_{\alpha,k}$.

Theorem 3.2. Let the singular kernel $K(x,y) = r^{-2}\, f(x,\theta)$ be defined for all $x,y \in R^2$, $x \neq y$ and satisfy inequality (3.3.1). Then the singular operator

$$v(x) = \int\limits_{R^2} K(x,y)\, u(y)\, dy \tag{3.3.8}$$

transforms any function $u \in A_{\alpha,k}$ into a function $v \in A'_{\alpha,k}$.

We can establish the fact that Function (3.3.8) satisfies condition (3.3.5) by introducing into the reasoning, proving theorem 3.1, a Hoelder constant equal to $A(x^2+1)^{-k/2}$; this is admissible, because in that reasoning the point x is fixed.

There remains to prove that for sufficiently large $\mid x \mid$ inequality (3.3.7) is satisfied. We will denote the terms on the right hand side of (3.2.6) by $v_1(x)$ and $v_2(x)$. By inequality (3.3.5)

$$\mid v_2(x) \mid \le \frac{A'}{(x^2+1)^{k/2}} \int\limits_{r<1} \frac{dy}{r^{2-\alpha}} = \frac{A''}{(x^2+1)^{k/2}} \, ; \quad A', A'' = const. \qquad (3.3.9)$$

Moreover, by inequality (3.3.6)

$$\mid v_1(x) \mid \le B' \int\limits_{r>1} r^{-2}(r^2+1)^{-k/2} \, dy \, , \quad B' = const.$$

We introduce now the polar coordinates r, θ with centre at the point x. Then

$$\mid v_1(x) \mid \le B' \int\limits_{-\pi}^{\pi} d\theta \int\limits_{1}^{\infty} r^{-1}(r^2+x^2+1-2r\mid x\mid \cos\theta)^{-k/2} \, dr \, ,$$

or, with the notation $\sigma = 2r\mid x\mid /(r^2+x^2+1)$,

$$\mid v_1(x) \mid \le 2B' \int\limits_{1}^{\infty} \frac{dr}{r(r^2+x^2+1)^{k/2}} \int\limits_{0}^{\pi} \frac{d\theta}{(1-\sigma\cos\theta)^{k/2}} \, . \qquad (3.3.10)$$

Furthermore,

$$\int\limits_{0}^{\pi} \frac{d\theta}{(1-\sigma\cos\theta)^{k/2}} = \int\limits_{0}^{\pi/2} \frac{d\theta}{(1-\sigma\cos\theta)^{k/2}} + \int\limits_{0}^{\pi/2} \frac{d\theta}{(1+\sigma\cos\theta)^{k/2}} \, . \qquad (3.3.11)$$

The second term on the right hand side of (3.3.11) does not exceed $\pi/2$. If $k < 1$, then, since $\sigma < 1$, also the first term is bounded, and we arrive at the estimate for the first integral in (3.3.10):

$$\int\limits_{1}^{\infty} \frac{dr}{r(r^2+x^2+1)^{k/2}} = \frac{dt}{(x^2+1)^{k/2}} \int\limits_{(x^2+1)^{-1/2}}^{\infty} \frac{dt}{t(t^2+1)^{k/2}} \le C\frac{ln\,(x^2+1)}{(x^2+1)^{k/2}} \, .$$

$$(3.3.12)$$

Inequalities (3.3.9) - (3.3.12) prove theorem 3.2 for the case $k > 1$.

Now let $1 < k < 2$. Setting $t = tg\,\theta/2$, $a^2 = (1 - \sigma)/(1 + \sigma)$ we find

$$\int_0^{\pi/2} \frac{d\theta}{(1 - \sigma\cos\theta)^{k/2}} < C_1 \int_0^1 \frac{dt}{(a^2 + t^2)^{k/2}} < \frac{C_2}{a^{k/2}} \int_0^\infty \frac{dz}{(1 + z^2)^{k/2}} <$$

$$< C_3(1 - \sigma)^{-\varepsilon}, \quad \varepsilon = (k - 1)/2.$$

Integral (3.3.11) is estimated by $\pi/2 + C(1-\sigma)^{-\varepsilon}$. We substite this result into (3.3.10). The expression estimating $v_1(x)$, subdivided into two parts one of which is estimated by (3.3.12), the other, exactly apart from a constant factor, has the form

$$I = \int_1^\infty \frac{dr}{r(r^2 + x^2 + 1)^{k/2}(1 - \sigma)^\varepsilon} = \int_1^\infty \frac{dr}{r(r^2 + x^2 + 1)^{1/2}[1 + (r - \mid x \mid)^2]^\varepsilon}.$$

We apply now Hoelder's inequality with an arbitrary index p.

$$I \leq \left\{ \int_1^\infty \frac{dr}{r(r^2 + x^2 + 1)^{p'/2}} \right\}^{1/p'} \cdot \left\{ \int_1^\infty \frac{dr}{r[1 + (r - \mid x \mid)^2]^{p\varepsilon}} \right\}^{1/p}. \qquad (3.3.13)$$

The first integral on the right hand side is estimated by (3.3.12) in which one must replace k by p':

$$\left\{ \int_1^\infty \frac{dr}{r(r^2 + x^2 + 1)^{p'/2}} \right\}^{1/p'} \leq C \frac{[ln\,(x^2 + 1)]^{1/p'}}{(x^2 + 1)^{1/2}}. \qquad (3.3.14)$$

We subdivide the second integral in (3.3.13) into two parts $(1, \mid x \mid)$ and $(\mid x \mid, \infty)$; we denote them by I' and I'', repsectively. The number p will be chosen so close to unity that $2p\varepsilon < 1$, and make then in I'' the substitution $r = \mid x \mid (1 + t)$:

$$I'' = \int_0^1 \frac{dt}{(1 + t)(1 + x^2 t^2)^{p\varepsilon}} + \int_1^\infty \frac{dt}{(1 + t)(1 + x^2 t^2)^{p\varepsilon}} <$$

$$< \int_0^1 \frac{dt}{(x^2 t^2)^{p\varepsilon}} + \int_1^\infty \frac{dt}{t^{1 + 2p\varepsilon}\, x^{2p\varepsilon}} = O((1 + x^2)^{-p\varepsilon}). \qquad (3.3.15)$$

In the integral I', we make the substitution $r = (1-t) \mid x \mid$:

$$I' = \int_0^{1/2} \frac{dt}{(1-t)(1+x^2t^2)^{p\varepsilon}} + \int_{1/2}^{1-1/|x|} \frac{dt}{(1-t)(1+x^2t^2)^{p\varepsilon}}\,.$$

We now estimate the first integral:

$$\int_0^{1/2} \frac{dt}{(1-t)(1+x^2t^2)^{p\varepsilon}} < 2 \int_1^{1/2} \frac{dt}{(1+x^2t^2)^{p\varepsilon}} =$$

$$= \frac{2}{\mid x \mid} \left[\int_0^1 \frac{dz}{(1+z^2)^{p\varepsilon}} + \int_1^{|x|/2} \frac{dz}{(1+z^2)^{p\varepsilon}} \right] < \frac{2}{\mid x \mid} \left[1 + \int_1^{|x|/2} z^{-2p\varepsilon}\, dz \right] =$$

$$= O(\mid x \mid^{-2p\varepsilon}) = O((1+x^2)^{-p\varepsilon})\,.$$

$$(3.3.16)$$

The estimate for the second integral is some what simpler:

$$\int_{1/2}^{1-1/|x|} \frac{dt}{(1-t)(1+x^2t^2)^{p\varepsilon}} < (1+x^2/4)^{-p\varepsilon} \int_{1/2}^{1-1/|x|} \frac{dt}{(1-t)} = \qquad (3.3.17)$$

$$= O[(x^2+1)^{-p\varepsilon}\, ln\,(x^2+1)]\,.$$

We now obtain from (3.3.12) - (3.3.17)

$$\mid v_1(x) \mid \le C \frac{ln\,(x^2+1)}{(x^2+1)^{k/2}}\,. \qquad (3.3.18)$$

The assertion of the theorem follows from Inequalities (3.3.9)- (3.3.18).

There remains to consider the case $k = 1$. The first integral on the right hand side in (3.3.11) then equals

$$\int_0^{\pi/2} \frac{d\theta}{\sqrt{1-\sigma\cos\theta}} = \int_0^1 \frac{dt}{\sqrt{(1-t^2)(1-\sigma t)}} < \int_0^1 \frac{dt}{\sqrt{(1-t)(1-\sigma t)}} =$$

$$= \frac{1}{2\sqrt{\sigma}}\, ln\, \frac{1+\sqrt{\sigma}}{1-\sqrt{\sigma}}\,.$$

The last expression is bounded as $\sigma \to 0$; is has $O(|\, 1 - \sigma\, |^{\eta})$, where η is any positive number, for $\sigma \to 1$. The further reasoning is as before.

Corollary 3.1. If $u \in A_{\alpha,k}$, then $v \in A_{\alpha,k'}$, for any k', $0 < k' < k$.

Corollary 3.2. If $u \in A_{\alpha,k}$ and the singular kernel

$$K_n(x,y) = r^{-2}\, f_n(x,\theta)$$

satisfies the conditions of theorem 3.2, then the function $v_n(x)$, determined by

$$v_0(x) = u(x), \quad v_n(x) = \int\limits_{R^2} K_n(x,y)\, v_{n-1}(y)\, dy\,,$$

belongs to the class $A_{\alpha,k'}$ for any $k' < k$.

§4. Differentiation of integrals with weak singularities

1^0. Let Γ be a domain in the two-dimensional plane and

$$w(x) = \int\limits_{\Gamma} u(y)\, \frac{\varphi(x,\theta)}{r}\, dy\,, \quad x \in \Gamma\,. \tag{3.4.1}$$

We will assume that the function $\varphi(x,\theta)$ is continuous and bounded together with its first derivatives with respect to the Cartesian coordinates of the point x and θ, when x covers the closed domain Γ, and θ is the circumference of radius unity with centre at the origin of coordinates. In addition, we will assume that $u(y)$ satisfies a Hoelder condition and that at infinity (if the domain Γ is infinite) $u(x) = O(|\, x\, |^{l})$, $l = const > 1$. We will show that under these conditions the first derivative of $w(x)$ is determined by

$$\frac{\partial}{\partial x_k} \int\limits_{\Gamma} u(y)\, \frac{\varphi(x,\theta)}{r}\, dy = \int\limits_{\Gamma} u(y)\, \frac{\partial}{\partial x_k}\left[\frac{\varphi(x,\theta)}{r}\right] dy -$$
$$- u(x) \int\limits_{-\pi}^{\pi} \varphi(x,\theta)\, cos(r, x_k)\, d\theta\,, \quad k = 1, 2\,, \tag{3.4.2}$$

where the first integral on the right hand side is singular.

Let $x \in \Gamma$ and $\varepsilon > 0$ be a number which is smaller than the distance from the point x to the boundary of the domain Γ. We surround the point x

by a circle with radius $r < \varepsilon$ and set $\Gamma \setminus (r < \varepsilon) = \Gamma_\varepsilon$. We will now construct the derivative

$$\frac{\partial}{\partial x_k} \int_{\Gamma_\varepsilon} u(y)\, \frac{\varphi(x,\theta)}{r}\, dy = \int_{\Gamma_\varepsilon} u(y)\, \frac{\partial}{\partial x_k} \left[\frac{\varphi(x,\theta)}{r}\right] dy -$$

$$- \int_{r=\varepsilon} u(y)\, \varphi(x,\theta)\, cos(r, x_k)\, d\theta\,.$$

Let D_k' denote the derivative with respect to x_k, computed under the assumption that r and θ do not depend on x, D_k'' the derivative with respect to x_k, computed under the assumption that only r and θ depend on x. Then the last formula becomes

$$\frac{\partial}{\partial x_k} \int_{\Gamma_\varepsilon} u(y)\, \frac{\varphi(x,\theta)}{r}\, dy = \int_{\Gamma_\varepsilon} u(y)\, D_k'' \left[\frac{\varphi(x,\theta)}{r}\right] dy +$$

$$+ \int_{\Gamma_\varepsilon} D_k'[\varphi(x,\theta)]\, \frac{u(y)}{r}\, dy - \int_{r=\varepsilon} u(y)\, \varphi(x,\theta)\, cos(r, x_k)\, d\theta\,. \tag{3.4.3}$$

If $\varepsilon \to 0$, then the sum of the second and thrid integrals on the right hand side of (3.4.3) tend uniformly to the limit

$$\int_{\Gamma} D_k'[\varphi(x,\theta)]\, \frac{u(y)}{r}\, dy - u(x) \int_{-\pi}^{\pi} \varphi(x,\theta)\, cos(r, x_k)\, d\theta\,.$$

We will consider next the first integral. Its kernel, may be given the form $D_k''[r^{-1}\, \varphi(x,\theta)] = r^{-2}\, f_k(x,\theta)$. Simple manipulatons yield

$$f_1(x,\theta) = \frac{\partial}{\partial\theta}\left[\varphi(x,\theta)\, sin\,\theta\right],\quad f_2(x,\theta) = -\frac{\partial}{\partial\theta}\left[\varphi(x,\theta)\, cos\,\theta\right],$$

and, clearly, both functions $f_k(x,\theta)$ satisfy condition (3.2.4). In that case, the first integral on the right hand side of (3.4.3) tends uniformly for $\varepsilon \to 0$ to the singular integral

$$\int_{\Gamma} \frac{f_k(x,\theta)}{r^2}\, u(y)\, dy\,,$$

hence it is seen that for $\varepsilon \to 0$ the right hand side of (3.4.3) tends to

$$\int\limits_{\Gamma} \frac{\partial}{\partial x_k}\left[\frac{\varphi(x,\theta)}{r}\right] u(y)\, dy - u(x) \int\limits_{-\pi}^{\pi} \varphi(x,\theta)\cos(\Gamma, x_k)\, d\theta\,,\ \ k = 1,2\,,$$

uniformly in any closed sub-region of Γ. At the same time,

$$\lim_{\epsilon \to 0} \int\limits_{\Gamma_\epsilon} \frac{\varphi(x.\theta)}{r}\, u(y)\, dy = w(x)$$

uniformly in the same sub-region, hence follows the correctness of (3.4.2). Note that $cos(r, x_1) = cos\,\theta$ and $cos(r, x_2) = sin\,\theta$.

§5. Singular integrals in the space of square summable functions

1^0. We will consider in this section singular integrals, extended over the entire plane R^2, with densities in the class $L_2(R^2)$. A study of double (and, in general, many-dimensional) singular integrals in a space of square summable functions requires a certain generalization of the concept of a singular integral; we will discuss this aspect below. In this context, an important role play the properties of Fourier transforms in R^2, which we will now discuss; a detailed study of these questions can be found in [16], and likewise in the literature which we will cite.

The Fourier transform of the function $u(x)$, specified and summable in the entire space R^m of arbitrary dimensionality m, is give by the integral

$$\forall \xi \in R^m\,,\ \tilde{u}(\xi) = (2\pi)^{-m/2} \int\limits_{R^m} u(x)\, e^{-i(x,\xi)}\, dx\,.$$

At times, the definition of the Fourier transform changes somewhat; the factor $(2\pi)^{-m/2}$ is omitted and one writes

$$\tilde{u}(\xi) = \int\limits_{R^m} u(x)\, e^{-i(x,\xi)}\, dx\,.$$

In fact, such a definition proves to be more convenient for our purposes, hence we will use it exclusively in what follows. Besides, since we will only be interested in the case $m = 2$, we will use the following definition.

We will call Fourier transform of the function $u(x)$, given and summable throughout the plane R^2, the integral

$$\tilde{u}(\xi) = \int_{R^2} u(x)\, e^{-i(x,\xi)}\, dx\,. \tag{3.5.1}$$

Fourier transform will likewise be called the operator, generated by (3.5.1). Obviously, we can use for this operator the notation

$$\tilde{u}(\xi) = (\mathcal{F}u)(\xi) = \mathcal{F}_{x\to\xi}u\,. \tag{3.5.2}$$

In particular, if the function $u(x)$ has a compact support, then it will vanish outside some circle $|x| < N$, when

$$\tilde{u}(\xi) = \int_{|x|<N} u(x)\, e^{-i(x,\xi)}\, dx\,.$$

We will now assume that the function $u(x)$, haviung a compact support, is square summable on it. Such functions belong to the space $L_2(R^2)$ and form in it a dense manifold. For those functions, Plancherl's equality

$$\|\tilde{u}\|_{L_2(R^2)} = 2\pi\, \|u\|_{L_2(R^2)}\,. \tag{3.5.3}$$

holds true. Thus, a Fourier transform is an operator, defined on a manifold, with density in $L_2(R^2)$ and bounded on this manifold. However, this operator is then extended continuously throughout the space $L_2(R^2)$. Under such conditions, one can defined the Fourier transform of functions from $L_2(R^2)$ by the formula

$$\tilde{u}(\xi) = \mathcal{F}_{x\to\xi}u = \lim_{N\to\infty} \int_{|x|<N} u(x)\, e^{-i(x,\xi)}\, dx\,, \tag{3.5.4}$$

Where the limit is performed in the sense of convergence in $L_2(R^2)$. In that case, there remains, of course, true also the relation (3.5.3) which, in particular, yields the existence of the operator, inverse to the Fourier transform operator which is naturally denoted by $\mathcal{F}^{-1}_{\xi\to x}$; obviously, its norm is $1/2\pi$. The inverse Fourier transform is determined by the simple formula

$$\mathcal{F}^{-1}_{\xi\to x}\tilde{u} = (2\pi)^{-2}\,(\mathcal{F}\tilde{u})(x) = \frac{1}{(2\pi)^2}\int_{R^2} \tilde{u}(\xi)\, e^{i(x,\xi)}\, d\xi\,. \tag{3.5.5}$$

2^0. One calls the convolution of the two functions $g(x)$ and $h(x)$, summable in R^2, the integral

$$(g * h)(x) = \int_{R^2} g(x - y)\, h(y)\, dy\,;$$

obviouly, $g * h = h * g$. It is readily shown that

$$\mathcal{F}(g * h) = \mathcal{F}g\,\mathcal{F}h\,. \tag{3.5.6}$$

The concept of convolution and formula (3.5.6) are easily extended to certain classes of functions which are not necessarily summable in R^2. In particular, singular integrals with constant characteristics, extended over R^2, can be approached formally under the concept of convolution; such an integral has the form

$$\int_{R^2} u(y) r^{-2}\, f(\theta)\, dy = \int_{R^2} K(x - y)\, u(y)\, dy\,;$$

$$K(x - y) = \mid y - x \mid^{-2}\, f\!\left(\frac{y - x}{\mid y - x \mid}\right). \tag{3.5.7}$$

It represents formally a convolution of the functions $K(x)$ and $u(x)$.

The Fourier transform of a singular kernel is defined by a formula which differs somewhat from (3.5.4):

$$\mathcal{F}_{x \to \xi} K = \lim_{\varepsilon \to 0, N \to \infty} \int_{\varepsilon < |x| < N} K(x)\, e^{-i(x,\xi)}\, dx\,, \tag{3.5.8}$$

note that the analogous formula

$$\mathcal{F}_{x \to \xi} u = \lim_{\varepsilon \to 0, N \to \infty} \int_{\varepsilon < |x| < N} u(x)\, e^{-i(x,\xi)}\, dx$$

is true also for the functions $u(x)$ described in 1^0.

In the sequel, an essential role is played by the Fourier transform for the particular case of the singular kernel $K(x - y) = r^{-2}\, e^{in\theta}$, where n is a positive or negative integer. The value $n = 0$ is excluded, because the characteristic $f(\theta) = 1$ does not satisfy the condition of (3.2.4).

3^0. **Theorem 3.3.** Let the characteristic $f(\theta)$ be constant, bounded and measurable on the segment $0 \leq \theta \leq 2\pi$, and the density of the singular integral $u \in A_{\alpha,k}$, $1 < k < 2$. Let, in addition,

$$v(x) = \int_{R^2} u(y)\, r^{-2}\, f(\theta)\, dy\,. \tag{3.5.9}$$

Then exists the function $\Phi_f(\nu)$, independent of $u(x)$ and continuous for $0 \leq \nu \leq 2\pi$, such that

$$\tilde{v}(\xi) = \mathcal{F}_{x \to \xi} v = \Phi_f(\nu)\, \tilde{u}(\xi)\,; \quad \tilde{u}(\xi) = \mathcal{F}_{x \to \xi} u\,, \tag{3.5.10}$$

with ν the polar angle of the point ξ with respect to the origin of coordinates.

We note that under the conditions of the theorem the function $f(\theta)$ is square summable on the segment $(0, 2\pi)$ and therefore expandable in a Fourier series which converges in the norm $L_2(0, 2\pi)$.

We will consider to start with the particular case $f(\theta) = e^{in\theta}$, where n is a non-zero natural number. Then $K(x - y) = r^{-2}\, e^{in\theta}$ and $K(x) = |x|^{-2} e^{in\theta}$; in this case, one must understand by θ the polar angle of the origin of coordinates relative to the point x. We will introduce polar coordinates with pole at the origin of coordinates and with the polar axis directed along the x_1 axis (Fig.4). Let (t, ω) and (ρ, ν) be the polar coordinates of the points x and ξ, respectively. Obviously, $\theta = \omega + \pi$. We find

$$\mathcal{F}_{x \to \xi} K = \lim_{\varepsilon \to 0, N \to \infty} \int_{\varepsilon}^{N} \frac{dt}{t} \left\{ \int_{0}^{2\pi} e^{in(\omega+\pi)} \cdot e^{-it\rho\cos(\omega-\nu)}\, d\omega \right\}\,.$$

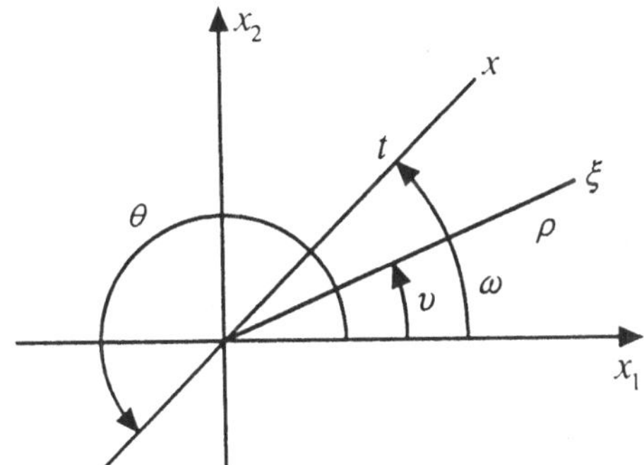

Fig. 4

We will assume that $\rho > 0$. With the substitution $t\rho = t'$ and notation $\varepsilon' = \varepsilon\rho$, $N' = N\rho$ we find

$$\mathcal{F}_{x\to\xi}K = (-1)^n \lim_{\varepsilon\to 0, N\to\infty} \int_{\varepsilon'}^{N'} \frac{dt'}{t'} \left\{ \int_{-\pi}^{\pi} e^{in\omega - it'\cos(\omega-\nu)}\, d\omega \right\}.$$

We may omit the dashes with t', ε', N' and write

$$\mathcal{F}_{x\to\xi}K = (-1)^n \lim_{\varepsilon\to 0, N\to\infty} \int_{\varepsilon}^{N} \frac{dt}{t} \left\{ \int_{-\pi}^{\pi} e^{in\omega - it\cos(\omega-\nu)}\, d\omega \right\}.$$

By a simple transformation, we may reduce the integral to the form

$$2i^{-n}\, e^{in\nu} \int_{0}^{\pi} \cos(n\omega - t\sin\omega)\, d\omega = 2\pi i^{-n}\, e^{in\nu}\, I_n(t);$$

the last equality follows from the known formula of the theory of Bessel functions ([2]. p.29)

$$I_n(t) = \frac{1}{\pi} \int_{0}^{\pi} \cos(n\omega - t\sin\omega)\, d\omega .$$

The integral of interest has now the form

$$2\pi i^n\, e^{in\nu} \lim_{\varepsilon\to 0, N\to\infty} \int_{\varepsilon}^{N} \frac{I_n(t)}{t}\, dt .$$

It follows from well known properties of Bessel functions that the ration $I_n(t)/t$ for natural n is summable on $(0, \infty)$, when the last expression becomes

$$2\pi i^n\, e^{in\nu} \int_{0}^{\infty} \frac{I_n(t)}{t}\, dt .$$

We will use the formula ([2], p. 428)

$$\int_{0}^{\infty} \frac{I_\nu(t)\, dt}{t^{\nu - \mu + 1}} = \frac{\Gamma(\mu/2)}{2^{\nu - \mu + 1}\, \Gamma(\nu + 1 - \mu/2)},$$

which holds true, if $0 < Re(\mu) < Re(\nu + 1/2)$. Setting $\mu = \nu = n$, we obtain

$$\int\limits_0^\infty \frac{I_n(t)}{t}\, dt = \frac{1}{n}$$

and, finally, for natural n,

$$\mathcal{F}_{x\to\xi}K = \frac{2\pi i^n}{n}\, e^{in\nu}\,, \quad K(x-y) = r^{-2}\, e^{in\theta}\,. \tag{3.5.11}$$

5^0. Now let $f(\theta) = e^{in\theta}$, where n is a negative integer. As before, we find that now

$$\mathcal{F}_{x\to\xi}K = 2\pi i^{|n|}\, e^{in\nu} \lim_{\varepsilon\to 0,N} \int\limits_\varepsilon^N \frac{I_n(t)}{t}\, dt = \frac{2\pi i^{|n|}}{|n|}\, e^{in\nu}\,, \quad n \le -1\,. \tag{3.5.12}$$

Obviously, this formula remains also true for positive integers n; we find now the following, more general formula: if $K(x-y) = r^{-2}\, e^{in\theta}$ and n is any non-zero integer, then

$$\mathcal{F}_{x\to\xi}K = \frac{2\pi i^{|n|}}{|n|}\, e^{in\nu}\,. \tag{3.5.13}$$

6^0. The Fourier transform - linear operator. It follows from (3.5.13) that, if

$$K(x-y) = r^{-2} \sum_{k=-n}^{n} a_k\, e^{ik\theta}\,, \quad a_0 = 0\,,$$

then

$$\mathcal{F}_{x\to\xi} = 2\pi \sum_{k=-n}^{n} \frac{i^{|k|}}{|k|}\, a_k\, e^{ik\nu}\,. \tag{3.5.14}$$

7^0. Let a characteristic be bounded and expand it in the infinite, uniformly convergent Fourier series

$$f(\theta) = \sum_{k=-\infty}^{+\infty} a_k\, e^{ik\theta}\,, \quad a_0 = 0\,. \tag{3.5.15}$$

We will introduce the notation

$$f_n(\theta) = \sum_{k=-n}^{n} a_k\, e^{ik\theta}\,, \quad \Phi_{f_n}(\nu) = 2\pi \sum_{k=-n}^{n} \frac{i^{|k|}}{|k|}\, a_k\, e^{ik\nu}\,,$$

$$v_n(x) = \int\limits_{R^2} u(y)\, r^{-2}\, f_n(\theta)\, dy\,, \tag{3.5.16}$$

when

$$\mathcal{F}_{x\to\xi}\, v_n = \Phi_{f_n}(\nu)\, \mathcal{F}_{x\to\xi} u\,. \tag{3.5.17}$$

From the reasoning of the proof of theorem 3.1, we see that $v_n(x) \longrightarrow v(x)$ uniformly, and likewise, in the metric of $L_2(R^2)$ ($v(x)$ is the singular integral (3.2.3)), $\mathcal{F}_{x\to\xi} v_n \longrightarrow \mathcal{F}_{x\to\xi} v$ in $L_2(R^2)$. Moreover, there exists the limit

$$\Phi_f(\nu) = \lim_{n\to\infty} \Phi_{f_n}(\nu) = 2\pi \sum_{k=-\infty}^{+\infty} \frac{i^{|k|}}{|\,k\,|}\, a_k\, e^{ik\nu}\,. \tag{3.5.18}$$

Taking the limit $n \to \infty$ in (3.5.17), we find identity (3.5.10).

8^0. Formula (3.5.10) yields a simple formula for the singular integral with constant characteristic, which satisfies the conditions 3.3:

$$v(x) = \mathcal{F}_{\xi\to x}^{-1}\, \Phi_f(\nu)\, \mathcal{F}_{x\to\xi} u\,. \tag{3.5.19}$$

Moreover, the last formula allows us to extend the concept of singular integral to the case of density from $L_2(R^2)$ and characteristic $f \in L_2(0, 2\pi)$. In fact, we see from (3.5.18) that for such characteristics the series (3.5.18) converges absolutely and uniformly, hence the function $\Phi_f(\nu)$ is bounded:

$$|\,\Phi_f(\nu)\,| \le 2\pi \sum_{k=-\infty}^{+\infty}{}' \frac{|\,a_k\,|}{|\,k\,|} \le 2\pi \Big(\sum_{k=-\infty, k\neq 0}^{+\infty} |\,a_k\,|^2 \Big)^{1/2} \Big(\sum_{k=-\infty, k\neq 0}^{+\infty} \frac{1}{|\,k\,|^2} \Big)^{1/2} =$$

$$= \sqrt{\frac{2\pi^3}{3}}\, \|f\|_{L_2(0,2\pi)}\,.$$

As we have noted above, $\|\mathcal{F}\| = 2\pi$, $\|\mathcal{F}^{-1}\| = (2\pi)^{-1}$. Hence it is clear that the right hand side of (3.5.19) meakes sense and

$$\|v\|_{L_2(R^2)} \le \sqrt{\frac{2\pi^3}{3}}\, \|f\|_{L_2(0,2\pi)} \cdot \|u\|_{L_2(R^2)}\,. \tag{3.5.20}$$

This yeilds

Theorem 3.4. The singular integral (3.5.8) the characteristic $f(\theta)$ of which is constant and belongs to the class $L_2(0, 2\pi)$, generates an operator, bounded iun $L_2(R^2)$; its norm does not exceed the quantity $\sqrt{\frac{2\pi^3}{3}}\, \|f\|_{L_2(0,2\pi)}$.

9^0. In conclusion, we will study the singular integral with the variable characteristic

$$f(x,\theta) = \sum_{k=-\infty}^{+\infty}{}' a_k(x)\,e^{ik\theta}$$

(where the dash indicates the absence of a free term). We will assume that for any fixed $x \in R^2$, it is square summable with respect to θ, when

$$\sup_{x \in R^2} \int_0^{2\pi} |\,f(x,\theta)\,|^2\,d\theta = C_0^2 < \infty. \qquad (3.5.21)$$

Note that then

$$\sup_{x \in R} \sum_{k=-\infty}^{+\infty}{}' a_k(x) = C_0^2/2\pi. \qquad (3.5.22)$$

If $u \in L_2(R^2)$, then it is natural to define the singular integral with the characteristic $f(x,\theta)$ by the formula

$$u(x) = \sum_{k=-\infty}^{+\infty}{}' a_k(x) \int_{R^2} e^{ik\theta}\, r^{-2}\, u(y)\,dy. \qquad (3.5.23)$$

We will prove that this series converges in the norm of $L_2(R^2)$. We will set

$$\int_{R^2} u(y)\, r^{-2}\, e^{ik\theta}\, dy = v_k(x).$$

By Cauchy's inequality

$$|\,v(x)\,|^2 \le \frac{C_0^2}{2\pi} \sum_{k=-\infty}^{+\infty}{}' |\,v_k(x)\,|^2\,.$$

Integrating over R^2 and taking into consideration (3.5.19) where in the this case one must set $f(\theta) = e^{ik\theta}$, we obtain

$$\|v\| = \sqrt{\frac{2\pi^3}{3}}\,\|f\|_{L_2(0,2\pi)}\,C_0\|u\|\,. \qquad (3.5.24)$$

This yields

Theorem 3.5. The singular integral with a variable characteristic, satisfying inequality (3.5.21) generates in $L_2(R^2)$ a bounded operator. The norm of this operator does not exceed the number $C_0\sqrt{\frac{2\pi^3}{3}}$.

We will make one more comment. By analogy with (3.5.18), we define a function $\Phi_f(x,\nu)$ by the formula

$$\Phi_f(x,\nu) = 2\pi \sum_{k=-\infty}^{+\infty} \frac{i^{|k|}}{|k|}\, a_k(x)\, e^{ik\nu}\,. \tag{3.5.25}$$

Obviously,

$$a_k(x)\int_{R^2} e^{ik\theta}\, r^{-2}\, u(y)\, dy = 2\pi\, \mathcal{F}^{-1}_{\xi\to x}\left(a_k(x)\frac{i^{|k|}}{|k|}e^{ik\nu}\,\tilde{u}(\xi)\right)$$

and (3.5.23), defining a singular integral with variable characteristic, assumes the form

$$v(x) = \int_{R^2} u(y)\, r^{-2}\, f(x,\theta)\, dy = \mathcal{F}^{-1}_{\xi\to x}\, \Phi_f(x,\nu)\, \mathcal{F}_{x\to\xi} u\,. \tag{3.5.26}$$

This form can serve for the definition of a singular integral with constant as well as with variable characteristic.

We note yet a simple, but important fact. We consider the operator

$$(Au)(x) = a_0(x)\, u(x) + \int_{R^2} u(y)\, r^{-2}\, f(x,\theta)\, dy\,. \tag{3.5.27}$$

The integral on the right hand side can be given the form (3.5.26). Moreover,

$$a_0(x)\, u(x) = a_0(x)\,\mathcal{F}^{-1}_{\xi\to x}\,\mathcal{F}_{x\to\xi} u = \mathcal{F}^{-1}_{\xi\to x}\, a_0(x)\,\mathcal{F}_{x\to\xi}\, u\,.$$

Substituting this expression into (3.5.12) and setting

$$a_0(x) + \Phi_f(x,\nu) = \Phi_A(x,\nu) \tag{3.5.28}$$

we find

$$(Au)(x) = \mathcal{F}^{-1}_{\xi\to x}\,\Phi_A(x,\nu)\,\mathcal{F}_{x\to\xi} u\,. \tag{3.5.29}$$

§6. Symbol and regularization

1^0. We shall call a general singular operator in the space $L_2(R^2)$ an operator A which acts in accordance with the formula

$$(Au)(x) = a_0(x)\, u(x) + \int\limits_{R^2} u(y)\, r^{-2}\, f(x,\theta)\, dy + (Tu)(x), \qquad (3.6.1)$$

where $a_0(x)$ is a bounded, measurable function, T an operator which is compact in $L_2(R^2)$, and the integral on the right hand side is singular with the characteristic $f(x,\theta)$, satisfying (3.2.4). We will dispose of the term "general" and simply call the operator of (3.6.1). a singular operator.

In the sequel, we will consider a narrower class of singular operators. We will assume that the coefficient $a_0(x)$ satisfies (1.1.11), and that the characteristic $f(x,\theta)$ satiusfies the condition

$$\mid f(x,\theta) - f(y,\theta) \mid \le Cr^{\lambda}[(1+\mid x\mid^2)(1+\mid y\mid^2)] \qquad (3.6.2)$$

in analogy with (1.1.11), and has a generalized first derivative with respect to θ the norm of which in $L_2(0,2\pi)$ is bounded independently of x. It follows from the results of §5 that a singular operator of this narrower class is bounded in $L_2(R^2)$. We note yet that the Fourier coefficients $a_k(x)$ of the characteristic $f(x,\theta)$ have the estimate

$$\mid a_k(x)\mid \le \frac{c}{\mid k\mid}\,,\quad c = const. \qquad (3.6.3)$$

We will denote this narrower class of singular operators by $\mathcal{R}$.

Lemma 3.1. The singular operators of the class $\mathcal{R}$ form a ring in $L_2(R^2)$.

Let $A, B \in \mathcal{R}$, where A is defined by (3.6.1) and B by the analogous formula

$$(Bu)(x) = b_0(x)\, u(x) + \int\limits_{R^2} u(y)\, r^{-2}\, g(x,\theta)\, dy + (T'u)(x).$$

Obviously, $(A + B) \in \mathcal{R}$, and there remains to consider the product AB.

The singular integral in (3.6.1) may be expanded according to (3.5.23). Denote by h the operator

$$(hu)(x) = \frac{1}{2\pi i} \int\limits_{R^2} u(y)\, r^{-2}\, e^{i\theta}\, dy\,.$$

To the characteristic $f_1(\theta)$ corresponds the function $\Phi_{f_1}(\nu) = e^{i\nu}$, and, by (3.5.19)

$$(hu)(x) = \mathcal{F}^{-1}_{\xi \to x} \, e^{i\nu} \, \mathcal{F}_{x \to \xi} \, u \, ;$$

it is readily seen that for any integer n

$$(h^n u) = \mathcal{F}^{-1}_{\xi \to x} \, e^{in\nu} \, \mathcal{F}_{x \to \xi} \, u$$

and, as a consequence of the last formula, $\|h^n\| = 1$. Likewise it is clear that, for $n \neq 0$, h^n is a singular operatior with the characteristic $(2\pi)^{-1} \, | \, n \, | \, i^{-|n|} \, e^{in\theta}$, and that $h^0 = I$, where I is the identity operator.

Now, using (3.5.23), one can represent A and B in the form

$$A = \sum_{n=-\infty}^{+\infty} \hat{a}_n(x) \, h^n + T := \hat{a}_0(x) I + A_0 + T \, ,$$

$$B = \sum_{n=-\infty}^{+\infty} \hat{b}_n(x) \, h^n + T' := \hat{b}_0(x) I + A_0 + T' \, ,$$

$$\hat{a}_0(x) = a_0(x) \, , \ \hat{b}_0(x) = b_0(x) \, ;$$

$$n \neq 0 \, , \ \hat{a}_n(x) = \frac{2\pi i^{|n|}}{|\, n \,|} \, a_n(x) \, , \ \hat{b}_n(x) = \frac{2\pi i^{|n|}}{|\, n \,|} \, b_n(x) \, .$$

From these results follows

$$AB = \hat{a}_0(x) \hat{b}_0(x) \, I + \hat{a}_0(x) B + A_0 \hat{b} + A_0 B_0 + T'' \, , \qquad (3.6.4)$$

where T'' is a compact operator. We will study the remaining terms of (3.6.4). The first of them is an operator of multiplication on the function $\hat{a}_0(x) \hat{b}_0(x)$, which, obviously, satisfies (1.1.11). The second term is the singular integral with the characteristic $\hat{a}_0(x) g(x, \theta)$ satisfies the conditions, imposed on the characteristic of an operator of the class $\mathcal{R}$. The third term may be transformed as follows:

$$(A_0 \hat{b}_0 u)(x) = \int\limits_{R^2} u(y) r^{-2} \, f(x, \theta) \, \hat{b}_0(y) \, dy =$$

$$= \int\limits_{R^2} u(y) r^{-2} \, f(x, \theta) \, \hat{b}_0(x) \, dy + \int\limits_{R^2} u(y) r^{-2} \, f(x, \theta) \, [\hat{b}_0(y) - \hat{b}_0(x)] \, dy \, .$$

Here, the second integral on the right hand side represents, as has been proved in 5^0, §1, chapter 1, an operator, compact in $L_2(R^2)$; the first integral on the right hand side is singular; its characteristic $\hat{b}_0(x)\,f(x,\theta)$ satisfies the requirement, imposed above on the characteristics of operators of the class $\mathcal{R}$.

There remains to consider the fourth term on the right hand side of (3.6.4). We have

$$(A_0 B_0 u)(x) = \sum_{k=-\infty}^{\infty}{}' A_0(\hat{b}_k(y)h^k u)(x) =$$

$$= \sum_{k=-\infty}^{\infty}{}' (A_0(\hat{b}_k(y) - \hat{b}_k(x))h^k u)(x) + \sum_{k=-\infty}^{\infty}{}' ((\hat{b}_k(x)A_0)h^k u)(x). \qquad (3.6.5)$$

Each term of the first series is a compact operator. The norm of the k–th term of this series has the estimate $2C\|A_0\|k^{-2}$; this series converges according to the norm, and its sum likewise is a compact operator. The second series in (3.6.5) is readily reduced to the form

$$\sum_{k=-\infty}^{\infty}{}' \hat{a}_k(x)\, h^k (\mathcal{F}_{\xi\to x}^{-1} \Phi_g(x,\nu)\, \mathcal{F}_{x\to\xi} u) =$$
$$= \mathcal{F}_{\xi\to x}^{-1} \Phi_f(x,\nu)\, \mathcal{F}_{x\to\xi} \mathcal{F}_{\xi\to x}^{-1} \Phi_g(x,\nu)\, \mathcal{F}_{x\to\xi} u = \qquad (3.6.6)$$
$$= \mathcal{F}_{\xi\to x}^{-1} \Phi_f(x,\nu)\, \Phi_g(x,\nu)\, \mathcal{F}_{x\to\xi} u\,.$$

We will expand the product $\Phi_f(x,\nu)\,\Phi_g(x,\nu)$ in the Fourier series

$$\Phi_f(x,\nu)\,\Phi_g(x,\nu) = \hat{c}_0 + \sum_{k=-\infty}^{\infty}{}' c_k(x)\, e^{ik\nu}\,.$$

Substituting it into (3.6.6), we find that the second series in (3.6.5) is the sum of the non-integral term $\hat{c}_0(x)\,u(x)$ and the singular integral

$$\mathcal{F}_{\xi\to x}^{-1} \sum_{k=-\infty}^{\infty}{}' \hat{c}_k(x)\, e^{ik\nu}\, \mathcal{F}_{x\to\xi}\, u\,.$$

Gathering the results, we verify that the product AB represents the sum of three operators: the non-integral operator

$$[\hat{a}_0(x)\,\hat{b}_0(x) + \hat{c}_0(x)]\, I\,,$$

a singular operator and a compact operator, and that all of them satisfy the conditions, defining the class $\mathcal{R}$. Hence the lemma is proved.

3^0. We can present the product AB in a comparatively simple form. As we have seen during the manipulation of 2^0 of this section

$$AB = \hat{a}_0(x)\,\hat{b}_0(x) + \hat{a}_0(x)\,B_0 + \hat{b}_0(x)\,A_0 + $$
$$+ \mathcal{F}^{-1}_{\xi\to x}\,\phi_f(x,\nu)\,\Phi_g(x,\nu)\,\mathcal{F}_{x\to\xi} + T\,, \tag{3.6.7}$$

where T is a compact operator. Moreover, $\hat{a}_0(x)B_0$ is a singular integral with charcateristic $\hat{a}_0(x)\,g(x,\nu)$ and, by (3.5.26),

$$\hat{a}_0(x)B_0 = \mathcal{F}^{-1}_{\xi\to x}\,\hat{a}_0(x)\,\Phi_g(x,\nu)\,\mathcal{F}_{x\to\xi}\,.$$

Analogously, one has

$$\hat{b}_0(x)A_0 = \mathcal{F}^{-1}_{\xi\to x}\,\hat{b}_0(x)\,\Phi_f(x,\nu)\,\mathcal{F}_{x\to\xi}\,.$$

Finally, it is obvious that

$$\hat{a}_0(x)\,\hat{b}_0(x)\,I = \hat{a}_0(x)\,\hat{b}_0(x)\,\mathcal{F}^{-1}_{\xi\to x}\,\mathcal{F}_{x\to\xi} = \mathcal{F}^{-1}_{\xi\to x}\,\hat{a}_0(x)\,\hat{b}_0(x)\,\mathcal{F}_{x\to\xi}\,.$$

Substituting this into (3.6.4) and using the notation (3.5.28) and its analogues, we find

$$AB = \mathcal{F}^{-1}_{\xi\to x}\,\Phi_A(x,\nu)\,\Phi_B(x,\nu)\,\mathcal{F}_{x\to\xi} + T\,. \tag{3.6.8}$$

It is seen from (3.6.8) that the products AB and BA differ only by compact terms. We can formulate the last property in the following manner: Multiplication of singular operators is exactly commutative apart from a completely continuous term. This product is strictly commutative, if both multiplied operators have the form

$$cu(x) + \int\limits_{R^2} u(y)\,r^{-2}\,f(\theta)\,dy\,,$$

where $c = const$ and the characteristic $f(\theta)$ is constant.

4^0. We introduce now into the consideration the ring $\mathcal{T}$ of the functions $\Phi(x,\theta)$, defined on the manifold $R^2 \times [0,2\pi]$ and possessing the following properties: For any fixed $x \in R^2$, they are 2π periodic with respect to ν and belong to the class $W_2^2(0,2\pi)$; they are bounded together with their first derivatives independently of x and ν; the Fourier coefficients of the function Φ satisfy the inequality (1.1.11). It is not difficult to see that the coefficients have the order $O(n^{-2})$.

5^0. Let A be an operator of the ring $\mathcal{R}$; the function $\Phi_A(x,\nu)$ constructed above (cf. (3.5.28)) belongs to the ring $\mathcal{T}$; to every operator of the ring $\mathcal{R}$ corresponds one such function, where the sum and product of operators from $\mathcal{R}$ correspond to the sum and product of functions from $\mathcal{T}$. Conversely, to every function from $\mathcal{T}$ corresponds an infinite manifold of operators from $\mathcal{R}$, which differ by their compact items. In correspondence with the definition, given in §2, chapter 1, one may take $\mathcal{T}$ as ring of symbols of the ring $\mathcal{R}$ and determine the symbol of an operator from the ring $\mathcal{R}$ by the formula

$$Smb\, A = \Phi_A(x,\nu), \qquad\qquad (3.6.9)$$

where the function $\Phi_f(x,\nu)$ is defined by (3.5.28). This definition has the following implications: a) The symbol of any compact operator from $\mathcal{R}$ equals zero, and conversely; b) the symbol of a singular operator with the characteristic $f(x,\theta)$ equals $\Phi_f(x,\nu)$ (cf. (3.5.25)); c) the symbol of the operator of multiplication on a function equals this function.

6^0. We will say that the symbol of an operator from the ring $\mathcal{R}$ is degnerate, if it vanishes even if for any pair of values $x \in R^2$, $\nu \in [0,2\pi]$. We call non-degenrate the opposite case (when the lower boundary of the value of the module of the symbol is positive). Obviously, an element $\Phi \in \mathcal{T}$ is inverse in $\mathcal{T}$ if and only if it is degenerate. In this case, the inverse element is $\Phi^{-1}(x,\nu)$. Let Φ and Φ^{-1} by symbols of the operators A and R, respectively. Since $\Phi\Phi^{-1} = \Phi^{-1}\Phi = 1$, then $RA = I + T$, $AR = I + T_1$, where T and T_1 are compact operators. Hence an operator from $\mathcal{R}$ admits regularization if and only if its symbol does not degenerate. Then (cf. chapter 1) Noether's theorem is true for the operator A: The number of zeroes of adjoint operators A and A^* is finite; both these operators are normally soluble; the index of the operator A is finite and does not depend on A its compact term.

7^0. We say that a singular operator of the form

$$a(x)\,u(x) + \int_{R^2} f(x,\theta) r^{-2}\, u(y)\, dy$$

is simplest. Then the general singular operator is a sum of two operators - of a simplest and a compact operator. The symbols of these terms are $\Phi_A(x,\nu)$ and zero, respectively. Our definition is correct, if there does not exist an operator, which differs from zero and belongs simultaneously to both stated classes. The truth of the last assertion follows from

Lemma 3.2. If $A \in \mathcal{R}$ is compact, then its symbol $\Phi_A(x, \nu)$, defined by (33.5.28), is identicalkly zero.

We will assume that the opposite is true. It may be assumed that the symbol is identically constant: $\Phi_A(x, \nu) = const$. In such a case $A = aI + T$. Such an operator is compact if and only if $a = 0$. In this case, the lemma is proved.

Now let the symbol $\Phi_A(x, \nu)$ differ from being identically constant. Being an element of the ring $\mathcal{T}$, the symbol is continuous; since it differs from being identically constant, and the manifold of pairs (x, ν) is infinite, then the manifold of the values of the symbol $\Phi_A(x, \nu)$ likewise is infinite.

We will consider the operator $B = \lambda I - A$, where λ is an arbitrary complex number. The symbol of this operator is $\Phi_B(x, \nu) = \lambda - \Phi_A(x, \nu)$. If λ equals one of the values of the symbol $\Phi_A(x, \nu)$, then the corresponding symbol $\Phi_B(x, \nu)$ degenerates and does not have an inverse element in the ring of symbols, and then the operator B does not admit regularization. However, if the operator A is compact, then for any $\lambda = 0$ the operator B has the obvious regularizator $\lambda^{-1}I$ in the ring $\mathcal{R}$. This contradiction proves the lemma.

§7. The index of a singular operator

1^0. We will call the symbol of a singular operator constant, if this symbol does not depend on the pole x. Obviously, a symbol is constant, if the characteristic and the coefficient for the external integral term are constant.

Theorem 3.6. If A is a singular operator with symbol $\Phi_A(x, \nu)$, then the adjoint operator A^* is likewise singular and its symbol is $\Phi_{A^*}(x, \nu) = \overline{\Phi_A(x, \nu)}$, where the bar above signifies the conjugate complex quantity.

Proof. The symbol of a singular operator does not depend on the compact term T, hence we may assume that the operator A is simplest.

We consider first the case of the simple symbol: $\Phi_A = \Phi_A(\nu)$. By (3.5.29), $A = \mathcal{F}^{-1}_{\xi \to x} \Phi_A(\nu) \mathcal{F}_{x \to \xi}$. Moreover, we see from (3.5.29) that $\mathcal{F}^*_{x \to \xi} = (2\pi)^2 \mathcal{F}^{-1}_{\xi \to x}$, and, consequuently, $(\mathcal{F}^{-1}_{x \to \xi})^* = (2\pi)^{-2} \mathcal{F}_{\xi \to x}$. Now

$$(u, A^*v) = (Au, v) = (\mathcal{F}^{-1}_{\xi \to x} \Phi_A(\nu) \mathcal{F}_{x \to \xi} u, v) =$$

$$= (2\pi)^{-2} (\Phi_A(\nu) \mathcal{F}_{x \to \xi} u, \mathcal{F}_{x \to \xi} v) = (2\pi)^{-2} (\mathcal{F}_{x \to \xi} u, \overline{\Phi_A(\nu)} \times \mathcal{F}_{x \to \xi} v) =$$

$$= (u, \mathcal{F}^{-1}_{\xi \to x} \overline{\Phi_A(\nu)} \mathcal{F}_{x \to \xi} v).$$

The last equality indicates that $A^* = \mathcal{F}^{-1}_{\xi \to x} \overline{\Phi_A(\nu)} \mathcal{F}_{x \to \xi} v$ and, therefore, that A is the simplest operator with symbol $\Phi_{A^*}(\nu) = \overline{\Phi_A(\nu)}$.

Next, we will consider the genral case. We expand the symbol $\Phi_A(x, \nu)$ in the Fourier series

$$\Phi_A(x, \nu) = \sum_{n=-\infty}^{+\infty}{}' \hat{a}_n(x)\, e^{in\nu} \, . \tag{3.7.1}$$

In correspondence to this, we obtain the expansion of the operator A

$$A = \sum_{n=-\infty}^{+\infty} \hat{a}_n\, h^n + T \, .$$

The operator h^n has the constant symbol $e^{in\nu}$; by what has been proved, the symbol $(h^n)^*$ is $e^{in\nu}$, therefore

$$A^*u = \sum_{n=-\infty}^{+\infty} h^{-n}(\bar{\hat{a}}_n u) = \overline{\hat{a}_0(x)}\, u(x) + \sum_{n=-\infty}^{+\infty}{}' \frac{|\,n\,|}{i|n|}\, \overline{\hat{a}_n(x)} \int_{R^2} u(y) r^{-2} e^{-in\theta}\, dy +$$

$$+ \sum_{n=-\infty}^{+\infty}{}' \frac{|\,n\,|}{i|n|} \int_{R^2} [\,\overline{a_n(y)} - \overline{a_n(x)}\,]\, u(y) r^{-2}\, e^{-in\theta}\, dy \, .$$

The first sum is a singular operator with symbol $\overline{\Phi_A(x, \nu)} - \overline{\hat{a}_0(x)}$, and the second an operator, compact in $L_2(R^n)$. Thus, theorem 3.6 has been proved.

2^0. Let $\Phi(x, \nu) \in T$, λ be a complex parameter and

$$(A_\lambda u)(x) = u(x) - \lambda \mathcal{F}_{\xi \to x}^{-1}\, \Phi(x, \nu)\, \mathcal{F}_{x \to \xi}\, u \, . \tag{3.7.2}$$

Clearly, $Smb\, A_\lambda = 1 - \lambda \Phi(x, \nu)$. In the complex plane of the parameter λ, we separate out the manifold σ of those values of λ for which $Smb\, A_\lambda$ degenrates; obviously, the manifold σ is closed. The manifold σ, additional to Δ, is the union of a finite or enumerable manifold of the open domains: $\Delta = \cup \Delta_j$.

Lemma 3.3. In each of the domains Δ_j, the index of the operator A_λ remains constant.

Proof. Let δ_j be a finite, closed subregion of the domain Δ_j. In this subregion $inf\, |\, 1 - \lambda \Phi(x, \nu)\, | > 0$, the symbol of the operator A_λ does not degenerate. The operator has the regularizator R_λ with the symbol $[1 - \lambda \Phi(x, \nu)]^{-1}$.

We now replace λ by $\lambda + \mu$; we choose μ so small in modulus that $|\mu| \cdot \|R_\lambda\| \cdot \|B\| < 1$, where B is any fixed singular operator with symbol $\Phi(x, \nu)$. By theorem 1.14, $Ind\, A_\lambda = Ind\, A_{\lambda + \mu}$; the assertion of lemma 3.3 follows now from Borel's lemma.

Lemma 3.4. In the notation of lemma 3.3. if $\sup\limits_{x} |\Phi(x, \nu)| < q < 1$, then $Ind\, A_\lambda = 0$.

Proof. We will consider the operator A_λ. It does not degenerate within the circle $|\lambda| < q^{-1}$ of the symbols of the operator A_λ:

$$inf\, |1 - \lambda\Phi(x, \nu)| \geq 1 - |\lambda|\, q > 0$$

and, by lemma 3.3, in this circle, $Ind\, A_\lambda$ is constant. However, for $\lambda = 0$, $Ind\, A_\lambda = Ind\, I = 0$, and the lemma is proved.

Lemma 3.5. Let for any fixed value of $t \in [0, 1]$ the function $\Phi(x, \nu; t) \in \mathcal{T}$, where $inf\, |\Phi(x, \nu; t)| > 0$, and let there be executed uniformly with respect to x and t the relation

$$\|\Phi(x, \nu; t + \tau) - \Phi(x, \nu; t)\| \to 0\, , \tau \to 0\, . \tag{3.7.3}$$

Moreover, let A_t be a singular operator with symbol $\Phi(x, \nu; t)$. Then $Ind\, A_t$ does not depend on t.

Proof. It follows from the results of §5 that the operator A_t is bounded in $L_2(R^2)$ independently of t, $\|A_{i+\tau} - A_t\|_{L_2(R^2)} \to 0\, , \tau \to 0$; if C_t is a simplest operator with symbol $[\Phi(x, \nu; t)]^{-1}$, then

$$\|C_t\|_{L_2(R^2)} \leq M = const\, .$$

We select τ so small that $\|A_{t+\tau} - A_t\|_{L_2(R^2)} < M^{-1}$. By theorem 1.14 $Ind\, A_{t+\tau} = Ind\, A$. It follows now from Borel's lemma that $Ind\, A_t$ does not depend on t.

3^0. **Theorem 3.7.** If the symbol $\Phi_A(x, \nu)$ does not degenerate, then $Ind\, A = 0$.

Proof. Expand the symbol $\Phi_A(x, \nu)$ in the series (3.7.1) and let

$$\tilde{\Phi}(x, \nu) = \sum_{k=-n}^{n} \hat{a}_k(x)\, e^{ik\nu}\, .$$

As has been noted in 4^0, §6, if $k > 0$, then $|\hat{a}_k(x)| \leq Ck^{-2}$, hence one may select n so large that the inequality

$$|a_n(x)| < \varepsilon/3\, , \; |\Phi_A(x, \nu) - \tilde{\Phi}(x, \nu)| < \varepsilon/3\, ,$$

is fulfilled, where ε is an arbitrarily small positive number. We will now set

$$\Phi_0(x,\nu) = \frac{2}{3}\varepsilon e^{in\nu} + \tilde{\Phi}(x,\nu);$$

$\Phi_0(x,\nu)$ is a trigonometric polynomial of order n with the following properties: The modulus of the coefficient of $e^{in\nu}$, equal to $1\varepsilon/3 + \hat{a}_n(x)$, is larger than $\varepsilon/3$; $| \Phi_A(x,\nu) - \Phi_0(x,\nu) | < \varepsilon$; the coefficients $\hat{a}_k(x)$ satisfy (1.1.11) and moreover are continuous on the Riemannian sphere; $inf \; | \Phi_0(x,\nu) | > 0$.

We can expand the function $\Phi_0(x,\nu)$, which may be considered to be the symbol of some simplest singular operator A_0, in a product of the form

$$\Phi_0(x,\nu) = \left[\frac{2\varepsilon}{3} + \hat{a}_n(x)\right] e^{-in\nu} \prod_{k=1}^{2n} [a_k(x) - e^{i\nu}].$$

On the barred plane $\bar{R}^2$, the roots $a_k(x)$ are continuous, however, in the general case, they are multi-valued. All of them differ in modulus from unity, because $inf \; | \Phi_0(x,\nu) | > 0$. Hence, on the barred plane $\bar{R}^2$, everyone of the roots $a_k(x)$ satisfies either the inequality $| a_k(x) | > 1$ or the inequality $| a_k(x) | < 1$. We will denote the roots of the first group by $a'_k(x)$, $k = 1, 2, ..., s'$, the roots of the second group by $a''_k(x)$, $k = 1, 2, ..., s''$. Now let

$$\prod_{k=1}^{s'} (a'_k(x) - e^{i\nu}) = \Phi'_0(x,\nu); \quad \prod_{k=1}^{s''} (a''_k(x) - e^{i\nu}) = \Phi''_0(x,\nu).$$

Let the point x describe on $\bar{R}^2$ a closed contour. The symbol $\Phi_0(x,\nu)$ is single-valued and continuous, hence for such a circuit its roots may experience only some rearrangement. Then, since the symbol above does not degenerate, its roots may not lie on the unit circle and will be rearranged between the roots of each group. Hence the triginometric polynomials $\Phi'(x,\nu)$ and $\Phi''(x,\nu)$ are single-valued as functions of x. We will now introduce the function

$$\Phi(x,\nu;t) = [2\varepsilon/3 + \hat{a}(x)]e^{-in\nu} \prod_{k=1}^{s'} [(1-t)a'_k(x) - e^{i\nu}] \prod_{k=1}^{s''} [a''_k(x) - (1-t)e^{i\nu}] =$$

$$= [2\varepsilon/3 + \hat{a}_n(x)](1-t)^{s'} e^{-in\nu} \Phi'\left(x, \frac{e^{i\nu}}{1-t}\right) \Phi''(x, (1-t)e^{i\nu}),$$

and denote by A_t the simplest singular operator with symbol $\Phi(x,\nu;t)$; for $t = 0$, we obtain the operator A_0, introduced above. The operator A_t satisfies the conditions of lemma 3.5, hence $Ind\,A_0 = Ind\,A_1$. However

$$A_1 = (.1)^{s''}[2\varepsilon_1/3 + \hat{a}(x)] \prod_{k=1}^{s''} (a_k''(x))h^{s'-n}\,.$$

By theorem 1.13, $Ind\,A_1 = Ind\,h^{s'-n} = (s' - n)Ind\,h = 0$, and then also $Ind\,A_0 = 0$.

We now select ε so small that $\sup |\,(\Phi - \Phi_0)/\Phi_0\,| < 1$. To a decomposition of the symbol $\Phi = \Phi_0[1 - (\Phi_0 - \Phi)/\Phi_0]$ cooresponds a decomposition of the operator $A = A_0(I - B) + T$, where T is a compact operator, and B a singular operator with the symbol $(\Phi_0 - \Phi)/\Phi_0$. Applying in succession theorems 1.12, 1.13 and lemma 3.3, we find that $Ind\,A = 0$.

§8. Singular integrals on a smooth surface without edge

1^0. Let Γ be a closed, sufficiently smooth surface in the three-dimensional Euclidean space R^3. We draw in R^3 about an arbitrary point $z \in \Gamma$ as centre a sphere of sufficiently small radius which, in general, may depend on z. We denote by Γ_z that part of Γ which lies inside that sphere. By Borel's lemma, we may select a finite number of points z_j, $j = 1, 2, ..., m$, such that the corresponding manifold $\Gamma_j = \Gamma_{z_j}$ forms a cover for Γ. If, as we have already stressed, the radiuses of the spheres are sufficiently small, then one may map each of the parts Γ_j suffiently smoothly and mutually single-valuedly on to some domain D_j of the plane R^2; we will denote the points of this plane by $x = (x_1, x_2)$. We denote the two-dimensional vector function, relaizing the mapping of Γ_j on to D_j, by $\omega_j(x)$.

Let the functions $\varphi_j(x)$, $j = 1, 2, ..., m$, have the properties

$$a)\; \varphi_j \in C^{(\infty)}(R^2);\; b)\; supp\,\varphi_j \subset D_j;\; c)\; \varphi(x) \geq 0;\; d)\; \sum_{j=1}^{m} \varphi_j(x) = 1\,.$$

We may define the Sobolev space $W_p^s(\Gamma)$ as the manifold of functions which are defined on Γ and have the finite norm

$$\|u\|_{W_p^s(\Gamma)} := \Big[\sum_{|a|=0}^{s} \sum_{j=1}^{m} \|u^{(a)}(\varphi_j)\|_{L_p(D_j)}^p\Big]^{1/p}\,. \tag{3.8.1}$$

In particular,

$$\|u\|_{W_p^\circ(\Gamma)} := \Big[\sum_{j=1}^{m} \|u(\varphi_j)\|_{L_p(D_j)}^p\Big]^{1/p}. \tag{3.8.2}$$

In the sequel, we will be interested in the value $p = 2$.

3^0. We will produce at the arbitrary point $z \in \Gamma$ a tangent plane to Γ on which we single out unit vectors with origin at the point of touch z. These vectors are tangent to Γ at the point z; their ends form a plane. We will refer to the manifold, comprising the points of the surface z and the unit vector, tangent to Γ at this point, as the element of the surface Γ.

4^0. A definition of the singular integral on a Lyapunov surface has been presented in §2 of this chapter; we will now give another definition which is due to R.T.Cili [56]; in essence, these definitions are equivalent, but the second one is somewhat more convenient for our investigations.

The operator A is called singular in $L_2(\Gamma)$, if

a) For any, sufficiently smooth functions $\varphi(z)$ and $\psi(z)$, the supports of which do not intersect, the operator $\varphi A\psi$ is compact in $L_2(\Gamma)$;

$$(\varphi A\psi u)(z) = \varphi(z)(A\psi u)(z).$$

b) If the supports of the functions $\varphi(z)$ and $\psi(z)$ lie in one and the same domain Γ_j of the cover of the surface Γ, then $\varphi A\psi = \varphi_j A_j\psi_j + T_j$, where T_j is an operator, compact in $L_2(\Gamma)$, $\varphi_j = \varphi(\omega_j(x))$, $\psi_j = \psi(\omega_j(x))$ and A_j is a simplest operator of the form

$$(A_j u)(x) = a_j(x)\,u(x) + \int_{R^2} r^{-2}\,f(x,\theta)\,u(y)\,dy. \tag{3.8.3}$$

The function of the linear element of the surface Γ is called thew symbol of the operator A; for $z \in \Gamma$, it coincides with the symbol of the operator (3.8.3). It may be shown that the symbol defined in this manner has the usual properties: The sum and product of the operators correspond to the sum and product of the symbols; the symbol of any operator, compact in $L_2(\Gamma)$, equals zero.

It has been shown in [56] that the symbol of a singular operator does not depend on the choice of the cover and and representations $\omega_j(x)$.

Note. A linear element of a surface Γ, to be denoted by τ, is the union of two vectors $z \in \Gamma$ and $\theta \in S_z$, where S_z is the unit circle with centre at z, lying in the tangent plane to Γ at the point z *. In the general case, it is impossible to specify these vectors independently, hence, in the general case, the symbol of a singular operator on a surface cannot be considered as a function of two independent vectors z and θ. Exceptions are presented by those surfaces on which one can introduce a right-handed coordinate system; for example, such a surface is the torus. However, if z covers one of the domains Γ_j, then θ may be defined independently of z. Hence we may, in general, denote the symbol of a singular operator on a surface by $\Phi_A(\tau)$; however, if $z \in \Gamma_j$ with fixed f, then one can also use the notation $\Phi_A(z, \theta)$.

A study of the operator A reduces to one of the operator A_j (cf. (3.8.3)), hence follows

Theorem 3.8. The singular operator A is bounded in $L_2(\Gamma)$, if the operator A_j is subjected to the conditions of 1^0, §6: The coefficients $a_j(x)$ satisfy conditions (1.1.11), and the characteristics $f_j(x, \theta)$ fulfill conditions (3.6.2) and have a first, generalized derivative with respect to θ the norm of which is bounded independently of x in $L_2(0, 2\pi)$.

Theorem 3.9. If $\Phi(\tau)$ is an arbitrary function of the linear element of a sufficiently smooth surface Γ, and if this function, as a function of θ, belongs for fixed z to the space $W_2^2(0, 2\pi)$, where $\|\Phi\|_{W_2^2(0,2\pi)}$ is bounded independently of z, but with respect to z satisfies a Hoelder condition with a constants independent of θ, then there exists an operator with symbol $\Phi(\tau)$ which is singular on Γ.

A proof of this theorem is given in [54].

Theorem 3.10. The index of a singular operator on a sufficiently smooth surface without edge is zero.

This theorem has been proved in [25] and in [54].

§9. Matrix singular operators of systems of singular equations

1^0. Let $\mathbf{u} = (u_1, u_2, ..., u_n)$ a vector the components of which are functions, defined on a suffiently smooth two-dimensional surface Γ; if $\Gamma = \bar{R}^2$, then one can employ the stereographic projection which maps the extended plane $\bar{R}^2$ into a sphere. Moreover, let A be a matrix of order $n \times n$, the elements A_{jk} of which are singular operators. We will call the matrix singular operator, for which we employ the same character A, an operator which maps

---

* The contemporary treatment of the concept of the symbol of an operator on a manifold is based on the theory of tangent atification of surfaces in Euclidean spaces [44,56].

a given vector **u** into another vector **v** according to the formula

$$v_j = \sum_{k=1}^{n} A_{jk} u_k \, ; \; j = 1, 2, ..., n; \tag{3.9.1}$$

we will assume that this operator acts out of the vector space $L_2(\Gamma)$ into the same space. We note that, if in (3.9.1), the vector **v** is given, and the vector **u** is subject to definition, then (3.9.1) maps a system of singular integral equations.

We will call the matrix of the singular operator A simplest, if all the operators $A_j k$ are simplest.

Let $\Phi_{jk}(\tau)$ be the symbol of the operator A_{jk}. It is readily seen that we may take as symbol of the matrix singular operator A the matrix of symbols ("symbolic matrix")

$$\Phi_A(\tau) = \begin{pmatrix} \Phi_{11}(\tau) & \Phi_{12}(\tau) & \cdots & \Phi_{1n}(\tau) \\ \Phi_{21}(\tau) & \Phi_{22}(\tau) & \cdots & \Phi_{2n}(\tau) \\ \cdots & \cdots & \cdots & \cdots \\ \Phi_{n1}(\tau) & \Phi_{n2}(\tau) & \cdots & \Phi_{nn}(\tau) \end{pmatrix} . \tag{3.9.2}$$

If for any values of τ the determinant of the ("symbolic") matrix (3.9.2) vanishes, then Symbol (3.9.2) is said to be degenerate; otherwise, when $inf \mid Det \, \Phi(\tau) \mid > 0$, the symbol is called non-degenerate.

If the symbol of a matrix, symmetric operator A does not degenerate, then this operator admits regularization. As regularizator (and besides two-sided one) serves the matrix singular operator with, the symbol $\Phi^{-1}(\tau)$. Hence Noether's theorem is true for a matrix singular operator with non-degenerate symbol.

2^0. It is readily shown that the operator A^*, adjoint to the patrix singular operator A, the symbol of which is given by (3.9.2), liekwise is a matrix singular operator, the symbol of which is the matrix, adjoint to the matrix (3.9.2)

$$\Phi_{A^*}(\tau) = \begin{pmatrix} \overline{\Phi_{11}(\tau)} & \overline{\Phi_{12}(\tau)} & \cdots & \overline{\Phi_{1n}(\tau)} \\ \overline{\Phi_{21}(\tau)} & \overline{\Phi_{22}(\tau)} & \cdots & \overline{\Phi_{2n}(\tau)} \\ \cdots & \cdots & \cdots & \cdots \\ \overline{\Phi_{n1}(\tau)} & \overline{\Phi_{n2}(\tau)} & \cdots & \overline{\Phi_{nn}(\tau)} \end{pmatrix} . \tag{3.9.3}$$

In contrast to the case of the scalar singular equations, considered in the preceding sections of this chapter, the index of a system of singular euqations may be different from zero; this subject has been the topic of significant effort (cf. [54]), nevertheless, apparently, methods for a practical evaluation of the index of an arbitrary matrix singular operator are unknown. On the other hand, great practical interest attaches to the case when the index a matrix singular operator vanishes. When calculating the index of a singular operator, we will always assume that its symbol does not degenerate.

A very simple case of such a type occurs, if $\Gamma = R^2$ and the symbolic matrix is constant, i.e., if it does not depend on x. Then the simplest singular operators with the symbolic matrices $\Phi_A^{-1}(\vartheta)$ and $\Phi_{A^*}^{-1}(\vartheta)$ are equivalent regularizators for A and A^*, respectively. By theorem 1.15, $Ind\,A \geq 0$ and $Ind\,A^* \geq 0$. However, $Ind\,A^* = -Ind\,A$, and hence, finally, $Ind\,A = 0$.

We will now present several of the less elementary theorems

Theorem 3.11. Let the symbolic matrix of the operator A has the form $\Phi_A(\tau) = I - \psi(\tau)$, where I is the unit matrix. If the moduli of all the characteristic numbers of the matrix ψ for any values of τ equal strictly unity, then $Ind\,A = 0$.

Proof. Let λ be an arbitrary complex number. We will consider the matrix singular operator A_λ with the symbol $I - \lambda\psi(\tau)$. In the complex plane of λ , we single out an open manifold of A on which the symbol $I - \lambda\psi(\tau)$ does not degenerate. Repeating the reasoning of lemma 3.3, we will verify that $Ind\,A_\lambda$ is constant in everyone of the connected regions in which disintegrates the open manifold Δ. It is apparent from the conditions of the theorem that the circle $\mid \lambda \mid \leq 1$ belongs entirely to one such region, hence $Ind\,A = Ind\,A_1 = Ind\,A_0$. However $Smb\,A_0 = I$, $A_0 = I + T$, where T is a compact operator; by theorems 1.6 and 1.7, $Ind\,A_0 = 0$.

Theorem 3.12. If the lower bounds of the moduli of the minors of the symbolic determinant of the operator

$$\delta_1 = \Phi_{11}, \ \delta_2 = \begin{pmatrix} \Phi_{11} & \Phi_{12} \\ \Phi_{21} & \Phi_{22} \end{pmatrix}, \ldots, \delta_n = \begin{pmatrix} \Phi_{11} & \Phi_{12} & \ldots & \Phi_{1n} \\ \Phi_{21} & \Phi_{22} & \ldots & \Phi_{2n} \\ \ldots & \ldots & \ldots & \ldots \\ \Phi_{n1} & \Phi_{n2} & \ldots & \Phi_{nn} \end{pmatrix}$$

are positive, then $Ind\,A = 0$.

Proof. We will separate out from the system (3.9.1) the first equation and consider it to be an equation with the known u_1. The symbol $\Phi_{11}(\tau)$ does not degenerate, hence the equation admits equivalent regularization, i.e., is may be reduced to an equivalent equation of the form

$$u_1 = \sum_{k=2}^{n} A_k^{(1)} u_k + \sum_{k=1}^{n} T_k^{(1)} u_k + g^{(1)}, \qquad (3.9.4)$$

where $g^{(1)}$ is some known function, $T_k^{(1)}$ a compact operator and $A_k^{(1)}$ are singular operators with symbols $\Phi_{1k}(\tau)/\Phi_{11}(\tau)$. We will substitute (3.9.4) into the remaining equations (3.9.1). This step leads us to a system of $n-1$ equations which contain only the unknowns $u_2, ..., u_n$ under the singular operator signs. The first of these equations contains the unknown u_2 under the sign of the singular operator the symbol of which equals δ_2/δ_1 and, consequently, it does not degenerate. However, then this equation reduces to an equivalent equation of the form

$$u_2 = \sum_{k=3}^{n} A_k^{(2)} u_k + \sum_{k=1}^{n} T_k^{(2)} u_k + g^{(2)}, \qquad (3.9.5)$$

where the significance of the notation is obvious. We will substitute (3.9.5) into the remaining equations. The first of the resulting equations contains the unknown u_3 under the sign of the singular integral with the symbolic determinant δ_3/δ_2. Continuing this process, we arrive eventually at the system

$$u_j(x) = \sum_{k=j+1}^{n} A_k^{(j)} u_k + \sum_{k-1}^{n} T_k^{(j)} u_k + g^{(j)}; \; j = 1, 2, ..., n, \qquad (3.9.6)$$

where for $j = n$ the first sum vanishes.

System (3.9.6) reduces in an obvious manner to an equivalent system with compact matrix operators; this means that system (3.9.1) admits equivalent regularization, and that $Ind\,A \geq 0$. However, the conditions of the theorem are also satisfied in an obvious manner for the adjoint system, hence $Ind\,A^* \geq 0$. Hence $Ind\,A = 0$.

Theorem 3.13. Let A be a matrix singular operator with symbol $\Phi(\tau)$, and let there exist in the complex ζ-plane a trajectory curve L which joins the point $\zeta = 0$ and $\zeta = \infty$ and does not contain common points with the manifold of the characteristic numbers of the matrix $\Phi(\tau)$, corresponding to all-possible values of the linear element τ. Then $Ind\,A = 0$.

Proof. Let $\zeta = (\lambda - 1)/\lambda$; under such a transformation, the trajectory L becomes some curve $\tilde{L}$ in the λ plane. We may assume that L does not pass through the point $\zeta = 1$, when the curve $\tilde{L}$ has finite length.

We will consider the singular system

$$u - \lambda(I - A)u = g(x), \qquad (3.9.7)$$

where $\lambda \in \tilde{L}$. We will show that the symbol of the operator on the left hand side of equation (3.9.7), equal to $\lambda[\Phi(\tau) - \zeta]$, does not degenerate.

In fact, if $\mid \lambda \mid$ is small, then the symbolic determinant is close to unity; consequently, we can find positive numbers η and q such that for $\mid \lambda \mid < \eta$ one has $\mid Det[(1 - \lambda)I_n + \lambda\Phi(\tau)] \mid > q$. Now let $\mid \lambda \mid \geq \eta$ then

$$\mid Det[(1 - \lambda)I_n + \lambda\Phi(\tau)] \mid = \mid Det\,\lambda(\zeta I_n - \Phi) \mid =$$
$$= \mid \lambda \mid^n \prod_{k=1}^n \mid \zeta - \zeta_k(\tau) \mid \geq q^n \prod_{k=1}^n \mid \zeta - \zeta_k(\tau) \mid,$$

where I_n is the unit matrix of order n, the $\tau_k(\tau)$ are the characteristic numbers of the matrix $\Phi(\tau)$. The manifold Γ is compact; it is not difficult to verify that so is the manifold of the corresponding linear elements , and likewise the manifold of the values of ζ for which $\mid \lambda \mid \geq q$.In the last product, every factor is bounded on the compact manifold on which is non-zero everywhere. Hence $inf \prod_{k=1} \mid \zeta - \zeta_k(\tau) \mid > 0$, and the assertion is proved.

It is now clear that in the vector space $L_2(\Gamma)$ the singular operator H_λ with symbol

$$[(1 - \lambda)I_n + \lambda\Phi(\tau)]^{-1}[I_n - \Phi(\tau)];$$

is bounded independently of λ; let $\|H_\lambda\| \leq C = const$. Setting $\lambda = 0$, we find that $\|I - A\| \leq C$, hence, if $\lambda_0 \in \tilde{L}$ and $\mid \lambda_0 \mid \leq 1/2C$, then the operator $[I - \lambda_0(I - A)]^{-1}$ exists, is defined on the entire space $L_2(\Gamma)$ and is bounded; likewise, it is obvious that the index of this operator is zero. We will now multiply both sides of equation (3.9.7) by the above operator. This step yields an equation which: a) is equivalent to (3.9.7) and has the same index; b) has the symbolic matrix

$$[(1 - \lambda_0)I_n + \lambda\Phi(\tau)]^{-1}[(1 - \lambda)I_n - \lambda\Phi(\tau)] =$$
$$= I_n - (\lambda - \lambda_0)[(1 - \lambda_0)I_n + \lambda\Phi(\tau)]^{-1}[I_n - \Phi].$$

Next, we will multiply both sides of the matrix singular equation, obtained in this manner, from the left by the operator $[I - (\lambda_1 - \lambda_0)H_{\lambda_0}]^{-1}$, $\lambda_1 \in \tilde{L}$, $\mid \lambda_1 - \lambda_0 \mid < 1/2C$. This step yields a new system which is likewise equivalent to system (3.9.7) and has the same index; the symbolic matrix of the new system is

$$I_n - (\lambda - \lambda_1)[(1 - \lambda_1)I_n + \lambda_1\Phi(\tau)]^{-1}[I_n - \Phi(\tau)].$$

By continuing this process, we arrive after k steps at a singular equation which satisfies condition a) above and has the symbolic matrix

$$I_k - (\lambda - \lambda_k)[(1 - \lambda_k)I_n + \lambda_k\Phi(\tau)]^{-1}[I_n - \Phi].$$

For sufficiently large k, we can set $\lambda_k = 1$; in the case of interest here , $\lambda = 1$ leads to a singular system the symbol of which equals the unit matrix. By design, the last system is equivalent to system (3.9.7), i.e., the operator A which is obtained from the operator (3.9.7) for $\lambda = 1$ admits equivalent regularization. By theorem 1.15, one has $Ind\, A \geq 0$.

Repeating our reasoning for the operator $I - \bar{\lambda}(I - A^*)$, adjoint to operator (3.9.7), we find that $Ind\, A^* \geq 0$ and, hence, $Ind\, A = 0$. Thus the theorem is proved.

Corollary 3.3. If the symbolic matrix of a matrix singular operator does not degenerate and is Hermitian, then the index of this operator vanishes.

CHAPTER 4
APPROXIMATE SOLUTION OF INTEGRAL EQUATIONS

Finite difference methods, widely employed for the solution of differential (or other operator) equations are also used to solve approximately integral equations; they include the methods of Ritz, Bubnov-Galerkin, least squares, collocation as well as a variety of these methods, linked to the application of so called finite elements. One has also been made of grid methods which, in the case of integral equations, assume the form of "methods of mechanical quadrature". The method of iteration has been applied for the solution of equations of the form $u - Au = f$, $\|A\| < 1$.

The method of mechanical quadrature is for integral equations one of the more important approximate methods. An essential scheme of this method is the utilization of quadratic and cubic formulas for the approximate evaluation of integrals. This last problematic acquires special significance for singular equations.

The problem of approximate solution of Fredholm integral equation is sufficiently well documented in many monographs; we note here the books [11,21,28], and likewise the paper [47,49]. Hence there exists no need to study such methods; we will limit our attention here to one- and two-dimensional singular integral equations as well as the Wiener-Hopf equations. We note that the problems of approximate solution of singular integral equations has been treated in separate chapters of the monographs [25,28,34,54]; the text [7] is devoted to that problem for the Wiener-Hopf equations. Questions of the errors of these approximate methods have been partially considered in the above monographs and in the book [53].

§1. Computation of singular integrals

1^0. One-dimensional singular integrals, extended over a sufficiently smooth closed contour, can be evaluated exactly in a wide enough class of cases. Such computations are based on the formulas (2.2.12) and (2.1.13). In particular, for example, if the density $P(\tau)$ is a polynomial in τ, then

$$\frac{1}{\pi i} \int_{\Gamma} \frac{P(\tau)}{t - \tau}\, d\tau = P(t)\,, \tag{4.1.1}$$

if, in addition, $P(0) = 0$, then

$$\frac{1}{\pi i} \int_{\Gamma} \frac{P(1/\tau)}{t - \tau}\, d\tau = -P(1/t)\,, \tag{4.1.2}$$

These formulas allow to compute Cauchy singular integrals the density of which are rational functions in τ. These formulas may be used in a number of cases of singular integrals, extended over infinite contours, if one can beforehand by use of some simple transformation reduce an infinite contour of integration to a finite, closed and sufficiently smooth contour. Likewise, it is obvious that a singular Cauchy integral, taken along a closed, sufficiently smooth contour, may be evaluated approximately, if the density of this integral can be approximated by a polynomial.

2^0. We will consider the two-dimensional operator h with the symbol $e^{i\nu}$:

$$(hu)(x) = \frac{1}{i} \int\limits_{R^2} u(y) r^{-2} e^{i\theta} \, dy \, .$$

One can construct a manifold of densities u for which this integral is readily evaluated. We have

$$(hu)(x) = \frac{1}{i} \int\limits_{R^2} r^{-2} (\cos\theta + i\sin\theta) \, u(y) \, dy \, .$$

Let

$$v(x) = \frac{1}{2\pi} \int\limits_{R^2} r^{-1} u(y) \, dy \, , \tag{4.1.3}$$

and assume that $u(y)$ fades out sufficiently quickly at infinity, so that this integral converges. By (3.4.2), differentiation of an integral with a weak singularity yields

$$(hv)(x) = \frac{2\pi}{i} \left(\frac{\partial v}{\partial x_1} + i \frac{\partial v}{\partial x_2} \right) \, . \tag{4.1.4}$$

We will introduce the three-dimensional Euclidean space R^3 with the coordinates x_1, x_2, x_3 and its half-space R^3_+ in which $x_3 > 0$, and consider the manifold of functions $v(x_1, x_2, x_3)$, harmonic in R^3_+, which fade away at infinity like $O((X_1^2 + X_2^2 + X_3^2)^{-1/2-\eta})$ with $\eta > 0$ and which are continuously differentiable in the closed half-space $\bar{R}^3_+ = (x_3 \geq 0)$. We will now set

$$u(x) = u(x_1, x_2) = \left. \frac{\partial v}{\partial x_3} \right|_{x_3=0} \, . \tag{4.1.5}$$

On the boundary x_3 of the half-space R^3_+, the function v satisfies the Neumann boundary condition $\partial v/\partial \nu = -u$; where ν is the external normal to R_+^3. The solution of this Neumann problem for the half-space is

$$v(x_1, x_2, x_3) = \frac{1}{2\pi} \int\limits_{R^2} \frac{u(y)}{\rho}\, dy \,;\ \rho^2 = (x_1 - y_1)^2 + (x_2 - y_2) + x_3 \,.$$

hence, by the already referred to formula (3.4.2),

$$\frac{\partial}{\partial x_k} \int\limits_{R^2} r^{-1} u(y)\, dy = 2\pi\, \frac{\partial v(x_1, x_2, 0)}{\partial x_k}\,,\ k = 1, 2;$$

if the function v is known, then the integral of interest is effectively evaluated.

We can use as function v the Kelvin transform of harmonic polynomials for the inversion, which transforms a sphere into the half-space R_+^3. Introduction of the same class of harmonic functions permits also the evaluation of the integrals $h^{-1}u$, $u = \frac{\partial v}{\partial x_3}\big|_{x_3=0}$. The evaluation of integrals $h^n u$, $n \neq \pm 1$ becomes more involved (cf. [26], §39).

3^0. In the general case, we may compute singular integrals by means of well known quadrature (cubature) formulas by beforehand reducing the integrals to ordinary non-singular or Lebesgue integrals.

Let Γ be a curve in the plane or on a surface in three-dimensional space, $K(x, y)$ the singular kernel and $u(y)$ a function which satisfies on Γ Hoelder condition. The singular integral with the kernel $K(x, y)$ can be transformed into the sum

$$\int\limits_{\Gamma} K(x, y)\, u(y)\, dy = u(x) \int\limits_{\Gamma} K(x, y)\, dy + \int\limits_{\Gamma} K(x, y)[u(y) - u(x)]\, dy \,. \quad (4.1.6)$$

The second integral on the right hand side is non-singular (in the general case, it is improper) and it may be evaluated by application of one or the other appropriate method. We mention here one method which is exposed in [32]. We will subdivide the surface Γ into curvilinear polygons Γ_k of small diameter. We will refer to their vertices, denoted y_{jk}, as nodes. Inside each polygon Γ_k, we select some point x_k, referred to as support. As approximate value of the above integral at the point x_k we take the sum

$$\sum_{j,l} K(x_k, y_{jl})[u(y_{jl}) - u(x_k)]\, |\, \Gamma_l\, |\,, \quad (4.1.7)$$

where $|\, \Gamma_l\, |$ is the area of the curvilinear polygon Γ_l. An extension of this method to dimension $m \neq 2$ is obvious.

As regards the first integral in (4.1.6), in a number of cases, its value will be known. Thus, if $K(x,y)$ is the Cauchy kernel $K(x,y) = dy/\pi i(x-y)$, then the relevant integral equals unity. If Γ is a circle with centre at x in R^2 and $K(x,y) = r^{-2}f(x,\theta)$, where f satisfies condition (3.2.4), then the integral of interest here vanishes.

In the general case, the integral in (4.1.6) may be evaluated approximately by letting

$$\int_\Gamma K(x,y)\,dy \approx \int_{\Gamma/\Gamma^\epsilon} K(x,y)\,dy\,, \qquad (4.1.8)$$

where Γ^ϵ is that part of the surface (or curve) which lies inside a sphere (or circle) of sufficiently small radius Γ with centre at the point x.

§2. The most important methods of solution of singular equations

1^0. We will apply the well known method of iteration to operator equations of the general form

$$u - \lambda Au = f\,, \qquad (4.2.1)$$

if the linear operator A is defined on some Banach space and bounded there, and the numerical parameter λ is such that $|\lambda|\,\|A\| < 1$. The exact solution of this equation is given by the series

$$u = \sum_{n=0}^\infty \lambda^n A^n f\,, \qquad (4.2.2)$$

in the capacity of an approximate solution, we may consider any partial sum of this series. Such an approximate solution is conveniently given the form

$$u^{(n)} = \sum_{k=0}^n \lambda^k f_k\,;\ f_0 = f\,,\ f_k = A(f_{k-1})\,. \qquad (4.2.3)$$

Since Fredholm operators as well as singular integral operators are bounded in corresponding spaces L_2 (and even in L_p), the method of iteration may be applied to Fredholm equations (and to equations with compact operators) as well as to singular equations, if they have the form (4.2.1) and the parameter λ is sufficiently small. The construction of the elements f_k in (4.2.3) essentially reduces to the evaluation of ordinary or singular integrals; in this context, the ideas of §2 may prove to be useful.

The range of applications of the iteration method may be somewhat extended, if use is made of the following observation by L.V.Kantorovich (cf.

[11]. §2, chapter 2). Let in equation (4.2.1) A be a Fredholm operator. As it is known, the Fredholm resolvent (cf., for example, [24],[33] regarding its concept and properties) is a meromorphic function in λ, the poles of which are the characteristic numbers of the corresponding Fredholm operator. Let λ_1 be the smallest in modulus of the characteristic numbers of such an operator. It is known that then the series (4.2.2) will converge for $\mid \lambda \mid < \mid \lambda_1 \mid$, and, generally speaking, it diverges for $\mid \lambda \mid \geq \mid \lambda_1 \mid$. We will assume that on the circle $\mid \lambda \mid = \mid \lambda_1 \mid$ there are no other characteristic numbers, so that the next such number is in modulus larger than λ_1. Without loss of generality, we may assume that $\lambda_1 = -1$. We will denote by p the multiplicity of the number $\lambda_1 = -1$ as pole of the resolvent. Then the product $(1 + \lambda)^p u$ is holomorphic in the circle $\mid \lambda \mid < \mid \lambda_2 \mid$ and, consequently, it may be expanded in this circle in a power series in λ:

$$(1 + \lambda)^p u = (1 + \lambda)^p \sum_{k=0}^{\infty} \lambda^k f_k = \sum_{\nu=0}^{p} \sum_{k=0}^{\infty} \lambda^{p+k} \binom{p}{\nu} f_k =$$

$$= \sum_{m=0}^{\infty} \lambda^m \sum_{\nu=0}^{min(p,m)} \binom{p}{\nu} f_{m-\nu} .$$

We can now take as approximate solution the expression

$$u^{(n)} = (1 + \lambda)^{-p} \sum_{m=0}^{\infty} \lambda^m \sum_{\nu=0}^{min(p,m)} \binom{p}{\nu} f_{m-\nu} . \tag{4.2.4}$$

This approximate formula is valid for $\mid \lambda \mid < \mid \lambda_2 \mid$; it requires evaluation of the same iterations f_m as formula (4.2.3). We not the case $p = 1$, $\lambda = 1$. As is readily seen, in this case

$$u^{(n)} = \sum_{m=0}^{n-1} f_m + \frac{1}{2} f_n . \tag{4.2.5}$$

This formula differs from (4.2.3) only by the fact that the term f_n is replaced by $f_n/2$.

    2^0. The method of Bubnov-Galerkin may b used to solve approximately Fredholm equations. This method is studied in detail in [27]; we will restrict ourselves here to a very simple summary. By means of this method, a sequence of finite-dimensional subspaces $\{H_n\}$ is generated which is complete in the

Hilbert space under consideration (for Fredholm equations, this is usually the corresponding space L_2); in H_n, the base is $(\varphi_{n1}, \varphi_{n2}, ..., \varphi_{nN})$, $N = N(n)$.

If the given equation has the form

$$u + Au = f, \qquad (4.2.6)$$

where A is a compact operator, then its approximate solution is constructed as element $u^{(n)} \in H_n$,

$$u^{(n)} = \sum_{k=1}^{N} a_k^{(n)} \varphi_{nk}, \qquad (4.2.7)$$

with coefficients which satisfy the linear algebraic system

$$\sum_{k=1}^{N} a_k^{(n)} (\varphi_{nk}, \varphi_{nj}) - (f, \varphi_{nj}), \; j = 1, 2, ..., N. \qquad (4.2.8)$$

It has been shown that, if the operator $(I + A)$ is boundedly invertible, then the system (4.2.8) is soluble for large n in a unique manner and $u^{(n)} \longrightarrow u$ in the metric of the given Hilbert space. The method of Bubnov-Galerkin may also be applied to singular integral equation, if their index is zero, and the equation is soluble in a unique manner; for this purpose, one must beforehand regularize the equation in an equivalent manner. However, it must be pointed out that the last operation, generally speaking, is not elementary.

3^0. The method of least squares may be applied to Fredholm as well as to singular integral equations. This method is studied in detail in [27]; we will summarize here its basic ideas.

Let in the equation

$$Au = f, \qquad (4.2.9)$$

A be a linear bounded and boundedly invertible operator in some Hilbert space H. As in 2^0, we will select a sequence of finite-dimensional subspaces $\{H_n\}$, which lie completely in H, and in each of these a base $(\varphi_{n1}, \varphi_{n2}, ..., \varphi_{nN})$. We will seek an approximate solution of (4.2.9) in the form (4.2.7); however, we will now determine the coefficients $a_k^{(n)}$ from the conditions $\|Au^{(n)} - f\| = min$ which lead to the system of linear algebraic equations

$$\sum_{k=1}^{N} (A\varphi_{nk}, A\varphi_{nj}) a_k^{(n)} = (f, A\varphi_{nj}), \; j = 1, 2, ..., N. \qquad (4.2.10)$$

This system has a unique solution for any n and the corresponding approximate solution is $u^{(n)} \to A^{-1} f$, $n \to \infty$.

The operators on the left hand sides of Fredholm as well as singular equations, under ordinary conditions (cf. chapters 2 and 3), are bounded in corresponding spaces L_2; the bounded invertibility of these operators imposes the following requirement: the homogeneous equation $Au = 0$ must have only the trivial solution $u = 0$; this condition is sufficient, if $Ind\,A = 0$.

Several aspects of the applicability of the method of least squares in the case when $Ind\,A \neq 0$ have been studied in chapter XVIII of [54].

4^0. We will investigate now the essence of the method of collocations, proposed at his time by L.V.Kantorovich in [9] (cf. also [10]). Let it be required to solve equation (4.2.9) in which the linear operator A acts on the Banach space X in the Banach space Y, where the elements of the manifold Y are functions continuous on some manifold K, compact in R^m; as usually, we denote here by R^m the Euclidean space of dimension m. We will assume that the operator A^{-1} exists, is defined throughout the space Y and is bounded in it.

We will select a sequence of finite-dimensional subspaces $\{X_n\}$, completely in X. Let $dim\,X_n = N(n) = N$ and $Y_n = AX_n$. By supposition, there exists the operator A^{-1} which is defined and continuous on the entire space Y. Hence $dim\,Y_n = dim\,X_n = N$. Moreover, if $\{\varphi_{nj}\}$, $j = 1, 2, ..., N$ is the base in X_n, and $\psi_{nj} = A\varphi_{nj}$, then $\{\psi_{nj}\}$ is the base in Y_n.

We will select in K points t_j, $j = 1, 2, ..., N$, the so-called nodes of the collocations. An approximate solution of equation (4.2.9) will be sought in the form (4.2.7), and we will determine the coefficients $a_k^{(n)}$ from the conditions that (4.2.9) is to be satisfied at the collocation nodes. This process leads to the algebraic equations

$$\sum_{k=1}^{N} a_k^{(n)}\,\psi_k(t_j) = f(t_j)\,. \tag{4.2.11}$$

The method of collocations can be applied to Fredholm as well as to singular integral equations. In the work of S. Proessdorf and his students, much attention has been given to applications of the method of collocations to one-dimensional singular equations; their results are studied in sufficient completeness in [34] and [54].

§3. On projection methods of solution of Wiener-Hopf equations

1^0. Let T be an operator in the Hilbert space H and $\{P_n\}$ an increasing manifold of projectors in H (numbered discretely $n = 0, 1, 2, ...$ or by continuous parameters $n \in [0, \infty]$), where for any $x \in H$ the relation $P_n x \to x$ for $n \to \infty$ is fulfilled.

We will say that T is invertible by the projection method according to the system of projectors $\{P_n\}$, if:

1. The operator $T_n = P_n T$ is invertible in the space $Im(T_n) = \{T_n y : y \in H\}$ for sufficiently large n;

2. there exists for any $y \in H$ the limit.

$$x = \lim_{n \to \infty} T_n^{-1} P_n y \,, \ Tx = y \,.$$

We will formulate now an assertion proved in [7].

Theorem 3.1. The operator T is invertible by the projection method according to the system $\{P_n\}$ if and only if for sufficiently large n the operators T_n are invertible and

$$\sup_{n \geq N} \|T_n^{-1}\| \leq c = const < \infty \,. \tag{4.3.1}$$

Corollary 3.1. Let T be invertible by the projection method according to the system $\{P_n\}$, then the estimate

$$\|x_n - P_n\| \leq c\|P_n T\| \cdot \|x - P_n x\| \,, \tag{4.3.2}$$

is true for the approximate solution $x_n = T_n^{-1} P_n y$, where $x \in H$ is the solution of the equation $Tx = y$.

Proof. It follows from the equations

$$P_n T P_n x_n = P_n y \,, \ P_n T x = P_n y \,, \ P_n x_n = x_n \,,$$

that

$$P_n T x_n = P_n y \,, \ P_n T P_n x + P_n T (I - P_n) x = P_n y \,.$$

We subtract the second equation from the first to obtain

$$P_n T (x_n - P_n x) = -P_n T (I - P_n) x \,. \tag{4.3.3}$$

Estimate (4.3.1) and equation (4.3.3) now yields (4.3.2).

We note that the union of all operators, invertible by the projection method according to the system $\{P_n\}$ does not contain all invertible operators. However, for any invertible operator, one can find a corresponding system of projectors $\{P_n\}$ according to which this operator is invertible by the projection method. A detailed study of general projection methods is contained in [7,34].

2^0. We have considered in §8 of chapter 2 the generalized Wiener-Hopf equation

$$T_p(a)u = v \, , \tag{4.3.4}$$

where $v = Pv$, P is an orthogonal projector in the Hilbert space H, A is a positive definite operator and $T_p(A)$ is the convolution of the operator PA on the subspace $Im(P)$.

We will now present several projection methods for the solution of equation (4.3.4).

Let $\{p_j\}$ be an orthonormal base in the space $Im(A^{1/2}P)$; then the system $\{A^{-1/2}p_j\}$ forms a base in the subspace $Im(P)$. In fact, if $v \in Im(P)$, then $A^{1/2}v \in Im(A^{1/2}P)$ and

$$A^{1/2}v = \sum_{j=1}^{\infty}(A^{1/2}v, p_j)p_j \, . \tag{4.3.5}$$

Hence

$$v = \sum_{j=1}^{\infty}(Av, A^{-1/2}p_j)A^{-1/2}p_j \tag{4.3.6}$$

and this series converges simultaneously with the series (4.3.5). Besides, if

$$v = \sum_{j=1}^{\infty}\xi_j(v)A^{-1/2}p_j \, ,$$

is any other expansion of v in a series according to the system $\{A^{-1/2}p_j\}$, then

$$A^{1/2}v = \sum_{j=1}^{\infty}\xi_j(v)p_j$$

and $\xi_j(v) = (Av, A^{-1/2}p_j)$, since $\{p_j\}$ is an orthonormal base. Thus, any element $v = Pv$ admits a single-valued representation in the form (4.3.6), i.e., $\{A^{-1/2}p_j\}$ is a base in $Im(P)$.

Next, let P_n be an orthogonal projector on the subspace, tight on the vectors $q_j = A^{-1/2}p_j$, $j = 1, 2, ..., n$. Obviously, $Im(P_n) \subset Im(P_{n+1})$, i.e., $\{P_n\}$ is an increasing family of projectors. Besides, $P_n v \to v$ for $n \to \infty$ for any $v \in Im(P)$.

Theorem 3.2. The solution if the Wiener-Hopf equation (4.3.4) with positive definite operator A cab be given the form

$$u = \sum_{j=1}^{\infty}(v, q_j)q_j \, , \quad v = Pv \, . \tag{4.3.7}$$

Proof. We will verify that

$$u_n = \sum_{j=1}^{\infty} (v, q_j) q_j \,, \quad v = Pv \,,$$

satisfies the equation

$$P_n T_n(A) u_n = P_n v \,. \tag{4.3.8}$$

For this purpose, we will prove the auxiliary

Lemma 3.1. Let P_0 be an orthogonal projector, then

$$Im(A^{1/2} P_0) = \{ h \in H : (h, g) = 0 \ \forall g : P_0 A^{1/2} g = 0 \} \,.$$

Proof. If $h \in Im(A^{1/2} P_0)$, then $h = A^{1/2} P_0 f$ and

$$(A^{1/2} P_0 f, g) = (P_0 f, A^{1/2} g) = (P_0 f, P_0 A^{1/2} g) + (P_0 f, (I - P_0) A^{1/2} g) = 0 \,.$$

Hence for all g such that $P_0 A^{1/2} g = 0$, we find $h \perp Ker(P_0 A^{1/2})$. The inverse conclusion is proved in an analogous manner. If $h \perp Ker(P_0 A^{1/2})$, then

$$(h, g) = 0 \,, \quad P_0 A^{1/2} g = 0$$

i.e.,

$$0 = (A^{-1/2} h, A^{1/2} g) \,, \quad A^{1/2} g \perp Im(P_0) \,.$$

Since the operator $A^{1/2}$ is invertible, then $A^{-1/2} h \in Im(P_0)$ and therefore $h \in Im(A^{1/2} P_0)$ which proves the lemma.

For $v = Pv$, we have

$$P_n P A u_n = P_n A^{1/2} \sum_{j=1}^{n} (A^{-1/2} v, A^{1/2} q_j) A^{1/2} q_j \,. \tag{4.3.9}$$

Let P_n^0 be an orthogonal projector on the subspace $Im(A^{1/2} P_n)$, then

$$P_n^0 A^{-1/2} v = \sum_{j=1}^{n} (A^{-1/2} v, A^{1/2} q_j) A^{1/2} q_j \in Im(A^{1/2} P_n) \,.$$

By lemma 3.1, we find that $H = Im(A^{1/2} P_n) \oplus Ker(P_n A^{1/2})$ is an expansion in the orthogonal sum of subspaces. Hence, by (4.3.9), we obtain

$$P_n P A u_n = P_n A^{1/2} (P_n^0 + (I - P_n^0)) A^{-1/2} v - P_n v \,,$$

i.e., u_n is a solution of equation (4.3.8). By theorem 3.1, we must still verify that the norm $\|u_n\|$ is uniformly bounded with respect to n.

We have the estimate

$$\|u_n\| = \left\|A^{-1/2}\sum_{j=1}^{n}(A^{-1/2}v, A^{1/2}q_j)A^{1/2}q_j\right\| \le$$

$$\le \|A^{-1/2}\|\,\|P_n^0 A^{-1/2}v\| \le \|A^{-1/2}\|^2\|Pv\|,$$

which complete the proof of theorem 3.2.

In the case of the Wiener-Hopf integral equation

$$u(t) - \int\limits_{0}^{\infty} k(t-s)u(s)\,ds = v(t),\ t > 0, \tag{4.3.10}$$

the projector P is determined by the equation

$$(Pu)(t) = \begin{cases} u(t),\, t \ge 0 \\ 0,\, t < 0 \end{cases},$$

and the operator

$$(Au)(t) = u(t) - \int\limits_{-\infty}^{\infty} k(t-s)u(s)\,ds,\ k \in L_1(R^1),$$

acts in the Hilbert space $L_2(R^1)$. Positive definiteness of the operator A is equivalent to the condition

$$1 - (Fk)(t) = 1 - \int\limits_{-\infty}^{\infty} k(s)e^{its}\,ds \ge \delta > 0$$

for almost all $t \in R^1$.

The obvious identity

$$1 - (Fk)(t) = \left[1 - \left(1 - \sqrt{1 - (Fk)(t)}\right)\right]^2$$

determines the operator

$$(A^{1/2}u)(t) = u(t) - \int\limits_{-\infty}^{\infty} K_{1/2}(t-s)\,u(s)\,ds\,,$$

$$K_{1/2}(\tau) = F^{-1}\big(1 - \sqrt{1 - Fk}\,\big)(\tau)\,.$$

A description of the subspace $Im(A^{1/2}P)$ and the choice in it of an orthonormal base are complicated in the general case of the problem. However, practical realization of the method, proposed in theorem 3.2, may be achieved in the following manner. Select a base $\{q_j\}$ (not necessarily an orthonormal one) in the space $Im(P)$. Then the system $\{q_j\}$ is orthogonalized (following to Gram Schmidt) with respect to the scalar product $(Au, v) = (u, v)_A$ in H, which will determine a norm, equivalent to the initial one. As a result, we obtain a system $\{q'_j\}$, satisfying the condition

$$(Aq'_j, q'_k) = (A^{1/2}q'_j, A^{1/2}q'_k) = \delta_{jk}\,,$$

where δ_{jk} is the Kronecker delta. It is readily shown that $\{A^{1/2}q'_j\}$ is an orthonormal base in $Im(A^{1/2}P)$. This process of orthogonalization permits us to construct consecutive approximations to the solution of the generalized Wiener-Hopf equation (4.3.4).

3^0. Certain special projection methods are applicable to equation (4.3.10) (cf. [7] where also this system of equations is treated).

We will denote by $P_\tau(0 < \tau < \infty)$ the projector, defined by

$$(P_\tau u)(t) = \begin{cases} u(t),\, 0 < t < \tau\,, \\ 0\,,\, t \notin (0, \tau). \end{cases}$$

Side by side with (4.3.10), we will consider the truncated equation

$$(T_\tau u)(t) \equiv (P_\tau A P_\tau u) = u(t) - \int\limits_{0}^{\infty} k(t-s)\,u(s)\,ds = v(t),\ 0 < t < \tau\,.$$

$$(4.3.11)$$

If, for sufficiently large $\tau \geq N > 0$, the operators T_τ are invertible in $P_\tau L_2(R^1_+)$ and $\sup\limits_{\tau \geq N} \|T_\tau^{-1}\| < +\infty$, then the operator $T_p(A)$, defined by the left hand side of equation (4.3.10), is invertible by the projection method with respect to the system of projectors $\{P_\tau\}$. By theorem 2.13 (cf. §8, chapter 2), we find

$$1 - (F\dot{k})(t) \neq 0, \ t \in \dot{R}^1, \ \kappa = \frac{1}{2\pi}\Big[arg(1 - Fk)\Big]_{-\infty}^{+\infty} = 0. \qquad (4.3.12)$$

It may be shown (cf. [7]) that this condition is sufficient for the invertibility of $T_p(A)$ by the projection method with respect to the system $\{P_\tau\}$. In that case, beginning with some $\tau > 0$, equation (4.3.11) has a unique solution $u_\tau(t)$ and the functions

$$\tilde{u}_\tau(t) = \begin{cases} u_\tau(t), \ 0 < t < \tau \\ 0, \ \tau < t < \infty \end{cases}.$$

converge for $\tau \to \infty$ with respect to the norm of the space $L_2(R_+^1)$ to the solution of equation (4.3.10).

The following method of solution of (4.3.10) is based on a special choice of the orthonormalized base $\{q_j'\}$ in the space $Im(P) = L_2(R_+^1)$. We will consider the operators

$$(Vu)(t) = u(t) - 2 \int\limits_{-\infty}^{t} e^{s-t}u(s)\,ds, \ 0 < t < \infty,$$

$$(V^{(-1)}u)(t) = u(t) - 2 \int\limits_{t}^{\infty} e^{t-s}u(s)\,ds,$$

which are mutually invertible. We will define the system of functions $\{q_j\}$, setting

$$q_0(t) = \sqrt{2}\,e^{-t}, \ q_{j+1}(t) = (Vq_j)(t), \ j = 0, 1, \ldots; \ t > 0.$$

Direct computations permit to establish a link between the functions $\{q_j\}$ and the normalized Laguerre polynomials $\{\Lambda_j\}$:

$$q_j(t) = \sqrt{2}\,e^{-t}\Lambda_j(2t), \ j = 0, 1, \ldots; \ t > 0.$$

As it is known, the sequence $\{\Lambda_j\}$ is simply determined from the relations

$$\int\limits_{0}^{\infty} e^{-t}\Lambda_j(t)\Lambda_k(t)\,dt = \delta_{jk}, \ (j, k = 0, 1, \ldots.),$$

where $\Lambda_0(t) = 1$, $\Lambda_j(t)$ are polynomials of degree j. Hence the system of functions $\{q_j\}$ form an orthonormal base in the space $L_2(R^1_+)$. Let P_n be an orthogonal projector on the subspace, tight on the first n functions of the system $\{q_j\}$:

$$(P_n u)(t) = \sum_{j=0}^{n-1} (u, q_j) q_j(t), \quad t > 0.$$

The equation

$$P_n T_k(A) P_n u = P_n v$$

is equivalent to the algebraic system of equations

$$\sum_{j=0}^{n-1} (T_p(A)q_j, q_k)(u, q_j) = (v, q_k), \quad k = 0, 1, ..., n-1, \tag{4.3.13}$$

relative to (u, q_j), $j = 0, 1, ..., n-1$. The elements of the matrix of this system depend only on differences of subscripts:

$$(T_k(A)q_{j+m}, q_{k+m}) = (T_k(A)V^m q_j, V^m q_k) = (V^{-m} T_p(A)V^m q_j, q_k) =$$
$$= (T_p(A)q_j, q_k),$$

since $V^{-m} T_p(A)V^m = T_p(A)$ and the operator V^{-1} is adjoint to the operator V (where V^m and V^{-m} are the degrees of the operators V and V^{-1}, respectively). These properties of the operators V and V^{-1} are established by direct inspection, - they called for the choice of the base $\{q_j\}$.

Using properties of Fourier transforms, we find

$$c_{k-j} = (T_p(A)q_k, q_j) = (APq_k, Pq_j) = \frac{1}{\pi} \int_{-\infty}^{\infty} \frac{1 - (Fk)(\lambda)}{1 + \lambda^2} \left(\frac{\lambda - i}{\lambda + i}\right)^{k-j} d\lambda.$$

For the positive definite operator A, we apply the method of 2^0. We will find the first approximation to the solution of equation (4.3.10). We have

$$q_0' = q_0, \quad q_1' = q_1 + a_{11}q_0, \quad a_{11} = -c_1 c_0^{-1},$$

when this first approximation assumes the form

$$\frac{1}{c_0}(Pv, q_0)q_0(t) + \frac{c_0}{p_1}\left(Pv, q_1 - \frac{c_1}{c_0}q_0\right)\left[q_1(t) - \frac{c_1}{c_0}q_0(t)\right],$$

where $b_1 = c_0^2 + |c_1|^2 - \overline{c_1 c_1} - c_1 c_1$.

PART II

PROBLEMS OF THE THEORY OF ELASTICITY AND
CRACKS MECHANICS

CHAPTER 5
THE INTEGRAL EQUATIONS OF CLASSICAL TWO–DIMENSIONAL PROBLEMS

The two–dimensional problems of the theory of elasticity form a class
of problems which lend themselves, in contrast to the majority of three–
dimensional problems, to analytical treatment and at the same time do not
have such intuitively clear answers as one-dimensional problems. This intermediate position between the one-dimensional and three-dimensional problems determines the special position of the two-dimensional problems of the
theory of elasticity.

This chapter deals in the following order with the so-called plane problems of the theory of elasticity: Plane strain, plane stress and generalized
plane stress.

§1. The plane problem of the theory of elasticity

1^0. A state of plane strain is defined by the displacement vector

$$\mathbf{U} = (u(x,y), v(x,y), 0). \tag{5.1.1}$$

One considers an elastic body Q, representing a massive long cylinder
$Q = \Omega \times [0, l]$, subject to the body forces.

$$\mathbf{F} = (F_x(x,y), F_y(x,y), 0) \tag{5.1.2}$$

and the, given on the side surface $\partial\Omega \times [0, l]$, surface forces

$$\mathbf{f} = (f_x(x, y), f_y(x, y), 0) \tag{5.1.3}$$

or displacements

$$\mathbf{u}_0 = (u_0, v_0, 0), \tag{5.1.4}$$

where $\mathbf{f}$ and $\mathbf{u}_0$ do not depend on z.

By (5.1.1), one has the strain tensor *

$$e_{xx} = \frac{\partial u}{\partial x}, \; e_{xy} = \frac{1}{2}\left(\frac{\partial v}{\partial x} + \frac{\partial u}{\partial y}\right), \; e_{yy} = \frac{\partial v}{\partial y},$$

$$e_{xz} = 0, \; e_{yz} = 0, \; e_{zz} = 0, \; e = e_{xx} + e_{yy} \tag{5.1.5}$$

and, in according to Hooke's law, the stress tensor

$$\sigma_{xx} = \lambda e + 2\mu e_{xx}, \; \sigma_{yy} = \lambda e + 2\mu e_{yy},$$

$$\sigma_{xy} = 2\mu e_{xy}, \; \sigma_{xz} = 0, \; \sigma_{yz} = 0. \tag{5.1.6}$$

System (5.1.5) and (5.1.6) is closed by the equilibrium equations

$$\frac{\partial\sigma_{xx}}{\partial x} + \frac{\partial\sigma_{yx}}{\partial y} + \mathbf{F}_x = 0;$$

$$\frac{\partial\sigma_{xy}}{\partial x} + \frac{\partial\sigma_{yy}}{\partial y} + \mathbf{F}_y = 0. \tag{5.1.7}$$

Written in terms of displacements, system (5.1.5)-(5.1.7) becomes

$$\mu\Delta\mathbf{u} + (\lambda + \mu)\frac{\partial}{\partial x}(div\,\mathbf{u}) + \mathbf{F}_x = 0,$$

$$\mu\Delta\mathbf{u} + (\lambda + \mu)\frac{\partial}{\partial y}(div\,\mathbf{u}) + \mathbf{F}_y = 0 \tag{5.1.8}$$

---

* Let's notice that here and hereafter we'll suppose a possibility of complete linearization (see,f.e. [56]). That is justified when lengthenings , displacements and angle turns are little in comparison with the unit and have the same order of littleness. In addition to that and under conditions of complete linearization we'll neglect the change of the body board due to the body deformations. In this case all tension tensors: "generalijed", of Euler, of Kirchhoff are identified.

with the boundary conditions

$$\mathbf{u}\,|_{\partial\Omega} = \mathbf{u}_0 \tag{5.1.9}$$

or

$$\sigma^{(n)}\,|_{\partial\Omega} = 2\mu\frac{\partial\mathbf{u}}{\partial n} + \lambda\mathbf{n}\,div\,\mathbf{u} + \mu\mathbf{n}\times rot\,u\,|_{\partial\Omega} \equiv B\mathbf{u}\,|_{\partial\Omega} = \mathbf{f}\,. \tag{5.1.10}$$

Note that under conditions of plane strain, the stress components σ_{zz}:

$$\sigma_{zz} = \lambda e\,, \tag{5.1.11}$$

which, however, is determined directly by the solution of problem (5.1.8)-(5.1.10).

2^0. The problem of plane stress [56]

$$\sigma_{zz} = \sigma_{xz} = \sigma_{yz} = 0\,,\ \sigma_{xx} = \sigma_{xx}(x,y)\,,$$

$$\sigma_{yy} = \sigma_{yy}(x,y)\,,\ \sigma_{xy} = \sigma_{yx} = \sigma_{yz}(x,y)$$

formulated in an analogous manner, leads to the surplus determined system. Its detailed formulation follows.

Consider an elastic region in the form of a finite, right cylinder with cross-section Ω. Let there act on the side surface of this cylinder forces which are perpendicular to its generators and depend only on the coordinates (x,y) in the section Ω. The ends of the cylinder are not loaded.

It is readily verified that the displacement vector $\mathbf{u} = (u,v,w)$ with the components

$$u = \frac{\lambda}{4\mu(2\mu+3\lambda)}\,a(z^2 - x^2) + \frac{\mu+\lambda}{\mu(2\mu+3\lambda)}\,bxy -$$

$$-\frac{\mu+\lambda}{4\mu(2\mu+3\mu)}\,ay^2 + \frac{2\mu+\lambda}{2\mu(2\mu+3\lambda)}\,cx - \frac{\varphi}{2\mu}\,,$$

$$v = \frac{\lambda}{4\mu(2\mu+3\lambda)}\,b(z^2 - x^2) - \frac{\mu+\lambda}{\mu(2\mu+3\lambda)}\,bx^2 + \tag{5.1.12}$$

$$+\frac{\mu+\lambda}{\mu(2\mu+3\mu)}\,axy + \frac{\mu+\lambda}{2\mu(2\mu+3\lambda)}\,cy + \frac{\psi}{2\mu}\,,$$

$$w = \frac{\lambda}{2\mu(2\mu+3\lambda)}\,(ax + by + 2c)z$$

satisfies the three-dimensional system of Lame equations, where $\varphi + i\psi$ is an analytic function of $x + iy$. Besides, the components σ_{12}, σ_{11} and σ_{22} depend only on x and y, while $\sigma_{13} = \sigma_{23} = \sigma_{33} = 0$, so that pale stress is achieved.

The boundary conditions on the side surface of the cylinder have the form

$$0 = \sigma_{11}n_1 + \sigma_{12}n_2 + f_x = byn_1 + cn_1 - \frac{\partial\varphi}{\partial n} + f_x ,$$
$$0 = \sigma_{21}n_1 + \sigma_{22}n_2 + f_y = axn_2 + cn_2 - \frac{\partial\varphi}{\partial s} + f_y ,$$
$$(5.1.13)$$

where $n_1 = cos(n, x)$, $n_2 = cos(n, y)$, n is the external normal to $\partial\Omega$ and s is the tangent to $\partial\Omega$. It will be assumed that the resultant vector of the forces and the resultant moment vanish, i.e.,

$$\int_{\partial\Omega} f_x \, ds = \int_{\partial\Omega} f_y \, ds = \int_{\partial\Omega} (xf_y - yf_x) \, ds = 0 . \qquad (5.1.14)$$

The choice of the coordinate system (x, y) may be made to ensure that

$$\int_{\Omega} x \, dx \, dy = \int_{\Omega} y \, dx \, dy = \int_{\Omega} xy \, dx \, dy = 0 . \qquad (5.1.15)$$

Using the Cauchy-Riemann equations

$$\frac{\partial\varphi}{\partial x} = \frac{\partial\psi}{\partial y} , \quad \frac{\partial\varphi}{\partial y} = -\frac{\partial\psi}{\partial x}$$

and (5.1.13), one finds

$$\frac{\partial\varphi}{\partial n} = \frac{\partial\varphi}{\partial x} n_1 + \frac{\partial\varphi}{\partial y} n_2 = \frac{\partial\psi}{\partial y} n_1 - \frac{\partial\psi}{\partial x} n_2 = byn_1 + cn_1 + f_x ,$$

$$\frac{\partial\varphi}{\partial s} = \frac{\partial\varphi}{\partial x} n_2 - \frac{\partial\varphi}{\partial y} n_1 = axn_2 - cn_2 - f_y .$$

Thus, one has for the function $H(\xi) = \varphi + i\psi$, analytic in the region Ω, with $\xi = x + iy$, the conditions

$$\frac{\partial H}{\partial s} = \frac{\partial\varphi}{\partial s} + i\frac{\partial\psi}{\partial s} = -(axn_2 + cn_2 + f_y) - i(byn_1 + cn_1 + f_x) , \qquad (5.1.16)$$

which are equivalent to (5.1.13). Define on the boundary of the region Ω, the function

$$G(\xi) = [(f_x + cn_1 + byn_1) - i(f_y + cn_2 + axn_2)] \frac{d\xi}{ds} ,$$

when (5.1.16) assumes the form

$$\frac{\partial H}{\partial s} = -iG\,\frac{d\xi}{ds}\,,\ \xi = x + iy\,,\ \xi \in \partial\Omega\,. \tag{5.1.17}$$

Consequently, $\partial H/\partial \xi = -iG$ for $\xi \in \partial\Omega$, hence the function G is analytically continued into Ω (assuming the boundary of Ω and the functions f_x, f_y to be smooth.).

The condition of analytic continuation of the function $G(\xi)$ may be given the form

$$\int_{\partial\Omega} \frac{G}{\xi - \eta}\,d\xi = 0\,,\ \eta \notin \bar\Omega\,,$$

or, using the formulas of Sokhotskii-Plemelj, the form

$$G(t) - \frac{1}{\pi i}\int_{\partial\Omega} \frac{G(\xi)\,d\xi}{\xi - t} = 0\,,\ t \in \partial\Omega\,. \tag{5.1.18}$$

If (5.1.18) is fulfilled, then follows from (5.1.17) the formula $\varphi + i\psi$

$$\varphi + i\psi = c_0 - \int_{\xi_0}^{\xi}\int_{\partial\Omega} \frac{G(t)\,dt}{t - \eta}\,d\eta\,,\ \xi_0, \xi \in \Omega\,. \tag{5.1.19}$$

In particular, for $f_x = pn_1$ and $f_y = pn_2$ (this is the case studied in the work of C. Cassisa and G. Fichera [86] and, likewise, W. Hayman [91]), one has

$$iG(\xi) = [(p + c + by)n_1 - i(p + c + ax)n_2](n_1 - in_2)\,,\ \xi \in \partial\Omega\,.$$

Thus, condition (5.1.18) ensures that on the boundary of $\partial\Omega$ a state of plane stress is attained.

Following [87], the boundary of a region Ω will be called singular * under the conditions: There exists a function $g = g_1 + ig_2 \not\equiv 0$, analytic in Ω and

---

* Don't confuse with the singular points of the contour.

continuous in $\bar{\Omega}$, such that

$$1. \; Re\left\{\frac{g(\xi)}{n_1 - in_2}\right\} = 0, \; \xi = x + iy \in \partial\Omega;$$

$$2. \; \int_{\partial\Omega} xg_1 n_2 \, ds = 0;$$

$$3. \; \int_{\partial\Omega} yg_2 n_1 \, ds = 0;$$

$$4. \; \int_{\partial\Omega} (g_1 n_2 - g_2 n_1) \, ds = 0.$$

Otherwise, the boundary $\partial\Omega$ will be called non-singular. It turns out that $\partial\Omega$ is non-singular if and only if $f_x = pn_1$ and $f_y = pn_2$ realize a state of plane stress.

W. Hayman [91] has obtained sufficient conditions on the boundary of a region Ω which ensure non-singularity of $\partial\Omega$. Besides, he has shown that, if a contour has a centre of symmetry, then there exist in an arbitrarily small neighbourhood of the contour analytic curves S_1 and S_2 which are singular and non-singular, respectively.

Consider now a cylinder the height of which is small compared to its cross-section dimensions, i.e., a plate. Let its free faces be loaded by surface forces $\mathbf{f} = (f_x(s, z), f_y(s, z), 0)$, and a body force $\mathbf{F} = (F_x(x, y), F_y(x, y), 0)$, which are symmetric with respect to the mean plane. Under such a loading, the plate will deform without bending.

Using

$$\sigma_{zz} = 0 \tag{5.1.20}$$

and mean $\sigma_{xx}, \sigma_{xy}, \sigma_{yy}, u, v, f_x, f_y, F_x, F_y$, defined by

$$\hat{\sigma}_{xx}(x, y) = \frac{1}{h} \int_{-h/2}^{h/2} \sigma_{xx}(x, y, z) \, dz \, , ..., \tag{5.1.21}$$

one arrives, using Cauchy's formula, Hooke's law and the equilibrium equations, at the system of equations

$$\hat{e}_{xx} = \frac{\partial \hat{u}}{\partial x} \, , \; \hat{e}_{yy} = \frac{\partial \hat{v}}{\partial x} \, , \; \hat{e}_{xy} = \frac{1}{2}\left(\frac{\partial \hat{u}}{\partial y} + \frac{\partial \hat{v}}{\partial x}\right), \tag{5.1.22}$$

$$\hat{\sigma}_{xx} = 2\mu\hat{e}_{xx} + \lambda^*(\hat{e}_{xx} + \hat{e}_{yy}), \ \hat{\sigma}_{xx} = 2\mu\hat{e}_{xy},$$
$$\hat{\sigma}_{yy} = 2\mu\hat{e}_{yy} + \lambda^*\hat{e}, \ \hat{e} = (\hat{e}_{xx} + \hat{e}_{yy}), \tag{5.1.23}$$

$$\frac{\partial\hat{\sigma}_{xx}}{\partial x} + \frac{\partial\hat{\sigma}_{xy}}{\partial y} + \hat{F}_x = 0, \ \frac{\partial\hat{\sigma}_{xy}}{\partial x} + \frac{\partial\hat{\sigma}_{yy}}{\partial y} + \hat{F}_y = 0 \tag{5.1.24}$$

and the boundary conditions

$$\hat{\sigma}^{(n)}\,|_{\partial\Omega} = \hat{\mathbf{f}}, \tag{5.1.25}$$

where

$$\lambda^* = \frac{2\mu\lambda}{2\mu + \lambda}. \tag{5.1.26}$$

Substitution of (5.1.23) into (5.1.24) yields

$$\mu\Delta\hat{u} + (\lambda^* + \mu)\frac{\partial}{\partial x}(div\ \hat{\mathbf{u}}) + \hat{F}_x = 0,$$
$$\mu\Delta\hat{v} + (\lambda^* + \mu)\frac{\partial}{\partial y}(div\ \hat{\mathbf{u}}) + \hat{F}_y = 0. \tag{5.1.27}$$

This system is supplemented by the boundary conditions

$$\hat{\sigma}^{(n)}\bigg|_{\partial\Omega} = B\hat{\mathbf{u}}\bigg|_{\partial\Omega} = \hat{\mathbf{f}} \tag{5.1.28}$$

or specification of displacements

$$\hat{\mathbf{u}}\bigg|_{\partial\Omega} = \hat{\mathbf{u}}_0. \tag{5.1.29}$$

It is now obvious that the problems of plane strain and generalized plane stress reduce to the same mathematical relations (5.1.16) - (5.1.18) or (5.1.27) - (5.1.29). Both problems are joined under the general title of plane problems of the theory of elasticity.

3^0. Consider the equivalent formulation of the plane problem in terms of stresses. For the sake of definiteness, consider the case of plane strain and not of generalized plane stress.

Let body forces be absent and on the boundary Ω be given the vector

$$\hat{\sigma}^{(n)}\bigg|_{\partial\Omega} = \mathbf{f}. \tag{5.1.30}$$

In order to solve the problem in terms of stresses, combine the equilibrium equations

$$\frac{\partial\sigma_{xx}}{\partial x} + \frac{\partial\sigma_{xy}}{\partial y} = 0, \ \frac{\partial\sigma_{xy}}{\partial x} + \frac{\partial\sigma_{yy}}{\partial y} = 0 \tag{5.1.31}$$

with the Beltrami-Mitchell equations

$$\Delta\sigma_{xx} + \frac{1}{1+\nu}\frac{\partial^2\sigma}{\partial x^2} = 0\,,$$

$$\Delta\sigma_{xy} + \frac{1}{1+\nu}\frac{\partial^2\sigma}{\partial x\,\partial y} = 0\,,$$

$$\Delta\sigma_{yy} + \frac{1}{1+\nu}\frac{\partial^2\sigma}{\partial y^2} = 0\,,$$

(5.1.32)

in the role of compatibility equations with

$$\sigma = \sigma_{xx} + \sigma_{yy} + \sigma_{zz}\,. \tag{5.1.33}$$

Since

$$e_{zz} = \frac{1}{E}\left[\sigma_{zz} - \nu(\sigma_{xx} + \sigma_{yy})\right] = 0\,, \tag{5.1.34}$$

one has

$$\sigma = (1+\nu)(\sigma_{xx} + \sigma_{yy})\,. \tag{5.1.35}$$

Then (5.1.32) assumes the form

$$\Delta\sigma_{xx} + \frac{\partial^2(\sigma_{xx} + \sigma_{yy})}{\partial x^2} = 0\,,$$

$$\Delta\sigma_{xy} + \frac{\partial^2(\sigma_{xx} + \sigma_{yy})}{\partial x\,\partial y} = 0\,,$$

$$\Delta\sigma_{yy} + \frac{\partial^2(\sigma_{xx} + \sigma_{yy})}{\partial y^2} = 0\,,$$

(5.1.36)

Results of B.E. Pobedrya [63] permit to formulate the problem (5.1.30), (5.1.31) and (5.1.36) in the following form of a standard boundary value problem of mathematical physics: Find the solution of equation (5.1.36) for the boundary conditions (5.1.30) and

$$R_1 \equiv \frac{\partial\sigma_{xx}}{\partial x} + \frac{\partial\sigma_{xy}}{\partial y}\bigg|_{\partial\Omega} = 0\,,$$

$$R_2 \equiv \frac{\partial\sigma_{xy}}{\partial x} + \frac{\partial\sigma_{yy}}{\partial y}\bigg|_{\partial\Omega} = 0\,.$$

(5.1.37)

It has been shown in [63] that the problems (5.1.30), (5.1.31), (5.1.36) and (5.1.30), (5.1.36) and (5.1.37) are equivalent.

§2. Complex representation

1^0. This section derives a number of relations which permit to express stresses and displacements for plane strain (and, likewise, for generalized plane stress) in terms of analytic functions of the complex variable $z = x + iy$, where x and y are Cartesian coordinates. A detailed derivation of these relations as well as a sufficiently complete bibliography is given in [53].

The differential equations of elastic plane deformation of homogeneous, isotropic media in the absence of body forces have the form

$$\frac{\partial \sigma_{xx}}{\partial x} + \frac{\partial \sigma_{xy}}{\partial y} = 0, \ \frac{\partial \sigma_{xy}}{\partial x} + \frac{\partial \sigma_{yy}}{\partial y} = 0, \tag{5.2.1}$$

$$\sigma_{xx} = \lambda e + 2\mu \frac{\partial u_x}{\partial x}, \ \sigma_{xy} = \mu \left(\frac{\partial u_x}{\partial y} + \frac{\partial u_y}{\partial x} \right),$$
$$\sigma_{yy} = \lambda e + 2\mu \frac{\partial u_y}{\partial y}, \ e = \frac{\partial u_x}{\partial x} + \frac{\partial u_y}{\partial y}, \tag{5.2.2}$$

where σ_{xx}, σ_{xy}, σ_{yy} are the components of the stress tensor, u_x and u_y are the components of the displacement vector. Equations (5.2.1) allow to express the stress components in terms of the so-called Airy function $W(x, y)$:

$$\sigma_{xx} = \frac{\partial^2 W}{\partial y^2}, \ \sigma_{xy} = -\frac{\partial^2 W}{\partial x\, \partial y}, \ \sigma_{yy} = \frac{\partial^2 W}{\partial x^2}, \tag{5.2.3}$$

hence follows from (5.2.2) that this function is bi-harmonic, i.e., it satisfies the bi-harmonic equation

$$\Delta^2 W = \frac{\partial^4 W}{\partial x^4} + 2 \frac{\partial^4 W}{\partial x^2\, \partial y^2} + \frac{\partial^4 W}{\partial y^4} = 0, \tag{5.2.4}$$

where Δ is the Laplace operator.

2^0. Let z denotes the complex variable $z = x + iy$. The bi-harmonic function $W(x, y)$ may be expressed in terms of two analytic functions of z, holomorphic in the neighbourhood of every point of a region Ω, occupied by an elastic medium:

$$W(x, y) = Re\left[z\varphi[z] + \chi(z) \right]. \tag{5.2.5}$$

The functions $\varphi(z)$ and $\psi(z) = \chi'(z)$ will be referred to as Goursat functions.

Next, the stresses and displacements as, well as the first derivatives of the Airy function will be expressed in terms of the Goursat functions:

$$\sigma_{xx} + \sigma_{yy} = 4Re\left[\varphi'(z) \right], \tag{5.2.6}$$

$$\sigma_{yy} - \sigma_{xx} + 2i\sigma_{xy} = 2[z\varphi''(z) + \psi'(z)], \tag{5.2.7}$$

$$2\mu(u_x + iu_y) = \kappa\varphi(z) - z\overline{\varphi'(z)} - \overline{\psi(z)}, \tag{5.2.8}$$

$$\kappa = \frac{\lambda + 3\mu}{\lambda + \mu} = 3 - 4\nu,$$

$$\frac{\partial W}{\partial x} + i\frac{\partial W}{\partial y} = \varphi(z) + z\overline{\varphi'(z)} - \overline{\psi(z)}. \tag{5.2.9}$$

where ν in (5.2.8) is the Poisson constant.

Whenever stresses are given, the Goursat functions are not a completely single-valued. In particular, the function $\varphi(z)$ is defined apart from a term $i\alpha z + \beta$, where α is a real and β a complex constant. The function $\psi(z)$ is defined to that accuracy by the first derivatives of the Airy function. For given displacements, the function $\varphi(z)$ is defined exactly apart from a constant, complex term.

3^0. In the case of a finite region Ω, the external forces, applied to its boundary, are in equilibrium, so that their resultant vector and resultant moment must vanish, hence one arrives at the following assertion regarding the behaviour of the Goursat functions: If the region Ω is simply connected, then these functions are single-valued in this region; if Ω is multiply connected, say, $(n + 1)$-ply connected, i.e., its boundary is given by $\Gamma = \bigcup_{k=0}^{n} \Gamma_k$, where the Γ_k are non-intersecting curves and Γ_0 bounds the region Ω outside, then

$$\varphi(z) = \sum_{k=1}^{n} B_k \, ln(z - z_k) + \varphi^*(z),$$
$$\psi(z) = -\kappa \sum_{k=1}^{n} B_k \, ln(z - z_k) + \psi^*(z). \tag{5.2.10}$$

In these formulas, the z_k denote fixed points inside the curves Γ_k, $\varphi^*(z)$ and $\psi^*(z)$ are holomorphic functions which are single-valued in Ω and

$$B_k = \frac{X_k + iY_k}{2\pi(I + \kappa)}, \tag{5.2.11}$$

where X_k and Y_k are the components of the resultant vector of the force applied to the curve Γ_k.

If the region Ω is infinite, but bounded by a finite contour and has connectivity n, then it is the outside of n non-intersecting curves Γ_k and formulas (5.2.10) and (5.2.11) also apply.

It is important for the following work to note that the function $\varphi'(z)$ is single-valued for regions of any connectivity.

4^0. Consider now two of the simplest boundary value problems of the theory of plane, elastic deformations.

Problem I: It has the boundary conditions

$$u_x\big|_\Gamma = g_1(s), \ u_y\big|_\Gamma = g_2(s), \tag{5.2.12}$$

where s is a parameter which determines the position of points on Γ, g_1 and g_2 are functions given on Γ.

Problem II: Let s be the length of the arc of the contour Γ, measured from some initial position. Its boundary conditions are

$$\sigma_{xx} \cos(n, x) + \tau_{xy} \cos(n, y) = X_n(s),$$
$$\sigma_{xy} \cos(n, x) + \sigma_{yy} \cos(n, y) = X_n(s), \tag{5.2.13}$$

where n is the normal to Γ which is external with respect to Ω, $X_n(s)$ and $Y_n(s)$ are functions, given on Γ which are the projections onto the coordinate axes of the external forces, applied at the point s to the contour Γ. Regarding this boundary value problem, it will be assumed that the external forces satisfy the conditions of statics.

5^0. The boundary conditions (5.2.12) and (5.2.13) are comparatively easily formulated in terms of the Goursat functions. Relations (5.2.8) and (5.2.9) permit to write the boundary conditions of the two problems of the last section in the form

$$\kappa\,\varphi(\zeta) - \zeta\,\overline{\varphi'(\zeta)} - \overline{\psi(\zeta)} = g(\zeta), \tag{5.2.14}$$

where $\kappa = 3 - 4\nu$ and $\kappa = -1$ for the problems I and II, respectively, $g(\zeta)$ is a complex-valued function given on Γ. The symbol ζ denotes here and below the complex coordinate of the point on Γ which is determined by s. Boundary condition (5.2.13) for problem II may be given the form

$$\varphi(\zeta) - \zeta\,\overline{\varphi'(\zeta)} + \overline{\psi(\zeta)} = f(\zeta) + C, \tag{5.2.15}$$

where

$$f(\zeta) = i \int_{\zeta_0}^{\zeta} (X_n + iY_n)\, ds,$$

and C is a complex constant which has different, in general complex values on different curves Γ_k; on one of these curves, the value of C may be chosen arbitrarily.

6^0. It frequently proves to be useful to map in advance conformally the region Ω onto a simpler region. Let $z = \omega(t)$ be this mapping function and introduce the notation

$$\varphi(z) = \varphi(\omega(t)) = \Phi(t), \ \psi(z) = \psi(\omega(t)) = \Psi(t).$$

Then (5.2.14) and (5.2.15) become

$$\kappa \, \Phi(t) - \frac{\omega(t)}{\overline{\omega'(t)}}\overline{\Phi'(t)} - \overline{\Psi(t)} = 2\mu(u_x + iu_y), \tag{5.2.16}$$

$$\Phi(t) - \frac{\omega(t)}{\overline{\omega'(t)}}\overline{\Phi'(t)} - \overline{\Psi(t)} = \frac{\partial W}{\partial x} + i\frac{\partial W}{\partial y}. \tag{5.2.17}$$

§3. Integral equations of N.I. Muskhelishvili

1^0. Formally, one can consider condition (5.2.15) as the particular case of condition (5.2.14) for $\kappa = -1$, hence we will start the derivation of the integral equation from (5.2.14).

Let the region Ω be simply connected, finite and bounded by a sufficiently smooth contour which, as above, will be denoted by Γ. Map Ω conformally on to the unit circle $\mid t \mid < 1$, denoting the circumference of this circle by γ, the variable point on γ by τ, and the mapping function by $\omega(t)$. If the contour $\Gamma \in C^{(k)}$, it is known [100] that in the circle $\mid t \mid \leq 1$ the mapping function $\omega(t) \in C^{(k-1)}$. Assume that $k \geq 3$. Formula (5.2.12) permits to write the boundary condition (5.2.14) in the form

$$\kappa \, \Phi(\tau) - \frac{\omega(\tau)}{\overline{\omega'(\tau)}} \, \overline{\Phi'(\tau)} - \overline{\Psi'(\tau)} = g(\omega(\tau)) := h(\tau). \tag{5.3.1}$$

Let the function $\Phi(t)$, $\Phi'(t)$, $\Psi(t)$ be holomorphic in $\mid t \mid < 1$ and continuous in $\mid t \mid \leq 1$.

Multiply equality (5.3.1) as well as its conjugate complex form by $d\tau/2\pi i(t - \tau)$, where $\mid t \mid < 1$, and integrate along γ to arrive at the two

          5. The integral equations of classical two–dimensional problems

integral relations

$$\kappa\Phi(t) - \frac{1}{2\pi i}\int_\gamma \frac{\omega(\tau)}{\overline{\omega'(\tau)}}\frac{\overline{\Phi'(\tau)}}{t-\tau}\,d\tau - \frac{1}{2\pi i}\int_\gamma \frac{\overline{\Psi(\tau)}}{t-\tau}\,d\tau =$$

$$= \frac{1}{2\pi i}\int_\gamma \frac{h'(\tau)}{t-\tau}\,d\tau := A(t),$$

$$-\Psi(t) - \frac{\kappa}{2\pi i}\int_\gamma \frac{\overline{\Phi(\tau)}}{t-\tau}\,d\tau - \frac{1}{2\pi i}\int_\gamma \frac{\overline{\omega(\tau)}}{\omega'(\tau)}\frac{\Phi'(\tau)}{t-\tau}\,d\tau =$$

$$= \frac{1}{2\pi i}\int_\gamma \frac{\overline{h(\tau)}}{t-\tau}\,d\tau := B(t).$$

$$(5.3.2)$$

Let $\omega(t)$ be an arbitrary function, holomorphic in $|\,t\,|< 1$ and continuous in $|\,t\,|\le 1$. It is readily shown that

$$\frac{1}{2\pi i}\int_\gamma \frac{\overline{\omega(\tau)}}{t-\tau}\,d\tau = \overline{\omega(0)},\qquad\qquad (5.3.3)$$

and, in particular,

$$\frac{1}{2\pi i}\int_\gamma \frac{\overline{\Phi(\tau)}}{t-\tau}\,d\tau = \overline{\Phi(0)},\quad \frac{1}{2\pi i}\int_\gamma \frac{\overline{\Psi(\tau)}}{t-\tau}\,d\tau = \overline{\Psi(0)}.$$

Since the function $\Psi(t)$ is defined only apart from a constant term, one can set $\Psi(0) = 0$. The first equality (5.3.2) contains now only the single unknown $\Phi(t)$; if it can be found, then the second equality (5.3.2) directly determines $\Psi(t)$.

Transform the first equation (5.3.2) into

$$\kappa\Phi(t) - \frac{1}{2\pi i}\int_\gamma \frac{\omega(\tau)-\omega(t)}{t-\tau}\frac{\overline{\Phi'(\tau)}}{\overline{\omega'(\tau)}}\,d\tau - \frac{\omega(t)}{2\pi i}\int_\gamma \frac{1}{t-\tau}\frac{\overline{\Phi'(\tau)}}{\overline{\omega'(\tau)}}\,d\tau = A(t).$$

By (5.3.3)

$$\frac{1}{2\pi i}\int_\gamma \frac{1}{t-\tau}\frac{\overline{\Phi'(\tau)}}{\overline{\omega'(\tau)}}\,d\tau = \frac{\overline{\Phi'(0)}}{\overline{\omega'(0)}}.$$

Substitute this result into the last equality and differentiate with respect to t:

$$\kappa \Phi'(t) + \frac{1}{2\pi i} \int_\gamma \frac{\partial}{\partial t}\left[\frac{\omega(\tau) - \omega(t)}{t - \tau}\right] \frac{\overline{\Phi'(\tau)}}{\overline{\omega'(\tau)}}\, d\tau - \frac{\overline{\Phi'(0)}}{\overline{\omega'(0)}}\, \omega'(t) = A'(t)\,.$$

With the transformation

$$\Phi'(t) = \frac{1}{\kappa} \frac{\overline{\Phi'(0)}}{\overline{\omega'(0)}}\, \omega'(t) + \vartheta(t) = A'(t)\,, \tag{5.3.4}$$

one finds for the new unknown $\vartheta(t)$ the integral equation

$$\vartheta(t) - \frac{1}{2\pi i \kappa} \int_\gamma \frac{\partial}{\partial t}\left[\frac{\omega(\tau) - \omega(t)}{t - \tau}\right] \frac{\overline{\vartheta(\tau)}}{\overline{\omega'(\tau)}}\, d\tau = \frac{1}{\kappa}\, A'(t)\,. \tag{5.3.5}$$

By the assumption $\Gamma \in C^k$, $k \geq 3$, the kernel of equation (5.3.5) is continuous. This equation was obtained by N.I. Muskhelishvili in 1931 [53]. It is clear from its derivation that it is true for $|\,t\,| < 1$. A simple limiting process shows that it is also true for $|\,t\,| = 1$. Separation of real and imaginary parts in the last equation yields a system of two Fredholm equations in the unknowns $Re\,\vartheta(t)$ and $Im\,\vartheta(t)$, hence it readily follows that Fredholm's alternative is true for equation (5.3.5). Hence the homogeneous equation (5.3.5) has only the trivial solution, and therefore the non-homogenous equation (5.3.5) has a unique solution.

In order to find $\Phi'(t)$, it is sufficient , as can be seen from (5.3.4), to determine the constant $l = \Phi'(0)/\omega'(0)$. Setting $t = 0$ in (5.3.4), one obtains the equation $\kappa l - \bar{l} = \kappa \vartheta(0)/\omega'(0)$. When solving problem I, then $\kappa > 1$ and the last equality determines uniquely l, and the problem is solved completely. For problem II, $\kappa = -1$ and the equation for l assumes the form $l + \bar{l} = \vartheta(0)/\omega'(0)$; this problem is soluble if and only if $Im\,(\vartheta(0)/\omega'(0)) = 0$. It has been sho that the last condition is equivalent the vanishing of the resultant moment of the forces acting on the contour Γ. If this condition is fulfilled, then problem II is soluble; the imaginary part of the number $\Phi'(0)/\omega'(0)$ then remains arbitrary.

Note one important peculiarity of the integral equation (5.3.5): If the mapping function $\omega(t)$ is rational, then the kernel

$$\frac{\partial}{\partial t}\left[\frac{\omega(\tau) - \omega(t)}{\tau - t}\right] \frac{1}{\omega'(\tau)}$$

degenerates, so that this equation can be solved in an elementary manner by reduction to a system of linear, algebraic equations.

These results show that exists a large enough class of simply connected regions, for which the plane problem of the theory of elasticity admits an elementary, exact solution. In the general case, the mapping function may be approximated by a rational (for example, polynomial) function, and an approximate solution may be found in an elementary manner.

2^0. Next, consider the case when the region Ω is simply connected and infinite, lying outside a finite contour. For the sake of simplicity, let it be assumed that in the case of problem II the external forces acting on the contour Γ are in static equilibrium. Then the functions $\Phi(t)$ and $\Psi(t)$ are holomorphic in the circle $\mid t \mid \leq 1$. As before, let $w(t)$ be a function which maps Ω conformally onto the circle $\mid t \mid < 1$ and assume that for this mapping the point $z = \infty$ corresponds to the point $t = 0$. Then

$$w(t) = \frac{a}{t} + w_0(t), \qquad (5.3.6)$$

where a is a constant and w_0 a function, holomorphic in $\mid t \mid < 1$. As in the case of a finite region, if $\Gamma \in C^k$, then $w_0 \in C^{(k-1)}$, and one must have $k \geq 3$.

As in 1^0, equations (5.3.2) apply. The second of them determines the function $\Psi(t)$, if $\Phi(t)$ is known.

As before, setting $\Psi(0) = 0$, the first of equations (5.3.2) reduces to

$$\kappa\Phi(t) - \frac{1}{2\pi i} \int_\gamma \frac{w(\tau)}{\overline{w'(\tau)}} \frac{\overline{\Phi'(\tau)}}{t - \tau}\, d\tau = A(t).$$

As in the case of a finite region, the last equation becomes

$$\kappa\Phi(t) + \frac{1}{2\pi i} \int_\gamma \frac{w(\tau) - w(t)}{\tau - t} \frac{\overline{\Phi'(\tau)}}{\overline{w'(\tau)}}\, d\tau - \frac{w(t)}{2\pi i} \int_\gamma \frac{1}{t - \tau} \frac{\overline{\Phi'(\tau)}}{\overline{w'(\tau)}}\, d\tau = A(t).$$

The function $\Phi'(t)/w'(t)$ is holomorphic in $\mid t \mid < 1$ and vanishes for $t = 0$. By (5.3.3), the second integral vanishes. Now, using (5.3.6) in the first integral, one finds $w(t)$

$$\frac{1}{2\pi i} \int_\gamma \frac{w(\tau) - w(t)}{\tau - t} \frac{\overline{\Phi'(\tau)}}{\overline{w'(\tau)}}\, d\tau = \frac{1}{2\pi i} \int_\gamma \frac{w_0(\tau) - w_0(t)}{\tau - t} \frac{\overline{\Phi'(\tau)}}{\overline{w'(\tau)}}\, d\tau +$$

$$+ \frac{a}{2\pi i} \int_\gamma \frac{1}{\tau - t} \frac{\overline{\tau\Phi'(\tau)}}{\overline{w'(\tau)}}\, d\tau .$$

According to (5.3.3), the second term on the right hand side vanishes, hence

$$\kappa\Phi(t) + \frac{1}{2\pi i}\int_\gamma \frac{\omega_0(\tau) - \omega_0(t)}{\tau - t}\,\frac{\overline{\Phi'(\tau)}}{\omega'(\tau)}\,d\tau = A(t).$$

Differentiating and allowing the point t to approach the circle γ, one arrives at the final integral equation

$$\kappa\Phi'(t) + \frac{1}{2\pi i}\int_\gamma \frac{\partial}{\partial t}\left[\frac{\omega_0(\tau) - \omega_0(t)}{\tau - t}\right]\frac{\overline{\Phi'(\tau)}}{\omega'(\tau)}\,d\tau = A'(t), \tag{5.3.7}$$

which is equivalent to two Fredholm equations. It has been shown that (5.3.7) has one and only one solution. It can be found by determining $\Psi(t)$ from the second equation (5.3.2).

§4. Generalization to multiply connected regions

1^0. Introduce now the Green function $G(x, y;\ \xi, \eta)$ of the region Ω for the first boundary value problem for the Laplace equation. Its basic properties are

$$G(x, y;\ \xi, \eta) = G_0(x, y;\ \xi, \eta) - (2\pi)^{-1}\ln r, \tag{5.4.1}$$

$$r^2 = (x - \xi)^2 + (y - \eta)^2,$$

where (x, y) and (ξ, η) are points of Ω, G_0 is the symmetric function of these points which is harmonic in Ω with respect to each of them; if one of these points approaches $\Gamma = \partial\Omega$, then $G_0 = (2\pi)^{-1}\ln r$ and, consequently, $G = 0$.

In order to simplify the notation, consider G and G_0 as (not analytic) functions of the complex variables $z = x + iy$ and $\zeta = \xi + i\eta$ and write (5.4.1) in the form $G(z, \zeta) = G_0(z, \zeta) - \ln|\zeta - z|/2\pi$. For some fixed point $a \in \Omega$, let

$$H(z, \zeta) = \int_a^z \left(-\frac{\partial G}{\partial y}\,dx + \frac{\partial G}{\partial x}\,dy\right). \tag{5.4.2}$$

The function H is harmonic conjugate to G with respect to z; it is multi-valued in Ω and one of its branches vanishes for $z = a$. The function

$$M(z, \zeta) = G(z, \zeta) + iH(z, \zeta) \tag{5.4.3}$$

is referred to as the complex Green function of the domain Ω. This function is analytic, but multi-valued with respect to the variable z. Its multi-valuedness, on the one hand, is determined by the fact that it contains the term $-ln(\zeta-z)$. On the other hand, as will be shown immediately, in the case of multiply connected regions, it undergoes some increment for a circuit of each of the internal curves, constituting the boundary of Ω. Let Γ_j, $j = 1, 2, ..., n$ be the closed curves bounding Ω inside and let z travel around the curve Γ_j, while $\zeta \in \Omega$ remains fixed inside Ω. As the contour Γ_j is circuited in a positive direction (leaving Ω on the left), the function $H(z, \zeta)$ undergoes the increment $- \beta_j(\zeta)$. It is readily seen from (5.4.2) that

$$\beta_j(\zeta) = - \int\limits_{\Gamma_j} \left(- \frac{\partial G}{\partial y} \, dx + \frac{\partial G}{\partial x} \, dy \right) = \int\limits_{\Gamma_j} \frac{\partial G(z, \zeta)}{\partial \nu_z} \, ds \, , \qquad (5.4.4)$$

where ν_z is the normal to Γ_j at the point z, external with respect to Ω and ds is the arc element on Γ_j. It is seen from (5.4.4) that $\beta_j(\zeta)$ is harmonic in Ω with respect to ξ and η, equal to unity on Γ_j and vanishes on the remaining parts of the boundary of Ω. Hence follows the representation of the complex Green function

$$M(z, \zeta) = M_0(z, \zeta) + \sum_{j=1}^{n} \beta_j(\zeta) \, ln(z - z_j) - (2\pi)^{-1} \, ln(\zeta - z) \, , \qquad (5.4.5)$$

where $M_0(z, \zeta)$ is a single-valued function of z and ζ, holomorphic with respect to z in Ω, and z_j is a point fixed inside Γ_j.

2^0. In (5.4.5), let $\zeta \in \Gamma$ and $z \in \Omega$, and the normal ν to Γ at the point ζ be outward with respect to Ω. Differentiating (5.4.5) with respect to ν, one obtains the function, which will be called the Schwartz kernel of the region Ω

$$T(z, \zeta) = \frac{\partial M(z, \zeta)}{\partial \nu} = \frac{\partial M_0(z, \zeta)}{\partial \nu} + \sum_{j=1}^{n} \alpha_j(\zeta) \, ln(z - z_j)$$

$$- \frac{1}{2\pi(\zeta - z)} \frac{\partial \zeta}{\partial \nu} \, ; \; \alpha_j(\zeta) = \frac{\partial \beta_j(\zeta)}{\partial \nu} \, . \qquad (5.4.6)$$

The Schwartz kernel is and analytic (multi-valued, if Ω is multiply connected) function of z in the region Ω; as a function of ζ, it is single-valued on Γ. Its real part is the normal derivative of the Green function. Hence, if $u(\zeta)$ is the real part of the single-valued, continuous function of $\zeta \in \Gamma$, then the integral

$$\Phi(z) = \int\limits_{\Gamma} T(z, \zeta) \, u(\zeta) \, d\sigma \, , \qquad (5.4.7)$$

where $d\sigma$ is the element of the arc length on Γ, is an analytic, generally speaking, multi-valued function of $z \in \Omega$, the real part of which coincides with $u(\zeta)$ on Γ, and one branch of the imaginary part of which vanishes for $z = a$. In particular, of the function $R(z)$ is holomorphic and single-valued in Ω and continuous in the closed region $\bar{\Omega} = \Omega \bigcup \Gamma$, then, as it is readily verified,

$$\frac{1}{2} \int_\Gamma R(\zeta)\, T(z,\zeta)\, d\sigma = R(z) - \frac{1}{2} R(\alpha),\qquad (5.4.8)$$

$$\frac{1}{2} \int_\Gamma \overline{R(\zeta)}\, T(z,\zeta)\, d\sigma = \frac{1}{2}\, \overline{R(\alpha)}\,.$$

In the more special case, when $R(z) \equiv 1$, one finds now

$$\int_\Gamma T(z,\zeta)\, d\sigma = 1\,. \qquad (5.4.9)$$

3^0. Let the multiply connected region Ω be finite. Multiply equation (5.2.14), as well as its adjoint equation by $T(z,\zeta)\, d\sigma$ and integrate along Γ. Since it is natural to assume that the displacement is single-valued, the function $g(t)$ for problem I will likewise be single-valued, and, as is seen from (5.2.10), so is the functions $\varphi'(z)$ and $\kappa\varphi(z) - \overline{\psi(z)}$. Using (5.4.8), one obtains the equations

$$\kappa\,\varphi(z) - \frac{1}{2} \int_\Gamma \zeta\, T(z,\zeta)\, \overline{\varphi'(\zeta)}\, d\sigma - \frac{1}{2}\big[\kappa\varphi(a) + \overline{\psi(a)}\big] =$$
$$= \frac{1}{2} \int_\Gamma g(\zeta)\, T(z,\zeta)\, d\sigma := A(z)\,, \qquad (5.4.10)$$

$$\psi(z) + \frac{1}{2} \int_\Gamma \bar{\zeta}\, T(z,\zeta)\, \varphi'(\zeta)\, d\sigma - \frac{1}{2}\big[\kappa\overline{\varphi(a)} + \psi(a)\big] =$$
$$= -\frac{1}{2} \int_\Gamma \overline{g(\zeta)}\, T(z,\zeta)\, d\sigma := B(z)\,. \qquad (5.4.11)$$

Impose now on the unknown function the condition

$$\kappa\,\varphi(a) + \overline{\psi(a)} = 0\,, \qquad (5.4.12)$$

and find from (5.4.11) the function $\psi(z)$. If $\varphi'(z)$ is known, then only equation (5.4.10) need be studied. Taking account of (5.4.8), it may be rewritten in the form

$$\kappa\,\varphi(z) - \frac{1}{2}\int\limits_{\Gamma}(\zeta - z)T(z,\zeta)\overline{\varphi'(\zeta)}\,d\sigma - \frac{z}{2}\overline{\varphi'(a)} = A(z)\,.$$

Let $\varphi'(z) = \vartheta(z) + l$, where l is the constant determined by

$$2\kappa l - \bar{l} - \varphi'(a) = 0\,. \tag{5.4.13}$$

Using (5.4.8), one finds

$$\vartheta(z) - \frac{1}{2\kappa}\int\limits_{\Gamma}\frac{\partial}{\partial z}\left[(\zeta - z)\,T(z,\zeta)\right]\overline{\vartheta(\zeta)}\,d\sigma = \frac{1}{\kappa}\,A'(z)\,, \tag{5.4.14}$$

which may still be simplified. It follows from (5.4.6) that

$$\frac{\partial}{\partial z}\left[(\zeta - z)\,T(z,\zeta)\right] = -\sum_{j=1}^{n}\alpha_j(\zeta)\,ln(z - z_j) + K(z,\zeta)\,,$$

where $K(z,\zeta)$ is a single-valued, continuous function of $z \in \overline{\Omega}$ and $\zeta \in \Gamma$ which is holomorphic with respect to z. The function $\vartheta(z)$ is holomorphic in Ω; starting from (5.4.3), it is easily shown that

$$\int\limits_{\Gamma}\alpha_j(\zeta)\,\overline{\vartheta(\zeta)}\,d\sigma = 0 \tag{5.4.15}$$

and equation (5.4.14) becomes the simpler equation

$$\vartheta(z) - \frac{1}{2\kappa}\int\limits_{\Gamma}K(z,\zeta)\,\overline{\vartheta(\zeta)}\,d\sigma = \frac{1}{2\kappa}\,A'(z)\,. \tag{5.4.16}$$

Letting now the point z approach Γ, one arrives at an integral equation which is equivalent to a system of two Fredholm equations (cf. §3, 2^0). It may be shown that this equation has a unique solution.

  The constant l has still to be determined. $\overline{\varphi'(a)} = \overline{\vartheta(a)} + \bar{l}$. Substituting this result into (5.4.13), one find $\kappa l - \bar{l} = \vartheta(a)/2$. As in the case of simply connected regions, the last equation has for problem I a unique solution, and the first boundary value problem of the plane of elasticity is solved.

Similar results also apply to infinite regions. They will not be discussed here in detail.

4^0. Problem II will be considered next. In that case the constants B_k are determined by (5.2.11). The problem reduces to the determination of the functions $\varphi^*(z)$ and $\psi^*(z)$ in Ω (5.2.10). The boundary conditions for these functions follow readily from (5.2.10) and (5.2.15). The work proceeds as in 3^0 with one important exception: If one sets $\varphi'^*(z) = \vartheta(z) + l$, then one obtains for l the equation $l + \bar{l} = -\vartheta(a)/2$. Thus it is necessary and sufficient for solubility of problem II, if the quantity $\vartheta(a)$ is real. It has been shown that this requirement if equivalent to the statement: The resultant moment of the external forces, acting on the contour Γ must vanish.

5^0. The results of this section have been obtained in the years 1933-34 and presented in detail in [41] and [44]. The methods has been applied to doubly connected regions for which the conformal mapping onto a circular ring is known: for such regions, the Green function is known. Two such examples are presented in [44].

§5. The potential theory equations of N.I. Muskhelishvili

1^0. For the sake of simplicity, consider the case of a simply connected region Ω. A more general case has been studied in [53]. Let the contour Γ of Ω be a Lyapunov curve.

Consider the conjugate complex form of the boundary condition (5.2.14).

$$\kappa\,\overline{\varphi(\zeta)} - \bar{\zeta}\varphi'(\zeta) + \psi(\zeta) = \overline{g(\zeta)}. \tag{5.5.1}$$

Let z' be a point of the complex plane which lies outside the closed region $\overline{\Omega} = \Omega + \Gamma$. Multiply (5.5.1) by $d\zeta/(2\pi i(\zeta - z'))$ and integrate along Γ. Since the function $\psi(z)$ is holomorphic in Ω, the corresponding Cauchy integral vanishes, and one arrives at an equation which contains only the unknown function φ:

$$\frac{\kappa}{2\pi i}\int_{\Gamma}\frac{\overline{\varphi(\zeta)}}{\zeta - z'}\,d\zeta - \frac{1}{2\pi i}\int_{\Gamma}\frac{\overline{\zeta\varphi'(\zeta)}}{\zeta - z'}\,d\zeta = \frac{1}{2\pi i}\int_{\Gamma}\frac{\overline{g(\zeta)}}{\zeta - z'}\,d\zeta := G(z'), \tag{5.5.2}$$

which will now be transformed. Since the functions φ and φ' are holomorphic in Ω, one has

$$\frac{1}{2\pi i}\int_{\Gamma}\frac{\varphi(\zeta)}{\zeta - z'}\,d\zeta \equiv 0, \quad \frac{1}{2\pi i}\int_{\Gamma}\frac{\varphi'(\zeta)}{\zeta - z'}\,d\zeta \equiv 0.$$

Multiply the first of these equations by κ and step over to its conjugate complex form; and the second equation by $\overline{z'}$; add the results to (5.5.2) and let $\overline{z'} \to t$, $t \in \Gamma$. Using (2.1 11), setting $\zeta - t = re^{i\theta}$ and integrating by parts, retaining $\varphi'(\zeta)$, one arrives at the integral equation

$$\kappa\,\overline{\varphi(\zeta)} - \frac{1}{\pi} \int\limits_{\Gamma} \left[\kappa\,\overline{\varphi(\zeta)} + e^{-2i\theta}\varphi(\zeta) \right] \frac{\partial \theta}{\partial s}\, ds = -G(t)\,, \qquad (5.5.3)$$

where s is the arc length of the curve Γ.

It follows from this result that for a Lyapunov contour Γ. equation (5.5.3) is equivalent to a system of two Fredholm equations and that, consequently, Fredholm's alternative holds.

2^0. Consider equation (5.5.3). First of all, it will be shown that its solution analytically continues with the contour Γ into the region Ω. Retracing the derivation of (5.5.3), it can be verified that it may be presented in the form

$$\lim_{z' \to t} \left\{ \frac{\kappa}{2\pi i} \int\limits_{\Gamma} \frac{\overline{\varphi(\zeta)}}{\overline{\zeta} - \overline{z'}}\, d\overline{\zeta} - r\frac{\overline{z'}}{2\pi i} \int\limits_{\Gamma} \frac{\varphi'(\zeta)}{\zeta - z'}\, d\zeta - \right.$$
$$\left. - \frac{1}{2\pi i} \int\limits_{\Gamma} \frac{\kappa\,\overline{\varphi(\zeta)} - \overline{\zeta}\,\varphi'(\zeta) - g(\zeta)}{\zeta - z'}\, d\zeta \right\} = 0\,, \qquad (5.5.4)$$

where $\varphi'(\zeta)$, must be interpreted as the derivative of $\varphi(\zeta)$, taken along the contour Γ. Introduce the notation

$$\frac{1}{2\pi i} \int\limits_{\Gamma} \frac{\varphi(\zeta)}{\zeta - z'}\, d\zeta = i\Phi(z')\,,$$
$$\frac{1}{2\pi i} \int\limits_{\Gamma} \frac{\kappa\overline{\varphi(\zeta)} - \overline{\zeta}\varphi'(\zeta) - g(\zeta)}{\zeta - z'}\, d\zeta = -i\Psi(z')\,. \qquad (5.5.5)$$

The functions $\Phi(z')$ and $\Psi(z')$ are holomorphic in the region Ω, lying outside Γ, and they vanish for $z' = \infty$. In terms of these functions, relation (5.5.4) assumes the form

$$\kappa\,\overline{\Phi(t)} - \overline{t}\Phi'(t) - \Psi(t) = 0\,. \qquad (5.5.6)$$

Thus, the functions $\Phi(z')$ and $\Psi(z')$ solve the homogeneous, plane problem of the theory of elasticity for the infinite region Ω'. By the theorem on the uniqueness of the solution of such a problem, one has that $\Phi(z') = \Psi(z') = 0$. This identity shows that the function $\varphi(t)$, which satisfies (5.5.3), and likewise the corresponding function $\psi(t) = \kappa\,\varphi(t) - \overline{t}\varphi'(t) - \overline{g(t)}$ are continued analytically into the region Ω.

In the general case, equation (5.5.3) has no solution. In fact, for $g(t) \equiv 0$, this equation is homogenous; the corresponding Green function $\varphi(z)$ may then (cf. §2) be taken as $\varphi(z) = i\alpha z + \beta$ (problem II) or $\varphi(z) = \beta$ (problem I). It can be verified directly that these values satisfy equation (5.5.4), and, consequently, also equation (5.5.3).

It has been shown by D.I. Sherman [76] that equation (5.5.3) may be somewhat modified so that it is unconditionally soluble and its solution leads to that of the corresponding problem of the theory of elasticity.

§6. The equations of Lauricella–Sherman

1^0. One of the first papers in which a plane problem of the theory of elasticity was solved by reduction to a Fredholm equation was by J. Lauricella [94]. Lauricella considered there the problem of bending of a plate, supported at its edge and subject to forces, applied normally to its faces; as it is well known, such a problem is mathematically equivalent to problem II of the theory of elasticity. Lauricella's integral equation and its derivation were somewhat involved. D.I. Sherman [79] obtained Lauricella's equations in complex form and presented for them a new, simpler derivation. The equations of Lauricella-Sherman turned out to resemble closely the structure of the equations of N.I. Muskhelishvili, described in the preceding section. Sherman's derivation will be reproduced here for the case of a finite (generally speaking, multiply connected) region Ω bounded by a Lyapunov contour. It will be assumed that $\Gamma = \bigcup_{j=0}^{n} \Gamma_j$ where Γ_j are closed curves without common points and self-intersections, and that Γ_0 bounds the region Ω outside.

2^0. The boundary condition of problem II, expressed in terms of the Goursat functions, has the form

$$\varphi(\zeta) + \zeta\,\overline{\varphi'(\zeta)} + \overline{\psi(\zeta)} = f(\zeta), \qquad (5.6.1)$$

where the function $f(\zeta)$ is given on Γ. The Goursat functions contain (in the case of multiply connected regions) logarithmic terms (cf. (5.2.10)); for problem II, these terms are completely known (cf. (5.2.11)). Using these formulas, equation (5.6.1) may be reduced to the form

$$\varphi^*(\zeta) + \zeta\,\overline{\varphi'^*(\zeta)} + \overline{\psi^*(\zeta)} = f^*(\zeta), \qquad (5.6.2)$$

where φ^* and ψ^* are single-valued functions, holomorphic in Ω and f^* is a function, continuous on Γ, which satisfies the condition

$$Re \int_\Gamma f^*(\zeta)\,\overline{d\tau} = 0\,. \tag{5.6.3}$$

The conditions, imposed on $f^*(\zeta)$ signify that the external forces, applied to the contour Γ, are in static equilibrium.

The functions $\varphi^*(z)$ and $\psi^*(z)$ will be sought in the form of the integrals

$$\varphi^*(z) = \frac{1}{2\pi i} \int_\Gamma \frac{\omega(\zeta)}{\zeta - z}\, d\zeta\,, \ z \in \Omega\,,$$

$$\psi^*(z) = \frac{1}{2\pi i} \int_\Gamma \frac{\overline{\omega(\zeta)}}{\zeta - z}\, d\zeta - \frac{1}{2\pi i} \int_\Gamma \frac{\overline{\zeta}\,\omega(\zeta)}{(\zeta - z)^2}\, d\zeta + \frac{1}{2\pi i} \int_\Gamma \frac{\omega(\zeta)}{\zeta - z}\, \overline{d\zeta}\,, \tag{5.6.4}$$

where $\omega(\zeta)$ is an unknown, complex valued function of the point $\zeta \in \Gamma$. Substituting from (5.6.4) into (5.6.2) and performing some simple transformations, similar to those of the last section, one obtains the integral equation for the unknown $\omega(t)$:

$$\omega(t) + \frac{1}{2\pi i} \int_\Gamma \omega(t)\, d\ln \frac{\zeta - t}{\overline{\zeta} - \overline{t}} - \frac{1}{2\pi i} \int_\Gamma \overline{\omega(t)}\, d\frac{\zeta - t}{\overline{\zeta} - \overline{t}} = f^*(t)\,, \ t \in \Gamma\,, \tag{5.6.5}$$

which is equivalent to a system of two Fredholm equations.

It is readily verified that this equation has no solution for an arbitrarily given $f^*(t)$. In order to make it unconditionally soluble, proceed in the following manner.

Add to the left hand side of (5.6.5) the term

$$b\Big(\frac{1}{t - a} - \frac{1}{\overline{t} - \overline{a}} + \frac{t - a}{(\overline{t} - \overline{a})^2}\Big)\,; \ b = \frac{1}{\pi i}\, Re \int_\Gamma \frac{\omega(\zeta)}{(\zeta - a)^2}\, d\zeta\,, \tag{5.6.6}$$

where a is an arbitrarily fixed point in Ω, and set $\zeta - t = r\,e^{i\theta}$ to obtain the equation

$$\omega(t) + \frac{1}{\pi} \int_\Gamma \omega(\zeta)\, d\theta - \frac{1}{\pi} \int_\Gamma \overline{\omega(\zeta)}\, e^{2i\theta}\, d\theta +$$

$$+ \frac{1}{\pi i}\Big(\frac{1}{t - a} - \frac{1}{\overline{t} - \overline{a}} + \frac{t - a}{(\overline{t} - \overline{a})^2}\Big) Re \int_\Gamma \frac{\omega(\zeta)}{(\zeta - a)^2}\, d\zeta = f^*(t)\,, \tag{5.6.7}$$

which will now be investigated.

First of all, it will be shown that, if $f^*(t)$ satisfies (5.6.3), then any solution of (5.6.7) (if it exists) makes the number b vanish. Stepping through the transformations leading from (5.6.2) to (5.6.7) in reverse order, one finds

$$\varphi^*(t) + t\,\overline{\varphi'^*(t)} + \overline{\psi(t)} + b\left(\frac{1}{t-a} + \frac{1}{\bar{t}-\bar{a}} + \frac{t-a}{(\bar{t}-\bar{a})^2}\right) = f^*(t)\,.$$

Multiply this equation by $\overline{dt}$ and integrate along Γ to find, after some simple transformations,

$$\int_\Gamma \left[\varphi^*(t)\,\overline{dt} - \overline{\varphi^*(t)}\,dt\right] + b\int_\Gamma \left(\frac{\overline{dt}}{t-a} + \frac{dt}{\bar{t}-\bar{a}}\right) + 2\pi i b = \int_\Gamma f^*(t)\,\overline{dt}\,. \tag{5.6.8}$$

It is seen from the definition of the number b (cf. (5.6.6)) that it is purely imaginary, hence it follows that the real part of the expression on the left hand side of (5.6.8) equals $2\pi i b$. It also follows from (5.6.3) that its right hand side is purely imaginary, hence $b = 0$.

It will now be shown that equation (5.6.7) is unconditionally soluble. By the strength of Fredholm's alternative, it will be sufficient to prove that the corresponding homogeneous equation has only the trivial solution.

Setting $f^*(t) = 0$, one obtains the homogeneous equation

$$\omega_0(t) + \frac{1}{\pi}\int_\Gamma \omega_0(\zeta)\,d\theta - \frac{1}{\pi}\int_\Gamma \overline{\omega(\zeta)}\,e^{2i\theta}\,d\theta + b_0\left(\frac{1}{t-a} - \frac{1}{\bar{t}-\bar{a}} + \frac{t-a}{(\bar{t}-\bar{a})^2}\right) = 0\,,$$

$$b_0 = \frac{1}{\pi i}\,\mathrm{Re}\int_\Gamma \frac{\omega_0(\zeta)}{(\zeta-a)^2}\,d\zeta\,. \tag{5.6.9}$$

It follows from what has been proved above that $b_0 = 0$. In correspondence with (5.6.4), let

$$\varphi_0(z) = \frac{1}{2\pi i}\int_\Gamma \frac{\omega_0(\zeta)}{\zeta-z}\,d\zeta\,,$$

$$\psi_0(z) = \frac{1}{2\pi i}\int_\Gamma \frac{\overline{\omega_0(z)}}{\zeta-z}\,d\zeta - \frac{1}{2\pi i}\int_\Gamma \frac{\bar\zeta\,\omega_0(z)}{(\zeta-z)^2}\,d\zeta + \frac{1}{2\pi i}\int_\Gamma \frac{\omega_0(z)}{\zeta-z}\,d\bar\zeta\,. \tag{5.6.10}$$

Since $b_0 = 0$, it follows from (5.6.9) and (5.6.10) that

$$\varphi_0(z) + \zeta\overline{\varphi_0'(\zeta)} + \overline{\psi_0(\zeta)} = 0 \,.$$

This equation signifies that the functions $\varphi_0(z)$ and $\psi_0(z)$ solve for the region Ω the plane problem of the theory of elasticity, provided the boundary of this region is not acted upon by external forces. By the uniqueness theorem, $\varphi_0(z) = i\alpha z + \beta$ and $\psi_0(z) = -\overline{\beta}$, where α is a real and β a complex constant.

Since $b_0 = 0$, one has by (5.6.9) and (5.6.10) that $0 = b_0 = 2i\,Im[\varphi_0'(\alpha)] = 2i\alpha$. However, $\alpha = 0$ and $\varphi_0(z) = -\overline{\psi_0(z)} = \beta$. Using this last result and (5.6.10), one finds the identities.

$$\frac{1}{2\pi i}\int_\Gamma \frac{\omega_0(\zeta) - \beta}{\zeta - z}\, d\zeta = 0, \quad \frac{1}{2\pi i}\int_\Gamma \frac{\overline{\omega_0(\zeta)} - \overline{\zeta}\omega_0'(\zeta) + \overline{\beta}}{\zeta - z}\, d\zeta = 0,$$

which indicate that the functions $\omega_0(\zeta) - \beta$ and $\overline{\omega_0(\zeta)} - \overline{\zeta}\omega_0'(\zeta) + \overline{\beta}$ are the contour values of functions which are holomorphic in each of the regions, complementary to $\overline{\Omega}$, and vanish at infinity. Consider these functions in the complementary region which contains the point at infinity. Denote this region by Ω_0 and the functions, which are holomorphic there, by $i\delta(z)$ and $i\varepsilon(z)$, respectively, so that

$$i\delta(\zeta) = \omega_0(\zeta) - \beta \,;\; i\varepsilon(\zeta) = \overline{\omega_0(\zeta)} - \overline{\zeta}\omega_0'(\zeta) + \overline{\beta} \,;$$

where ζ is the variable point on the curve Γ_0, bounding the region Ω_0 outside. Excluding $\omega_0(\zeta)$ from the last equalities, one obtains

$$\overline{\delta(\zeta)} + \overline{\zeta}\delta'(\zeta) + \varepsilon(\zeta) + 2i\overline{\beta} = 0 \,,\; \zeta \in \Gamma_0 \,.$$

Thus, the functions $\delta(z)$ and $\varepsilon(z) + 2i\overline{\beta}$ solve the homogeneous problem of the theory of elasticity for the region Ω. By the uniqueness theorem, $\delta(z) = i\alpha'z + \beta'$, $\varepsilon(z) + 2i\overline{\beta} = -\overline{\beta'}$. However, $\delta(z)$ is holomorphic in Ω_0 and vanishes at infinity, hence $\alpha' = \beta' = 0$; since $\varepsilon(\infty) = 0$, one has $\beta = 0$. Thus $\delta(z) = \varepsilon(z) \equiv 0$, and, consequently, $\omega_0(\zeta) = 0$, as was to be proved.

§7. Plane strain in anisotropic media

This section is devoted to the results presented in [43].

1^0. In the general case, the elastic properties of anisotropic media are characterized by 81 elastic constants c_{jklm}; $j, k, l, m = 1, 2, 3$. Since they satisfy the conditions of symmetry $c_{jklm} = c_{kjlm} = c_{lmjk}$, there are altogether 21 independent constants. They form a symmetric, fourth order tensor; Hooke's law thus becomes

$$\sigma_{jk} = \sum_{l,m=1}^{3} c_{jklm} e_{lm} \, , \tag{5.7.1}$$

where σ_{jk} and e_{lm} are the components of the stress and strain tensors, respectively. The components of the strain tensor are expressed in terms of the components of the displacement vector $u = (u_1, u_2, u_3)$ by the relations

$$e_{jk} = \frac{1}{2} \left(\frac{\partial u_j}{\partial x_k} + \frac{\partial u_k}{\partial x_j} \right); \; j, k = 1, 2, 3 \, .$$

Let it be assumed that the medium under consideration is homogeneous, so that the elastic constants in a fixed coordinate system are constant. Let the elastic medium under consideration be in a state of plane strain, i.e., $u_3 = 0$ and $\frac{\partial u_1}{\partial x_3} = \frac{\partial u_2}{\partial x_3} = 0$. Then $e_{j3} = e_{3j} = 0$, $j, k = 1, 2, 3$. The equations of Hooke's law assume now the somewhat, simpler form:

$$\sigma_{jk} = \sum_{l,m=1}^{2} c_{jklm} e_{lm} \, ; \; j, k = 1, 2, 3 \, . \tag{5.7.2}$$

Now the strains e_{jk}, and with them the stresses σ_{jk}, do not depend on x_3. In the absence of body forces, the equilibrium equations become

$$\frac{\partial \sigma_{11}}{\partial x_1} + \frac{\partial \sigma_{12}}{\partial x_2} = 0 \, , \; \frac{\partial \sigma_{12}}{\partial x_1} + \frac{\partial \sigma_{22}}{\partial x_2} = 0 \, , \; \frac{\partial \sigma_{13}}{\partial x_1} + \frac{\partial \sigma_{23}}{\partial x_2} = 0 \, . \tag{5.7.3}$$

As before, the first two of these equations suggest the existence of a function $W(x_1, x_2)$ (Airy function):

$$\sigma_{11} = \frac{\partial^2 W}{\partial x_2^2} \, , \; \sigma_{12} = -\frac{\partial^2 W}{\partial x_1 \partial x_2} \, , \; \sigma_{22} = \frac{\partial^2 W}{\partial x_1^2} \, . \tag{5.7.4}$$

Replacing in (5.7.2) the stress components σ_{11}, σ_{12}, σ_{22} using (5.7.4), and solving the corresponding equations with respect to the strains e_{11}, e_{12}, e_{13}, one obtains

$$\frac{\partial u_j}{\partial x_k} + \frac{\partial u_k}{\partial x_j} = A_{jk1} \frac{\partial^2 W}{\partial x_1^2} + A_{jk2} \frac{\partial^2 W}{\partial x_1 \partial x_2} + A_{jk3} \frac{\partial^2 W}{\partial x_2^2} \, , \tag{5.7.5}$$

where the A_{jkl} are certain constants.

Differentiating this equation with respect to x_1 and x_2 and eliminating the displacements, one finds for the function W the fourth order, elliptic differential equation

$$L(W) = B_{40}\frac{\partial^4 W}{\partial x_1^4} + B_{31}\frac{\partial^4 W}{\partial x_1^3 \partial x_2} + B_{22}\frac{\partial^4 W}{\partial x_1^2 \partial x_2^2} +$$
$$+ B_{13}\frac{\partial^4 W}{\partial x_1 \partial x_2^3} + B_{04}\frac{\partial^4 W}{\partial x_2^4} = 0\,;\ B_{\alpha\beta} = const. \tag{5.7.6}$$

which was first obtained by S.G. Lekhnitskii [32].

Assume for the sake of simplicity that the plane region Ω , occupied by an anisotropic, elastic medium is finite and simply connected, and that for this region problem II is posed; naturally, the forces applied to the contour $\Gamma = \partial\Omega$ must be in static equilibrium. As a rule, one can determine from the given forces the boundary values of the derivatives $\partial W/\partial x_1$ and $\partial W/\partial x_2$. It will be shown below that one can find from the known contour values of these derivatives the function W inside Ω, and hence, using (5.7.4) and (5.7.5), the stresses and strains, respectively. Finally, the stresses σ_{j3} are obtained from (5.7.2). Thus, by solving the boundary value problem for the function W, one obtains all quantities characterizing the solution of the plane problem. However, by the strength of the third equation (5.7.3), the function W must still satisfy a certain, third order differential equation. In order to find this equation, differentiate with respect to x_1 equation (5.7.2) with subscripts $j = 1$, $k = 3$ and with respect to x_2 equation (5.7.2) with subscripts $j = 2$, $k = 3$. Add the results and eliminate the strains, using (5.7.5). These operations yield the equation:

$$a_1\frac{\partial^3 W}{\partial x_1^3} + a_2\frac{\partial^3 W}{\partial x_1^2 \partial x_2} + a_3\frac{\partial^3 W}{\partial x_1 \partial x_2^2} + a_4\frac{\partial^3 W}{\partial x_2^3} = 0\,, \tag{5.7.7}$$

the coefficients $a_1, ..., a_4$ are readily found. In the sequel, it will be assumed that these coefficients are zero, so that equation (5.7.7) is satisfied identically and need not be taken into consideration. It may be shown [42] that the coefficients a_k, vanish, if the elastic properties of the medium do not change on mirror reflection in any plane, normal to the x_3 axis.

2^0. The polynomial

$$f(r) = B_{40}r^4 + B_{31}r^3 + B_{22}r^2 + B_{13}r + B_{04} \tag{5.7.8}$$

has only complex roots, as follows from the elliptic property of equation (5.7.6) and has been proved in C.G. Lekhnitzkii's paper [32]. This polynomial factorizes into two quadratic polynomials

$$f(r) = (\alpha_1 r^2 + \alpha_2 r + \alpha_3)(\beta_1 r^2 + \beta_2 r + \beta_3)$$

and therefore

$$L(W) = \left(\alpha_1 \frac{\partial^2}{\partial x_1^2} + \alpha_2 \frac{\partial^2}{\partial x_1 \partial x_2} + \alpha_3 \frac{\partial^2}{\partial x_2^2}\right)$$
$$\left(\beta_1 \frac{\partial^2}{\partial x_1^2} + \beta_2 \frac{\partial^2}{\partial x_1 \partial x_2} + \beta_3 \frac{\partial^2}{\partial x_2^2}\right) W. \tag{5.7.9}$$

If the roots of the polynomial (5.7.8) equal in pairs, then a linear co-ordinate transformation reduces equation (5.7.6) to a bi-harmonic equation for which the corresponding boundary value problem has been studied in the earlier sections of this chapter. If the roots of the polynomial (5.7.8) are not equal in pairs, equation (5.7.6) may be simplified somewhat by the following procedure. Reduce the second quadratic factor in (5.7.9) by a linear coordinate transformation to its canonical form, and then get rid of, by a rotation of the coordinate axes the mixed product in the first quadratic factor of (5.7.9). Denoting the new coordinates again by x and y, one arrives at an equation of the form

$$\left(\frac{\partial^2}{\partial x^2} + k^2 \frac{\partial^2}{\partial y^2}\right) \Delta W = 0, \tag{5.7.10}$$

where k is a positive constant which does not equal unity and Δ is the Laplace operator.

3^0. Every expression of the form

$$W = V_0(x, y) + V_1(x, y/k), \tag{5.7.11}$$

where V_0 and V_1 are harmonic functions in their arguments, satisfies equation (5.7.10). It is readily seen that, conversely, every solution of this equation has the form (5.7.11). In order to prove this statement, let $\frac{\partial^2 W}{\partial x^2} + k^2 \frac{\partial^2 W}{\partial y^2} = \Psi(x, y)$, and then $\Delta \Psi = 0$. Construct a harmonic function $Q(x, y)$ such that $\partial^2 Q/\partial y^2 = \Psi$ and $V_0(x, y) = Q(x, y)/(k^2 - 1)$. Obviously,

$$\left(\frac{\partial^2}{\partial x^2} + k^2 \frac{\partial^2}{\partial y^2}\right)(W - V_0) = \Psi - \frac{\partial^2 Q}{\partial y^2} = 0$$

and, therefore, $W - V_0 = V_1(x, y/k)$, where V_1 is a harmonic function in terms of x and y/k. It is easily verified that for this function W, satisfying equation

(5.7.10), the functions V_0 and V_1 are determined exactly apart from the terms $axy + bx + cy + d$ and $-(axy + bx + cy + d)$; a, b, c, d - constants.

With $x + iy = z$, $x + iy/k = z_1$, $i = \sqrt{-1}$ one has

$$V_0(x,y) = Re(f(z)) = \frac{1}{2}\left[f(z) + \overline{f(z)}\right],$$

$$V_1(x,y/k) = Re(f_1(z_1)) = \frac{1}{2}\left[f_1(z_1) + \overline{f_1(z_1)}\right], \tag{5.7.12}$$

where $f(z)$ and $f_1(z_1)$ are holomorphic functions in their arguments. Now let $f'(z)/2 = \varphi(z)$, and $f_1'(z_1)/2 = \psi(z_1)$, then

$$\frac{\partial W}{\partial x} = \varphi(z) + \overline{\varphi(z)} + \psi(z_1) + \overline{\psi(z_1)},$$

$$-ik\frac{\partial W}{\partial y} = \psi(z_1) - \overline{\psi(z_1)} + k\varphi(z) - k\overline{\varphi(z)}. \tag{5.7.13}$$

Note that the derivatives $\partial W/\partial x$ and $\partial W/\partial y$ are continuous on Γ; as a rule, this indicates that the resultant vector of the external forces, acting on Γ vanishes. If the function W has been found, then, as follows from (5.7.12), the function $f(z)$ is defined exactly apart from a quadratic term of the form $-aiz^2/2 + (b - ic)z + d + id$ the function $\varphi(z)$ apart from a linear term $aiz/2 + (b - ic)/2 := \alpha iz + \beta$, where α is a real and β a complex constant. Fixing α and β in any manner, one obtains a single-valued function $\varphi(z)$, and, consequently, also $\psi(z)$. Denote by Ω^* the region of change of the coordinates x and y. This region is finite and simply connected, as is also Ω. Locate the origin of coordinates inside Ω^* and demand that

$$\varphi(0) = 0, \; Im(\varphi'(0)) = 0. \tag{5.7.14}$$

With these values, the constants α and β are determined uniquely. Assume now that the boundary $\Gamma = \partial\Omega$ us a curve of Class $C^{(2)}$; the same will be true for the boundary $L = \partial\Omega^*$. Let ξ and η be the coordinates on the curve L and $\zeta = \xi + i\eta$, $\zeta_1 = \xi + i\eta/k$. Denote by $g(\zeta)$ and $h(\zeta_1)$ the contour values of the derivatives $\partial W/\partial x$ and $\partial W/\partial y$, which are known, if problem II has been solved. Equations (5.7.13) now yield the for the functions $\varphi(z)$ and $\psi(z_1)$ the boundary conditions

$$\varphi(\zeta) + \overline{\varphi(\zeta)} + \psi(\zeta_1) - \overline{\psi(\zeta_1)} = g(\zeta),$$

$$\psi(\zeta_1) - \overline{\psi(\zeta_1)} + k\varphi(\zeta) - k\overline{\varphi(\zeta)} = -ikh(\zeta_1). \tag{5.7.15}$$

Multiply the first of these by $d\zeta/2\pi i(\zeta - z)$, the second by $d\zeta_1/2\pi i(\zeta - z_1)$ and integrate along L to arrive at

$$
\begin{aligned}
\varphi(z) - \frac{1}{2\pi i} \int_L \frac{\overline{\varphi(\zeta)}}{\zeta - z}\, d\zeta + \frac{1}{2\pi i} \int_L \frac{\psi(\zeta) - \overline{\psi(\zeta_1)}}{\zeta - z}\, d\zeta = \\
= \frac{1}{2\pi i} \int_L \frac{g(\zeta)}{\zeta - z}\, d\zeta := G(z);
\end{aligned}
\tag{5.7.16}
$$

$$
\begin{aligned}
\psi(z_1) - \frac{1}{2\pi i} \int_L \frac{\overline{\psi(\zeta_1)}}{\zeta_1 - z_1}\, d\zeta_1 + \frac{k}{2\pi i} \int_L \frac{\varphi(\zeta) - \overline{\varphi(\zeta)}}{\zeta_1 - z_1}\, d\zeta_1 = \\
= \frac{1}{2\pi i} \int_L -\frac{ikh(\zeta_1)}{\zeta_1 - z_1}\, d\zeta_1 := H(z).
\end{aligned}
\tag{5.7.17}
$$

Equations (5.7.14) and (5.7.16) yield

$$
A := Im\left\{ \frac{1}{2\pi i} \int_L \frac{\overline{\varphi(\zeta)}}{\zeta^2}\, d\zeta + \frac{1}{2\pi i} \int_L \frac{\psi(\zeta_1) + \overline{\psi(\zeta_1)}}{\zeta^2}\, d\zeta \right\} = Im\, G'(0),
\tag{5.7.18}
$$

$$
B := \frac{1}{2\pi i} \int_L \frac{\overline{\varphi(\zeta)}}{\zeta}\, d\zeta + \frac{1}{2\pi i} \int_L \frac{\psi(\zeta_1) + \overline{\psi(\zeta_1)}}{\zeta}\, d\zeta = G(0).
\tag{5.7.19}
$$

Multiplying (5.7.18) by iz and adding it to (5.7.19), and subtracting the result from (5.7.16), one is led to the equation

$$
\begin{aligned}
\varphi(z) + \frac{1}{2\pi i} \int_L \frac{\overline{\varphi(\zeta)}}{\zeta - z}\, d\zeta + \frac{1}{2\pi i} \int_L \frac{\psi(\zeta_1) + \overline{\psi(\zeta_1)}}{\zeta - z}\, d\zeta - \\
- Aiz - B = G(z) - iz\, Im(G'(0)) - G(0).
\end{aligned}
\tag{5.7.20}
$$

Next, isolate in (5.7.19) the real part and subtract it from (5.7.17) to obtain the new equation

$$
\begin{aligned}
\psi(z_1) - \frac{1}{2\pi i} \int_L \frac{\overline{\psi(\zeta_1)}}{\zeta_1 - z_1}\, d\zeta_1 + \frac{k}{2\pi i} \int_L \frac{\varphi(\zeta) - \overline{\varphi(\zeta)}}{\zeta_1 - z_1}\, d\zeta_1 - \\
- Re(B) = H(z_1) - Re(G(0)).
\end{aligned}
\tag{5.7.21}
$$

The function $\varphi(z)$, determined from the system of equations (5.7.20) and (5.7.21) replacing (5.7.16) and (5.7.17), automatically satisfies condition (5.7.14). However, this does not yet imply that by solving this system, the corresponding problem of the theory of elasticity is solved. For this purpose, it is necessary and sufficient that the functions $\varphi(z)$ and $\psi(z_1)$,constituting the solutions of the new system, also satisfy equations (5.7.18) and (5.7.19). It will be shown below that this will be so, if the external forces, applied to the contour Γ are in static equilibrium.

4^0. Let the point (x,y) approach from inside the region Ω^* the contour L. Understanding now by (x,y) the point on L, one finds from (5.7.20) and (5.7.21), using the formulas of Sokhotskii-Plemelj (2.1.11), the system of homogeneous, singular integral equations

$$
\varphi(z) + \frac{1}{2}\overline{\varphi(z)} + \frac{1}{2}\left[\psi(z_1) + \overline{\psi(z_1)}\right] + \frac{1}{2\pi i}\int_L \frac{\overline{\varphi(\zeta)}}{\zeta - z}\,d\zeta +
$$

$$
+ \frac{1}{2\pi i}\int_L \frac{\psi(\zeta_1) + \overline{\psi(\zeta_1)}}{\zeta - z}\,d\zeta - Aiz - B = G(z)-
$$

$$
- iz\,Im(G'(0)) - G(0),
$$

$$
\psi(z_1) - \frac{1}{2}\overline{\psi(z_1)} + \frac{k}{2}\left[\varphi(z) - \overline{\varphi(z)}\right] - \frac{1}{2}\int_L \frac{\overline{\psi(\zeta_1)}}{\zeta_1 - z_1}\,d\zeta_1 +
$$

$$
+ \frac{k}{2\pi i}\int_L \frac{\varphi(\zeta) - \overline{\varphi(\zeta)}}{\zeta_1 - z_1}\,d\zeta_1 - Re(B) =
$$

$$
= H(z_1) - Re(G(0)).
$$

(5.7.22)

It is useful to somewhat transform this system prior to finding its solution. Let the parametric equations of the contour L in the form $x = x(t)$ and $y = y(t)$, $0 \leq t \leq 2\pi$ be periodic functions of t which are continuous together with their first two derivatives, where $x'^2(t) + y'^2(t) > 0$. Let $x(t) + iy(t) = z(t)$, $x(t) + \frac{i}{k}y(t) = z_1(t)$. Moreover, set $\zeta = z(\tau)$, $\zeta_1 = z_1(\tau)$, $0 \leq \tau \leq 2\pi$. Using (2.2.10), one obtains the identities

$$
\frac{d\zeta}{\zeta - z} = \frac{1}{2}ctg\frac{\tau - t}{2}\,d\tau + P(t,\tau)\,d\tau,
$$

$$
\frac{d\zeta_1}{\zeta_1 - z_1} = \frac{1}{2}ctg\frac{\tau - t}{2}\,d\tau + Q(t,\tau)\,d\tau,
$$

where $P(t, \tau)$ and $Q(t, \tau)$ are bounded and measurable functions. Finally, let

$$\varphi(z(t)) = W_1(t) + iW_2(t), \quad \psi(z(t)) = W_3(t) + iW_4(t),$$

$$G(z(t)) - iz(t) Im(G'(0)) - G(0) = f_1(t) + if_2(t),$$

$$H(z_1(t)) - Re(G(0)) = f_3(t) + if_4(t),$$

where the functions W_j, f_j, $j = 1, ..., 4$, are real, and

$$\frac{1}{2\pi} \int\limits_0^{2\pi} \rho(\tau) \, ctg \frac{\tau - t}{2} \, d\tau = (P\rho)(t),$$

where ρ is an arbitrary function.

Rewrite system (5.7.22) in the new notation, separate real and imaginary parts and multiply the resulting equations by suitable factors to free them of denominators, to arrive at the system of four singular integral equations

$$\begin{aligned}
3W_1(t) - (PW_2)(t) + 2W_3(t) + T_1(W_1, ..., W_4) &= 2f_1(t); \\
-(PW_1)(t) + W_2(t) - 2(PW_3)(t) + T_2(W_1, ..., W_4) &= 2f_2(t); \\
2k(PW_2)(t) + W_3(t) + (PW_4) + T_3(W_1, ..., W_4) &= 2f_3(t); \\
2kW_2(t) + (PW_3)(t) + 3W_4(t) + T_4(W_1, ..., W_4) &= 2f_4(t);
\end{aligned} \tag{5.7.23}$$

where the system T_j, $j = 1, ..., 4$ are certain compact operators.

As it seen from (5.7.10), the symbol of the operator P is θ/i, where θ is the symbol of the Cauchy operator S. Recalling that the variable θ has only the values ± 1, the symbolic matrix of the system (5.7.23) is

$$\begin{pmatrix}
3 & -\theta/i & 2 & 0 \\
-\theta/i & 1 & -2\theta/i & 0 \\
0 & 2k\theta/i & 1 & \theta/i \\
0 & 2k & \theta/i & 3
\end{pmatrix}. \tag{5.7.24}$$

For $k \neq 1$, this matrix does not degenerate, since its determinant if $16(1 - k)$. The elements of this symbolic matrix do not depend on t, and it follows from (2.6.5) that the index of the system (5.7.23) is zero, hence, in turn, Fredholm's theorems and their consequence - Fredholm's alternative - are true for this system.

5^0. Next, it will b shown that system (5.7.22) is soluble for arbitrary right hand sides. For this purpose, by Fredholm's alternative, it will be sufficient to show that the corresponding homogeneous system, has only the trivial solution.

Let $\varphi_0(z)$ and $\psi_0(z_1)$ be solutions of the homogeneous system, corresponding to the system (5.7.22). Denote by A_0 and B_0 the values of the constants A and B in the homogeneous version of system (5.7.22). These solution of the homogeneous system are naturally analytic continuations into the region Ω^* and satisfy the system of equations

$$\varphi_0(z) + \frac{1}{2\pi i} \int_L \frac{\overline{\varphi_0(\zeta)}}{\zeta - z}\, d\zeta + \frac{1}{2\pi i} \int_L \frac{\psi_0(\zeta_1) + \overline{\psi_0(\zeta_1)}}{\zeta - z}\, d\zeta = A_0 i z + B_0 ,$$

$$\psi_0(z_1) - \frac{1}{2\pi i} \int_L \frac{\overline{\psi_0(\zeta_1)}}{\zeta_1 - z_1}\, d\zeta_1 + \frac{k}{2\pi i} \int_L \frac{\varphi_0(\zeta) - \overline{\varphi_0(\zeta)}}{\zeta_1 - z_1}\, d\zeta_1 = Re(B_0) .$$

$$(5.7.25)$$

Note (cf. 3^0) that $\varphi_0(0) = 0$, $Im(\varphi_0'(0)) = 0$; it then follows from the first equation (5.7.25) and from (5.7.18) and (5.7.19) that $A_0 = B_0 = 0$, hence it is seen that $\varphi_0(z)$ and $\psi_0(z_1)$ satisfy the simpler homogeneous system

$$\varphi_0(z) + \frac{1}{2\pi i} \int_L \frac{\overline{\varphi_0(\zeta)}}{\zeta - z}\, d\zeta + \frac{1}{2\pi i} \int_L \frac{\psi_0(\zeta_1) + \overline{\psi_0(\zeta_1)}}{\zeta - z}\, d\zeta = 0 ,$$

$$(5.7.26)$$

$$\psi_0(z_1) - \frac{1}{2\pi i} \int_L \frac{\overline{\psi_0(\zeta_1)}}{\zeta_1 - z_1}\, d\zeta_1 + \frac{k}{2\pi i} \int_L \frac{\varphi_0(\zeta) - \overline{\varphi_0(\zeta)}}{\zeta_1 - z_1}\, d\zeta_1 = 0 .$$

The function $W_0(x, y)$ which will satisfy in the region Ω^* equation (5.7.10) and the relations

$$\lambda_0 := \frac{\partial W_0}{\partial x} = \varphi_0(z) + \overline{\varphi_0(z)} + \psi_0(z_1) + \overline{\psi_0(z_1)} ;$$

$$-ik\mu_0 := -ik\frac{\partial W_0}{\partial y} = \psi_0(z_1) - \overline{\psi_0(z_1)} + k\left[\varphi_0(z) - \overline{\varphi_0(z)}\right] .$$

can be constructed with the aid of the functions $\varphi_0(z)$ and $\psi_0(z_1)$. Replacing the last relations z and z_1 by ζ and ζ_1, respectively, multiplying by the Cauchy kernel and integrating along L, one finds

$$\varphi_0(z) + \frac{1}{2\pi i} \int\limits_L \frac{\overline{\varphi_0(\zeta)}}{\zeta - z}\, d\zeta + \frac{1}{2\pi i} \int\limits_L \frac{\psi_0(\zeta_1) + \overline{\psi_0(\zeta_1)}}{\zeta - z}\, d\zeta =$$

$$= \frac{1}{2\pi i} \int\limits_L \frac{\lambda_0(\zeta)}{\zeta - z}\, d\zeta\,,$$

$$\psi_0(z_1) - \frac{1}{2\pi i} \int\limits_L \frac{\overline{\psi_0(\zeta_1)}}{\zeta_1 - z_1}\, d\zeta_1 + \frac{k}{2\pi i} \int\limits_L \frac{\varphi_0(\zeta) - \overline{\varphi_0(\zeta)}}{\zeta_1 - z_1}\, d\zeta_1 =$$

$$= \frac{1}{2\pi i} \int\limits_L \frac{ik\,\mu_0(\zeta_1)}{\zeta_1 - z_1}\, d\zeta_1\,.$$

A comparison of these equations with equations (5.7.26) yields the identities

$$(x,y) \in \Omega^*\,;\quad \frac{1}{2\pi i} \int\limits_L \frac{\lambda_0(\zeta)}{\zeta - z}\, d\zeta \equiv 0\,;\quad \frac{1}{2\pi i} \int\limits_L \frac{ik\,\mu_0(\zeta_1)}{\zeta_1 - z_1}\, d\zeta_1 \equiv 0\,.$$

This means that the functions $\lambda_0(z)$ and $\mu_0(z)$ are continued analytically outside the contour L and vanish at infinity. However, on L, these functions are real, hence they vanish identically. By the uniqueness theorem, $W_0(x,y) = const$ in Ω^* and, since $\varphi_0(0) = 0$, $Im(\varphi_0'(0)) = 0$, one has $\varphi_0(z) \equiv 0$ and $\psi_0(z_1) \equiv 0$.

The results above show that the singular system (5.7.23) or, what is the same thing, the system (5.7.22) has a unique solution.

6^0. There remains to prove that the functions $\varphi(z)$ and $\psi(z_1)$ which solve system (5.7.22) satisfy conditions (5.7.15) and (5.7.19). It will be shown that this is so provided

$$\int\limits_L \left[g(\zeta)\, dx + h(\zeta_1)\, dy\right] = 0\,, \tag{5.7.27}$$

which signifies that the resultant moment of the external forces, applied to the contour Γ, vanish. The functions $g(\zeta)$ and $h(\zeta_1)$ are the right hand sides of the boundary conditions (5.7.15).

Let $\varphi(z)$ and $\psi(z_1)$ be the solutions of system (5.7.22). These functions will be used to construct the function $W^*(x,y)$ which satisfies (5.7.10) and the boundary conditions

$$\varphi(\zeta) + \overline{\varphi(\zeta)} + \psi(\zeta_1) + \overline{\psi(\zeta_1)} = g^*(\zeta)\,,$$
$$\psi(\zeta_1) - \overline{\psi(\zeta_1)} + k\left[\varphi(\zeta) - \overline{\varphi(\zeta)}\right] = -ikh^*(\zeta_1)\,, \tag{5.7.28}$$

where $g^*(\zeta) = \partial W^*/\partial x\big|_L$, $h^*(\zeta_1) = \partial W^*/\partial y\big|_L$. Note that

$$\int_L g^*\, dx + h^*\, dy = 0\,. \tag{5.7.29}$$

Multiplication of (5.7.28) by the corresponding Cauchy kernel and integration yield

$$\varphi(z) + \frac{1}{2\pi i}\int_L \frac{\overline{\varphi(\zeta)}}{\zeta - z}\, d\zeta + \frac{1}{2\pi i}\int_L \frac{\psi(\zeta_1) + \overline{\psi(\zeta_1)}}{\zeta - z}\, d\zeta = \frac{1}{2\pi i}\int_L \frac{g^*(\zeta)}{\zeta - z}\, d\zeta\,,$$

$$\psi(z_1) - \frac{1}{2\pi i}\int_L \frac{\overline{\psi(\zeta_1)}}{\zeta_1 - z_1}\, d\zeta_1 + \frac{k}{2\pi i}\int_L \frac{\varphi(\zeta) - \overline{\varphi(\zeta)}}{\zeta_1 - z_1}\, d\zeta_1 = \frac{ik}{2\pi i}\int_L \frac{h^*(\zeta_1)}{\zeta_1 - z_1}\, d\zeta_1\,.$$

Comparing this result with (5.7.20) and (5.7.21), one finds

$$(x,y) \in \Omega^*\,,\quad \int_L \frac{g^*(\zeta) - g(\zeta) - i\zeta(A - Im(G'(0))) - (B - G(0))}{\zeta - z}\,, d\zeta \equiv 0\,,$$

$$\frac{1}{2\pi i}\int_L \frac{ik[h^*(\zeta_1) - h(\zeta_1)] - Re(B)}{\zeta_1 - z_1}\, d\zeta_1 \equiv 0\,. \tag{5.7.30}$$

The second of these equations shows that

$$ik[h^*(\zeta_1) - h(\zeta_1)] - Re(B)$$

analytically continues with respect to the variable z_1 throughout the outside of the contour L and the continued function vanishes at infinity. On the contour L, the real part of this function is constant, hence this function is identically zero and, consequently, $h^* \equiv h$ and $Re(B) = 0$. It now follows from (5.7.27) and (5.7.29) that

$$\int_L (g^* - g)\, dx = 0\,. \tag{5.7.31}$$

Consider now the first relation (5.7.30). It means that the function $g^*(\zeta) - g(\zeta) - i\zeta[Im\, G'(0) - A] - [G(0) - B]$ is analytically continued with respect to z into the entire outside region of the contour L and vanishes at infinity. Let the continued function be denoted by $\gamma(z)$.

Map conformally the outside of the contour L onto the interior of the circle $|s| < 1$, where s is a new complex variable. The mapping function has the form

$$z = \omega(s) = \sum_{n=-1}^{+\infty} \omega_n s^n \, ;$$

where the coefficient ω_{-1} may be assumed to be positive. If $z \in L$, then the modulus of the corresponding value of s equals unity, and one can set $s = e^{i\vartheta}$. Then, obviously,

$$\gamma(z) = \gamma(\omega(t)) = \sum_{n=+1}^{\infty} \gamma_n e^{in\vartheta} \, , \quad z = \sum_{n=-1}^{\infty} \omega_n e^{in\vartheta} \, , \quad g \quad g^* = \sum_{n=-\infty}^{\infty} g_n e^{in\vartheta} \, .$$

Since the difference $g - g^*$ is real, then $g_{-n} = \overline{g_n}$. Substitution of this expression for $\gamma(z)$ and comparison of the coefficients of the same powers of $e^{i\vartheta}$ verifies that $g_n = 0$, $n \leq -2$. Consequently, $g_n = 0$, $n \geq 2$. Equating now the coefficients of $e^{in\vartheta}$, $|n| \leq 1$, one finds that

$$g - g^* = i\omega_{-1}[Im(G'(0)) - A](e^{i\vartheta} - e^{-i\vartheta}) + const. \tag{5.7.32}$$

The identity (5.7.31) will be considered next. On the contour L

$$dx = \frac{1}{2}\left(dz + \overline{dz}\right) = \frac{1}{2}\left[\omega'(s)\,ds + \overline{\omega'(s)\,ds}\,\right].$$

hence, taking into consideration (5.7.32), equation (5.7.31) is transformed into

$$\int_L (g - g^*)dx = 2\pi[Im(G'(0)) - A_0](\omega_{-1} + Re\,\omega_1). \tag{5.7.33}$$

The sum $\omega_{-1} + Re\,\omega_1$ is non-zero. In fact, by the area theorem of Bieberbach [85], one has

$$\sum_{n=1}^{\infty} n\left|\frac{\omega_n}{\omega_{-1}}\right|^2 \leq 1$$

and the sum $\omega_{-1} + Re\,\omega_1 = 0$ is only possible, if $\omega_1 = -\omega_{-1}$ and $\omega_n = 0$, $n > 1$. However, this case is excluded, since then $z = \omega(s) = \omega_0 + \omega_{-1}(s^{-1} - s)$ and the inside of the circle $|s| < 1$ is mapped on the plane R^2 with a cut, and this contradicts the assumption according to which the tangent along L

hanges continuously. It follows now from (5.7.33) that $A_0 = Im\,G'(0)$, and $Im\,B = Im\,G(0)$. Since it has been shown earlier that $Re\,B = Re\,G(0)$, one has $B = G(0)$. Thus, by solving system (5.7.20) and (5.7.21), problem II has been solved for anisotropic media.

7^0. The method of this section may be extended to the case of infinite, multiply connected regions as well as to problem I.

Another method of solution of the plane problems for anisotropic elastic media was developed by D.I. Sherman [80]; in this method, the problems under consideration were reduced to Fredholm equations.

CHAPTER 6
POTENTIAL THEORY FOR BASIC THREE–DIMENSIONAL PROBLEMS

The classical results of S.G. Mikhlin [45], N.I. Muskhelishvili [54], N.P. Vekua [8], V.D. Kupradze [28], and also the monographs of V.Z. Parton and P.I. Perlin [58], T.V. Burchuladze and T.P. Gegelia [7] and others [69, 9], it might be said, have presented comprehensive answers to questions, linked to applications of integral equations to problems of the theory of elasticity.

However, within the last decade, many applications of the methods of integral equations to such complex problems of the theory of elasticity as the theories of cracks, sharp concentrators, contact problems, bending problems, etc. as well as the publication of a large number of different numerical schemes of solution with the aid of integral equations of the equations of elasticity theory [6, 5, 25, 37, 38, 69] have led to new results, systemization of the material, sharpening of the terminology, and in the parts, where this was not the case, to establishment of sound foundations. Some of these questions will be discussed in this chapter.

§1. The equilibrium equations in terms of displacements

By tradition, there exist two classical formulations of the problems of mechanics of deformable, solid media: In terms of displacements and in terms of stresses. A start will be made with displacements.

Apply to he relations linking strains to displacements

$$e_{ij} = \frac{1}{2}(u_{i,j} + u_{j,i}) \tag{6.1.1}$$

Hooke's law

$$\sigma_{ij} = \delta_{ij}\lambda e + 2\mu e_{ij}, \; e = e_{ii}, \tag{6.1.2}$$

to obtain the stress-strain relations

$$\sigma_{ij} = \lambda\sigma_{ij}\, div\, \mathbf{u} + \mu(u_{i,j} + u_{j,i})* \tag{6.1.3}$$

which, on substitution into the equilibrium equations

$$\sigma_{ij,j} + F_j = 0, \; i = 1, 2, 3, \tag{6.1.4}$$

---

* The summation convention for Cartesian tensors has been applied here.

yield the Lame equations

$$\mu\Delta u_i + (\lambda + \mu)(div\,\mathbf{u}),i + F_i = 0\,,\; i = 1,2,3\,. \qquad (6.1.5)$$

These equations may be written in the vectorial form

$$\begin{aligned}
L\mathbf{u} \equiv{}& \mu\Delta\mathbf{u} + (\lambda + \mu)grad\,div\,\mathbf{u} \equiv \\
\equiv{}& (\lambda + 2\mu)grad\,div\,\mathbf{u} - \mu\,rot\,rot\,\mathbf{u} \equiv \\
\equiv{}& 2\mu\Delta\mathbf{u} + \lambda\,grad\,div\,\mathbf{u} + \mu\,rot\,rot\,\mathbf{u} = -F\,.
\end{aligned} \qquad (6.1.6)$$

When dealing with a linearly elastic body occupying a region Ω, these equations will apply in that region, while on its boundary Ω the boundary conditions

$$\mathbf{u}\Big|_{\partial\Omega_1} = \mathbf{u}_0\,, \qquad (6.1.7)$$

$$\sigma^{(n)}\Big|_{\partial\Omega_2} = f\,, \qquad (6.1.8)$$

are applied, where $\partial\Omega = \partial\Omega_1 \bigcup \partial\Omega_2$; in particular, it can happen that $\partial\Omega_1$ (or $\partial\Omega_2$) are empty set. Condition (6.1.8) may be written on the form of a differential boundary operator in $\mathbf{u}$. In fact,

$$\sigma^{(n)} \equiv \sigma^{(x_i)}cos(n, x_i) = \sigma_{x_i x_j}\mathbf{k}_j\,cos(n, x_i) =$$

$$= \mathbf{k}_j cos(n, x_i)[\lambda\sigma_{ij}div\,\mathbf{u} + \mu(u_{i,j} + u_{j,i})] =$$

$$= 2\mu\frac{\partial\mathbf{u}}{\partial n} + \lambda\mathbf{n}\,div\,\mathbf{u} + \mu\mathbf{n} \times rot\,\mathbf{u} \equiv \sigma^{(n)}[\mathbf{u}]\,,$$

when (6.1.8) becomes

$$\sigma^{(n)}[\mathbf{u}]\Big|_{\partial\Omega_2} = f\,,$$

where $\sigma^{(n)}[\mathbf{u}]$ is a first order, differential operator acting on $\mathbf{u}$.

Direct verification confirms that the boundary value problem (6.1.5)-(6.1.8) is strongly elliptic [10].

Let E denote the specific strain energy

$$E = \frac{1}{2}\,\sigma_{ij}e_{ij} = \frac{1}{2}\left[\lambda e_{kk}^2 + 2\mu e_{ij}e_{ij}\right]. \qquad (6.1.9)$$

For the theoretical analysis of problems of the theory of elasticity, Betti's formulas, which are a variant of the Green formulas for the Lame equations, assume a central role. Consider a simple means for deriving these formulas.

Obtain , by (6.1.1), expressions for the strains $e_{ij}(\mathbf{u})$ in terms of the vector u, and then from (6.1.2), expressions for the stresses $\sigma_{ij}[\mathbf{u}]$.

Obtain in an analogous manner expressions for the $e_{ij}(\mathbf{v})$ and $\sigma_{ij}[\mathbf{v}]$ in terms of the vect $\mathbf{v}$. Integration of the expression

$$\int_\Omega \mathbf{u} L \mathbf{v} \, d\Omega ,$$

by parts yields

$$\int_\Omega \mathbf{u} L \mathbf{v} \, d\Omega = \int_\Gamma \mathbf{u} \sigma^{(n)}(\mathbf{v}) \, d\Gamma - \int_\Omega E(\mathbf{u}, \mathbf{v}) \, d\Omega , \ \Gamma = \partial\Omega , \qquad (6.1.10)$$

where

$$E(\mathbf{u}, \mathbf{v}) = \frac{1}{2} \left[\lambda e_{kk}(\mathbf{u}) e_{rr}(\mathbf{v}) + 2\mu e_{ij}(\mathbf{u}) e_{ij}(\mathbf{v}) \right] . \qquad (6.1.11)$$

Formula (6.1.10) and the symmetry of E now give

$$\int_\Omega [\mathbf{u} L \mathbf{v} - \mathbf{v} L \mathbf{u}] \, d\Omega = \int_\Gamma [\mathbf{u} \sigma^{(n)}(\mathbf{v}) - \mathbf{v} \sigma^{(n)}(\mathbf{u})] \, d\Gamma . \qquad (6.1.12)$$

Equations (6.1.10) and (6.1.12) are known as Betti's formula.

Note 6.1. Betti's formula may be generalized. Consider the bi-linear form.

$$\tilde{E}(\mathbf{u}, \mathbf{v}) = \mu u_{i,j} v_{i,j} + \alpha u_{j,i} v_{j,i} + \beta u_{i,i} v_{j,j} . \qquad (6.1.13)$$

Using integration by parts, one finds

$$(\alpha + \beta) \int_\Omega v_l u_{m,pq} \, d\Omega = \alpha \int_\Gamma v_l u_{m,p} n_q \, d\Gamma + \beta \int_\Gamma v_l u_{m,q} n_p \, d\Gamma -$$

$$- \alpha \int_\Omega v_{l,q} u_{m,p} \, d\Omega - \beta \int_\Omega v_{l,p} u_{m,q} \, d\Omega , \qquad (6.1.14)$$

and hence Betti's generalized formulas

$$\int_\Omega \mathbf{u} L \mathbf{v} \, d\Omega + \int_\Omega \tilde{E}(\mathbf{u}, \mathbf{v}) \, d\Omega = \int_\Gamma \mathbf{u} p^{(n)}(\mathbf{v}) \, d\Gamma , \qquad (6.1.15)$$

$$\int\limits_{\Omega} (\mathbf{u}L\mathbf{v} - \mathbf{v}L\mathbf{u})\, d\Omega = \int\limits_{\Gamma} (\mathbf{u}\mathbf{p}^{(n)}(\mathbf{v}) - \mathbf{v}\mathbf{p}^{(n)}(\mathbf{u}))\, d\Gamma\,, \qquad (6.1.16)$$

where $\mathbf{p}^{(n)}$ is the generalized stress vector

$$\mathbf{P}^{(n)}(\mathbf{u}) \equiv (\alpha + \beta)\frac{\partial \mathbf{u}}{\partial n} + \beta\mathbf{n}\, div\, \mathbf{u} + \alpha\mathbf{n} \times rot\, \mathbf{u}\,, \quad \alpha + \beta = \lambda + \mu\,, \quad (6.1.17)$$

which for $\alpha = \mu$ and $\beta = \lambda$ coincides with the true stress vector.

For

$$\alpha = \frac{\mu(\lambda + \mu)}{\lambda + 3\mu}\,;\; \beta = \frac{(\lambda + \mu)(\lambda + 2\mu)}{\lambda + 3\mu}$$

the operator $\mathbf{P}^{(n)}$ is referred to as the pseudo-stress operator [55] and is denoted by $\mathbf{S}^{(n)}$.

The formulation of problems of the theory of elasticity in terms of stresses will now be presented. Consider an elastic body, occupying the region Ω on the boundary of which the stress vector $\sigma^{(n)}$ is specified, i.e., $\partial\Omega \equiv \partial\Omega_2$ and

$$\sigma^{(n)}\Big|_{\partial\Omega} = f\,. \qquad (6.1.18)$$

Since it is known [14] that the solution of the problem of the theory of elasticity in this case involves satisfaction of the boundary conditions (6.1.18), the equilibrium equations

$$R_j \equiv \sigma_{ij,i} + F_j = 0\,, \; j = 1,2,3\,, \qquad (6.1.19)$$

and the Beltrami-Michell equations

$$\Delta\sigma_{11} + \frac{1}{1+\nu}\frac{\partial^2 \sigma}{\partial x_1^2} = -2\frac{\partial F_1}{\partial x_1} - \frac{1}{1-\nu}\, div\, \mathbf{F}\,,$$
$$\Delta\sigma_{23} + \frac{1}{1+\nu}\frac{\partial^2 \sigma}{\partial x_2 \partial x_3} = -\left(\frac{\partial F_3}{\partial x_2} + \frac{\partial F_2}{\partial x_3}\right)\,, \; 1 \to 2 \to 3 \to 1\,, \qquad * \qquad (6.1.20)$$

$$\sigma = \sigma_{11} + \sigma_{22} + \sigma_{33}\,, \; \nu = \frac{\lambda}{2(\lambda + \mu)}\,.$$

---

* This form of the equations implies that the remaining equations are obtained by cyclic permutation of the subscripts.

Because the presence of equations of different orders in equations (6.1.19) and (6.1.20) did not satisfy investigators, B.E. Pobedrya proposed a new formulation of problems of the theory of elasticity in terms of stresses. For the sake of simplicity, assume that, following [63], the body stress vector $\mathbf{F} \equiv 0$.

As it is known, for the solution of such problems, the equilibrium equations (6.1.19) and the boundary condition (6.1.18) must by fulfilled by the Saint Venant compatibility conditions which permit replacement of the displacement vector by strain components. Satisfaction of these conditions means that incompatibility tensor vanishes.

Saint Venant's incompatibility tensor [34]

$$\mathbf{M} = rot(rot\hat{Q})^* , \qquad (6.1.21)$$

is the rotor of the transposed rotor of the Green strain tensor $\hat{Q} = \{e_{\alpha,\beta}\}$. The components of this tensor may be written in the form

$$\eta_{ij} = \varepsilon_{ikl}\varepsilon_{jmn}e_{kn,lm} , \qquad (6.1.22)$$

where ε_{prq} is a Levi-Civita pseudo-tensor, which is defined as follows [34]: $\varepsilon_{prq} = 0$ if p, r, q; $\varepsilon_{prq} = 1$ if the subscripts (1,2,3), (2,3,1), (3,1,2) are cyclic, and $\varepsilon_{prq} = -1$, if they are acyclic. Since the Green strain tensor is symmetric, there six relations for the η_{ij}:

$$\eta_{11} = \frac{\partial^2 e_{33}}{\partial x_2^2} + \frac{\partial^2 e_{22}}{\partial x_3^2} - 2\frac{\partial^2 e_{23}}{\partial x_2 \partial x_3} , \; 1 \to 2 \to 3 \to 1 , \qquad (6.1.23)$$

$$\eta_{12} = \eta_{21} = \frac{\partial}{\partial x_3}\left(\frac{\partial e_{23}}{\partial x_1} + \frac{\partial e_{31}}{\partial x_2} - \frac{\partial e_{12}}{\partial x_3}\right) - \frac{\partial^2 e_{33}}{\partial x_1 \partial x_2} , \qquad (6.1.24)$$
$$1 \to 2 \to 3 \to 1 .$$

As has already been stated above, the Saint-Venant conditions have the form

$$\eta_{ij} = 0 . \qquad (6.1.25)$$

Equations (6.1.25), (6.1.10) and (6.1.18) represent a classical variant of the formulation of problems of theory of elasticity in terms of stresses. It follows from the vanishing of the tensor $\{\eta_{ij}\}$ that its deviator and its convolution

$$det\{\eta_{ij}\} = 0 , \qquad (6.1.26)$$

$$e_{kl,kl} - e_{kk,ll} = 0 \, , \tag{6.1.27}$$

hence their linear combinations also vanish:

$$\Delta e_{ij} + e_{kk,ij} - e_{ik,kj} - e_{jk,ki} + (1 + \xi)\delta_{ij}(e_{kl,kl} - e_{kk,ll}) = 0 \, . \tag{6.1.28}$$

Let ξ be a parameter the value of which will be defined later. Using Hooke's law to go to stresses, one finds the compatibility conditions in the form

$$
\begin{aligned}
S_{ij} \equiv \Delta\sigma_{ij} + \frac{\sigma_{kk,ij}}{1+\nu} &+ \frac{1 + \xi(1 - \nu)}{1+\nu}\,\delta_{ij}\sigma_{kk,ll} - \\
&- \sigma_{ik,kj} - \sigma_{jk,ki} + (1 + \xi)\delta_{ij}\sigma_{kl,kl} = 0 \, .
\end{aligned}
\tag{6.1.29}
$$

Obviously, by the strength of the preceding transformation's, $\{S_{ij}\}$ is likewise a tensor. Form its matrix divergence [34]. The divergence of a second order tensor is the vector with the components

$$div\,\hat{Q} = \frac{\partial e_{st}}{\partial x_s}\,\mathbf{K}_t \, , \tag{6.1.30}$$

hence

$$div\,S_{ij} = \xi\,grad\left[\sigma_{kl,kl} - \frac{1-\nu}{1+\nu}\,\Delta\sigma\right] . \tag{6.1.31}$$

Equation (6.1.29) yields

$$\delta_{ij}S_{ij} = (1 + 3\xi)\left[\sigma_{kl,kl} - \frac{1-\nu}{1+\nu}\,\Delta\sigma\right] . \tag{6.1.32}$$

Introduce, following [63], the two operators D and V

$$D\{a_{ij}\} \equiv \frac{1}{1 + 3\xi}\,grad\,\{\delta_{ij}a_{ij}\} - \frac{1}{\xi}\,div\,a_{ij} \, , \tag{6.1.33}$$

$$\{V\mathbf{b}\}_{ij} = b_{i,j} + b_{j,i} - (1 + \xi)\delta_{ij}\,div\mathbf{b} \, . \tag{6.1.34}$$

Using (6.1.29), (6.1.31)-(6.1.33), one finds

$$DS_{ij} = 0 \, . \tag{6.1.35}$$

In an analogous manner, by (6.1.19) and (6.1.34), one has

$$\delta_{ij}(V\mathbf{R})_{ij} = -(1 + 3\xi)\,div\mathbf{R} \, , \tag{6.1.36}$$

$$div\{(V\mathbf{R})_{ij}\} = \Delta\mathbf{R} - \xi\,grad\,div\mathbf{R}. \tag{6.1.37}$$

B.E. Pobedrya [63] proposed for a new formulation of problems of the theory of elasticity the system of equations

$$S_{ij}[\sigma] + a\{V\mathbf{R}\}_{ij} = 0 \tag{6.1.38}$$

with the boundary conditions

$$\sigma^{(n)}\Big|_{\partial\Omega} = f, \tag{6.1.39}$$

$$\mathbf{R}\Big|_{\partial\Omega} = 0. \tag{6.1.40}$$

The parameters a and ξ will be specified below. System (6.1.38) is a modification of the well known system of Beltrami-Michell equations (cf., for example, [55]). It is easily seen from the preceding manipulations, that the stress tensor solving problem (6.1.25), (6.1.19) and (6.1.18) solves also problem (6.1.38)-(6.1.40). Following [63], the converse will be proved.

Let σ_{ij} solve the last problem. Apply the operator D to (6.1.39) when one obtains in the region Ω the equations

$$D\{S_{ij}[\sigma] + a\{V\mathbf{R}\}_{ij}\} = -\frac{a}{\xi}\Delta\mathbf{R} = 0. \tag{6.1.41}$$

The relation $\Delta\mathbf{R} = 0$ together with the conditions $\mathbf{R}\big|_{\partial\Omega} = 0$ leads in Ω to

$$R_j = 0, \tag{6.1.42}$$

hence $V\mathbf{R} = 0$, and, by (6.1.38), $S_{ij} = 0$, etc. It has thus been established that problem (6.1.10), (6.1.18) and (6.1.25), and problem (6.1.38)-(6.1.40) are equivalent.

Note 6.2. Problem (6.1.38)-(6.1.40) introduces the transitional parameter ν as well as the parameters a and ξ. B.E. Pobedrya [63] has shown that:

1. The operator on the left hand side of the system is symmetric in

$$\xi = -\frac{1}{a(1+\nu)}; \tag{6.1.43}$$

2. Problem (6.1.38)-(6.1.40) is elliptic in the sense of Douglas-Nierenberg and soluble for all values of a with the exclusion of a discrete set of values.

Note 6.3. Next, note that the choice of the parameters a and ξ is essential for the organization of a numerical process of solution of problem (6.1.38)-(6.1.40), but it has no significance for the final solution. In fact, it follows from the above manipulations that σ_{ij}^1 and σ_{ij}^2 solving this problem for different values of these parameters, solve one and the same problem of the theory of elasticity, i.e., that they are identical.

§2. Fundamental solutions of the equations of the theory of elasticity

The preceding section was devoted to a study of the formulation of the boundary value problems of the theory of elasticity. For the transition to integral equations, the fundamental solutions play an important role.

Consider the concept of the δ-function, well known to mechanics. More obvious is the so-called sequential transition [2], when $\delta(x - x_0)$ is studied as the limit of a sequence $\varphi_n(x)$ so that $\int_{R^3} \varphi_n(x)\, dx = 1$ and for all $x \neq x_0$, $\varphi_n(x) \to 0$ as $n \to \infty$. In [11], an equivalent transition is employed. The δ-function is introduced as such a generalized function or, what is the same thing, such a linear functional, which for any finite function $\varphi(x)$ of $C^\infty(R^3)$ acts according to the rule

$$< \delta_{x_0}, \varphi > = \varphi(x_0)\,, \quad \delta_{x_0} = \delta(x - x_0)\,. \tag{6.2.1}$$

Let $L(\partial)$ be a system of differential operators with constant coefficients. By the fundamental solution of the system

$$L(\partial) = 0\,, \tag{6.2.2}$$

will be understood the matrix g for which

$$L(\partial)g = \delta \cdot I\,, \tag{6.2.3}$$

where I is the unit matrix and δ is the delta-function.

Next, consider the scheme for the construction of fundamental solutions, proposed by Hoermander [73, 74]:

1. Compute the symbolic determinant $det\, L(\partial)$ and find the fundamental solution of the equation

$$det\, L(\partial)F = \delta(x - y)\,; \tag{6.2.5}$$

2. Construct the operator matrix M from the algebraic addition of the transposed matrix L

$$ML = LM = det\, L \cdot I\,. \tag{6.2.6}$$

The matrix

$$g = MIF\,, \tag{6.2.7}$$

is wanted. In fact,

$$Lg = LMIF = \delta I\,.$$

Following [36], examples of the construction of fundamental matrices will now be presented. For the Cauchy-Riemann system

$$L \begin{pmatrix} u \\ v \end{pmatrix} \equiv \begin{pmatrix} \partial_1 & \partial_2 \\ \partial_2 & -\partial_1 \end{pmatrix} \begin{pmatrix} u \\ v \end{pmatrix} = 0 \,, \ det\, L = -\Delta \,, \tag{6.2.8}$$

one must introduce F in the form: $F = -1/2\pi \ln r$,

$$M = \begin{pmatrix} -\partial_1 & -\partial_2 \\ -\partial_2 & \partial_1 \end{pmatrix} \,, \ g = MFI = \frac{1}{2\pi r} \begin{pmatrix} e_1 & e_2 \\ e_2 & -e_1 \end{pmatrix} \,, \tag{6.2.9}$$

$$r = \mid x - y \mid \,, \ e_i = (x_i - y_i)/r \,, \ i = 1, 2 \,.$$

The fundamental vectors are

$$g^{(1)} = \frac{1}{2\pi r} \begin{pmatrix} e_1 \\ e_2 \end{pmatrix} \,, \ g^{(2)} = \frac{1}{2\pi r} \begin{pmatrix} e_2 \\ -e_1 \end{pmatrix} \,. \tag{6.2.10}$$

Another example is the three-dimensional Laplace equation

$$\Delta u = u_{,11} + u_{,22} + u_{,33} = 0 \,. \tag{6.2.11}$$

As before, one finds that its fundamental solution has the form

$$g = -\frac{1}{4\pi r} \,. \tag{6.2.12}$$

Consider now the construction of the fundamental matrix for the system of Lame equations

$$Lu \equiv \mu\Delta u + (\lambda + \mu)\, grad\, div\, u = 0 \,, \ u = (u_1, u_2, u_3) \,. \tag{6.2.13}$$

After computations analogous to those of example 1 above, one arrives at the matrix

$$\Gamma(x - y) = \begin{pmatrix} \Gamma_1^1 & \Gamma_1^2 & \Gamma_1^3 \\ \Gamma_2^1 & \Gamma_2^2 & \Gamma_2^4 \\ \Gamma_3^1 & \Gamma_3^2 & \Gamma_3^3 \end{pmatrix} \,, \tag{6.2.14}$$

where

$$\Gamma_j^k(x, y) = \frac{1}{8\pi\mu(\lambda + 2\mu)} [(\lambda + 3\mu)\delta_{kj} + (\lambda + \mu + e_k)e_j] \frac{1}{r} \,. \tag{6.2.15}$$

This matrix, well known in the literature of the theory of elasticity as the Kelvin-Somigliani matrix (cf., for example, [5]), yields the fundamental vectors

$$\Gamma^1 = \begin{pmatrix} \Gamma_1^1 \\ \Gamma_2^1 \\ \Gamma_3^1 \end{pmatrix}, \ \Gamma^2 = \begin{pmatrix} \Gamma_1^2 \\ \Gamma_2^2 \\ \Gamma_3^2 \end{pmatrix}, \ \Gamma^3 = \begin{pmatrix} \Gamma_1^3 \\ \Gamma_2^4 \\ \Gamma_3^3 \end{pmatrix}, \tag{6.2.16}$$

where

$$L\Gamma^1 = \begin{pmatrix} \delta(x-y) \\ 0 \\ 0 \end{pmatrix}, \ L\Gamma^2 = \begin{pmatrix} 0 \\ \delta(x-y) \\ 0 \end{pmatrix}, \ L\Gamma^3 = \begin{pmatrix} 0 \\ 0 \\ \delta(x-y) \end{pmatrix}. \tag{6.2.17}$$

In conclusion, the fundamental matrix of system (6.1.20) will be presented:

$$\Gamma = \begin{pmatrix} \Delta + a\partial_1^2 & a\partial_2^2 & a\partial_3^2 & 2a\partial_1\partial_2 & 2a\partial_1\partial_3 & 2a\partial_2\partial_3 \\ a\partial_1^2 & \Delta + a\partial_2^2 & a\partial_3^2 & 2a\partial_1\partial_2 & 2a\partial_1\partial_3 & 2a\partial_2\partial_3 \\ a\partial_1^2 & a\partial_2^2 & \Delta + a\partial_3^2 & 2a\partial_1\partial_2 & 2a\partial_1\partial_3 & 2a\partial_2\partial_3 \\ 0 & 0 & 0 & \Delta & 0 & 0 \\ 0 & 0 & 0 & 0 & \Delta & 0 \\ 0 & 0 & 0 & 0 & 0 & \Delta \end{pmatrix} G, \tag{6.2.18}$$

where $a = -(1+\nu)/(2+\nu)$ and G is the fundamental solution of the biharmonic equation $\Delta^2 G = \delta(x-y)$.

These fundamental solutions will be used in the following sections.

§3. Lichtenstein's boundary integral equations

In 1924, Lichtenstein proposed an elegant method for the solution of the boundary integral equations of the first boundary value problem of the theory of elasticity [96]:

$$Lu = 0, \ \mathbf{u}\Big|_{\partial\Omega} = \mathbf{u}_0. \tag{6.3.0}$$

Setting, for the sake of simplicity, the body forces equal to zero, he rewrote the Lame equations in the form

$$\Delta u_i = -\frac{\lambda+\mu}{\mu}\frac{\partial\theta}{\partial x_i}, \ \theta = div\,\mathbf{u}, \ i = 1,2,3, \tag{6.3.1}$$

hence, differentiating the i-th equation with respect to x_i and adding the results, one obtains

$$\Delta\theta = 0 \tag{6.3.2}$$

and hence

$$\Delta\left(u_i + \frac{\lambda+\mu}{2\mu}\, x_i\theta\right) = 0\,. \tag{6.3.3}$$

Introducing the Green function $G(x,y)$ for Laplace's equation, one finds from this equation, following [96], the integral equations

$$u_i + \frac{\lambda+\mu}{2\mu}\, x_i\theta = \frac{1}{4\pi}\int_{\partial\Omega} \frac{\partial}{\partial n_y}\, G(x,y)\, u_i(y)\, dS_y +$$
$$+ \frac{\lambda+\mu}{8\pi\mu}\int_{\partial\Omega} \frac{\partial}{\partial n_y}\, G(x,y)\, y_i\, \theta(y)\, dS_y\,, \quad \mathbf{u} = (u_1, u_2, u_3)\,. \tag{6.3.4}$$

Since

$$\theta(x) = \frac{1}{4\pi}\int_{\partial\Omega} \frac{\partial}{\partial n_y}\, G(x,y)\, \theta(y)\, dS_y \tag{6.3.5}$$

and

$$x_i\theta(x) = \frac{1}{4\pi}\int_{\partial\Omega} x_i\theta(y)\frac{\partial}{\partial n_y}\, G(x,y)\, dS_y\,,$$

one arrives at

$$u_i = \frac{1}{4\pi}\int_{\partial\Omega} \frac{\partial}{\partial n_y}\, G(x,y)\, u_i(y)\, dS_y +$$
$$+ \frac{\lambda+\mu}{8\pi\mu}\int_{\partial\Omega} (y_i - x_i)\theta(y)\frac{\partial}{\partial n_y}\, G(x,y)\, dS_y\,. \tag{6.3.6}$$

The first integral on the right hand side is a known function, to be denoted by $R_i(x)$.

Differentiating equations (6.3.6) with respect to x_i and adding the results, one finds

$$\theta(x) = \Lambda - \frac{3(\lambda+\mu)}{2\mu}\,\theta + \frac{\lambda+\mu}{8\pi\mu}\int_{\partial\Omega} (y_i - x_i)\frac{\partial^2 G(x,y)}{\partial x_i\partial n_y}\,\theta(y)\, dS_y\,, \tag{6.3.7}$$

where

$$\Lambda = \frac{\partial R_i}{\partial x_i}\,.$$

Recalling that

$$(y_i - x_i)\frac{\partial}{\partial x_i} = -\rho\frac{\partial}{\partial \rho}\,,\ \ \rho^2 = (y_i - x_i)(y_i - x_i)\,,$$

one find finally

$$\theta(x) = \frac{2\mu}{3\lambda + 5\mu}\Lambda - \frac{\lambda + \mu}{4\pi(3\lambda + 5\mu)}\int\limits_{\partial Q} \rho\theta(y)\frac{\partial^2 G(x,y)}{\partial\rho\,\partial n_y}\,dS_y\,, \qquad (6.3.8)$$

where the point $x \in \Omega$.

As $x \to x_0 \in \partial\Omega$, additional analysis of the kernel in (6.3.8) is required [96]. Introduce the local coordinate system $(\alpha_1, \alpha_2, \alpha_3)$ with centre at a point $y \in \partial\Omega$, so that α_3-axis is directed along the negative normal to $\partial\Omega$, and the α_1 and α_2 axes are tangent to the lines of curvature. The equation of the surface $\partial\Omega$, may be represented in the neighbourhood of the point y in the form

$$\alpha_3 = \frac{1}{2}\left(c_1\alpha_1^2 + c_2\alpha_2^2\right) + o\left(\alpha_1^2 + \alpha_2^2\right), \qquad (6.3.9)$$

where c_1 and c_2 are the principal curvatures at the point y.

Reference [55] presents a result by Levy on the representation of the harmonic functions of Green in the vicinity of a boundary

$$\frac{\partial G(\alpha,y)}{\partial n_y} = \frac{2\alpha_3}{\rho^3} - \frac{\alpha_1 + \alpha_2}{2\rho} - \frac{\alpha_1 - \alpha_2}{2\rho}\frac{\alpha_1^2 - \alpha_2^2}{(\rho + \alpha_3)^2} + Q(\alpha,y), \qquad (6.3.10)$$

where $\rho = (\alpha_1^2 + \alpha_2^2 + \alpha_3^2)^{1/2}$ and Q has near y partial first order derivatives which, for $x \to y$ are of order ρ^{-1}.

The results of the manipulations will now be presented, retaining only the principal terms involving ρ^{-1}:

$$\rho\frac{\partial}{\partial\rho}\frac{\partial G(\alpha,y)}{\partial n_y} = -\alpha_i\frac{\partial}{\partial\alpha_i}\frac{\partial G(\alpha,y)}{\partial n_y} = -\alpha_i\left\{\frac{\delta_{3i}2}{\rho^3} - \frac{6\alpha_3}{\rho^5}\alpha_i + ...\right\} =$$

$$= -\frac{2\alpha_3}{\rho^3} + \frac{6\alpha_3\rho^2}{\rho^5} + ... = \frac{4\alpha_3}{\rho^3} + ... = \qquad (6.3.11)$$

$$= 4\frac{\partial}{\partial n}\left(\frac{1}{\rho}\right)(1 + o(1))\,,\ \rho \to 0\,.$$

Proceeding in the expression

$$\int\limits_{\partial\Omega} \theta(y)\frac{\partial}{\partial n_y}\left(\frac{1}{\rho}\right)dS_y$$

to the limit as $x \to x_0$, one finds, on the basis of the well known limit theorems of potential theory,

$$2\pi\theta(x_0) + \int\limits_{\partial\Omega} \theta(y)\frac{\partial}{\partial n_y}\left(\frac{1}{\rho'}\right) dS_y \,, \tag{6.3.12}$$

where $\rho' = |\, x_0 - y \,|$.

Using (6.3.12), one obtains for the determination of $\theta(y)$ the integral equation

$$\theta(x_0) = \frac{2\mu}{5\lambda + 7\mu}\Delta(x_0) - \frac{\lambda+\mu}{4\pi(5\lambda+7\mu)}\int\limits_{\partial\Omega} \rho\theta(y)\frac{\partial^2 G}{\partial\rho\,\partial n_y}\, dS_y \,. \tag{6.3.13}$$

For sufficiently smooth surfaces $\partial\Omega$, for example, for surfaces which satisfy the Lyapunov conditions, by (6.3.11), the kernel in (6.3.13) has a weak singularity and, consequently, equation (6.3.13) is a Fredholm equation of the second kind.

If $\Lambda(x_0)$ is a sufficiently smooth function, then it is sufficiently simple to verify that the solution $\theta(x_o)$ of equation (6.3.13), on substitution in (6.3.8) and (6.3.6), determines a vector function u which satisfies the Lame equations and boundary conditions.

It will now be shown that equation (6.3.13) has the unique solution. Let $\Lambda(x_0) = 0$ and $\theta^0(x_0)$ be a non-trivial solution of the homogeneous equation (6.3.13). Substitute $\theta^0(x_0)$ into (6.3.8) and determine $\theta^0(x)$. Then one finds, using the homogeneous equations (6.3.6), u_i^0. On the basis of the statements above, the u_i^0 satisfy the homogeneous Lame equations and the homogeneous boundary conditions (6.3.0). At the same time, it has been verified that

$$div\,\mathbf{u}^0 = \theta^0 \,. \tag{6.3.14}$$

By assumption, θ^0 is non-zero, but this contradicts the uniqueness theorem for the first boundary value problem of the theory of elasticity. Hence the uniqueness of the solution of equations (6.3.13) for every smooth curve $\Lambda(x_0)$ has been proved.

§4. Solution of space problems by the potential method

Lichtenstein's integral equations, described in §3 and derived with the aid of Green's harmonic function is the simplest example of the application of the potential method to space problems of the theory of elasticity. A more general scheme of the potential method for three dimensional problems of the theory of elasticity will be studied in this section.

Employing the fundamental solution of the Lame equations, constructed in §2 of this chapter, -the Kelvin-Somigliani tensor-, the integral of a potential type

$$V[\mu] = \int_s \Gamma(x - y)\,\mu(y)\,d_y S\,, \quad S = \partial\Omega\,, \tag{6.4.1}$$

will be introduced which, by analogy, with the harmonic potential of the form

$$\int_s \frac{\mu(y)}{\mid x - y \mid}\,dyS\,,$$

will, be called potential of a simple layer, where $\Gamma(x - y)$ is defined in correspondence with formulas (6.2.14) and (6.2.15) as

$$\Gamma = \begin{pmatrix} \Gamma_1^1 & \Gamma_1^2 & \Gamma_1^3 \\ \Gamma_2^1 & \Gamma_2^2 & \Gamma_2^4 \\ \Gamma_3^1 & \Gamma_3^2 & \Gamma_3^3 \end{pmatrix}\,, \tag{6.4.2}$$

$$\Gamma_j^k(x,y) = \Gamma_k^j(x,y) = \Gamma_k^j(y,x) = \Gamma_j^k(y,x)\,,$$

$$\Gamma_j^k = -\frac{1}{8\pi\mu(\lambda + 2\mu)}\left[(\lambda + 3\mu)\delta_{kj} + (\lambda + \mu)\frac{(x_k - y_k)(x_j - y_j)}{\mid x - y \mid^2}\right]\frac{1}{\mid x - y \mid}\,.$$

Consider the vector (6.2.16)

$$\Gamma^{(j)} = \begin{pmatrix} \Gamma_1^j \\ \Gamma_2^j \\ \Gamma_3^j \end{pmatrix}\,.$$

Find the generalized stress vector $\mathbf{P}^{(n_y)}[\Gamma^j]$ and select the vector from the j-th column of the matrix P:

$$P = \begin{pmatrix} P_{11} & P_{12} & P_{13} \\ P_{21} & P_{22} & P_{23} \\ P_{31} & P_{32} & P_{33} \end{pmatrix}\,, \tag{6.4.3}$$

$$\mathbf{P}^{(n_y)}[\Gamma^j] = \left[-\frac{\alpha}{\mu} + \frac{(\alpha + \mu)(\lambda + \mu)}{2\mu(\lambda + 2\mu)}\right]\left[\nabla_y F \times \mathbf{n}_y\right] \times \mathbf{k}+$$

$$+\left\{\left[\frac{(\alpha+\mu)(\lambda+\mu)}{2\mu(\lambda+2\mu)}-\frac{\alpha}{\mu}\right]\mathbf{k}^j+\frac{3(\alpha+\mu)(\lambda+\mu)}{2\mu(\lambda+2\mu)}\,e_j\mathbf{r}_0\right\}\frac{\partial F}{\partial n_y}\,, \qquad (6.4.4)$$

$$F=-\frac{1}{4\pi}\frac{1}{r}\,,\ r=\mid x-y\mid,\ e_i=\frac{y_i-x_i}{r}\,,\ r_0=e_i\mathbf{k}^j\,.$$

For $\alpha=\mu$

$$\mathbf{P}^{(n_y)}=\sigma^{(n_y)}\,,$$

$$\sigma^{(n_y)}[\Gamma^j]=-\frac{\mu}{\lambda+2\mu}\left(\bar{\nabla}_y F\times\mathbf{n}_y\right)\times\mathbf{k}^k+$$

$$+\left\{\frac{\mu}{\lambda+2\mu}\mathbf{k}^j+\frac{3(\lambda+\mu)}{\lambda+2\mu}\,e_j\mathbf{r}_0\right\}\frac{\partial F}{\partial n_y}\,. \qquad (6.4.5)$$

For $\alpha=\mu(\lambda+\mu)/(\lambda+3\mu)$, one finds the operator of the pseudo-stress $\mathbf{S}^{(n_y)}$

$$\mathbf{S}^{(n_y)}[\Gamma^j]=\left\{\frac{2\mu}{\lambda+3\mu}\mathbf{k}^j+3\frac{(\lambda+\mu)}{\lambda+3\mu}\,e_j\mathbf{r}_0\right\}\frac{\partial F}{\partial n_y}\,. \qquad (6.4.6)$$

With the aid of Vectors (6.4.5) and (6.4.6), construct in analogy with the construction of the matrix (6.4.3) the matrices σ and S. Using these matrices, develop the corresponding potentials

$$\mathbf{W}_1[\nu_1]=\int_{\partial\Omega}S\nu_1\,d_y\partial\Omega\,, \qquad (6.4.7)$$

$$\mathbf{W}_2[\nu_2]=\int_{\partial\Omega}S\nu_2\,d_y\partial\Omega\,, \qquad (6.4.8)$$

which will be referred to as potentials of double layers of the first and second kinds, respectively. Note that here and in the sequel, for the sake of simplicity, it is assumed that the Lame equations are homogeneous.

Since for all $y\in\partial\Omega$ the vector $\Gamma^j(x,y)$ as well as the vector $\partial\Gamma^j(x,y)/\partial y_i$ satisfy with respect to x the Lame equations, then these equations are satisfied for $x\in\Omega$ by the potentials $\mathbf{V}[\mu]$, $\mathbf{W}_1[\nu_1]$ and $\mathbf{W}_2[\nu_2]$.

The following limit theorems of the Sokhotskii-Plemelj type hold for these potentials:

$$\lim_{\Omega \ni x \to x_0 \in \partial\Omega} \mathbf{W}_1[\nu_1] = \frac{1}{2}\nu_1(x_0) + \{\mathbf{W}_1[\nu_1]\},$$

$$\lim_{\Omega \ni x \to x_0 \in \partial\Omega} \mathbf{W}_2[\nu_2] = \frac{1}{2}\nu_2(x_0) + \{\mathbf{W}_2[\nu_2]\},$$

$$\lim_{\bar{\Omega} \not\ni x \to x_0 \in \partial\Omega} \mathbf{W}_1[\nu_1] = -\frac{1}{2}\nu_1(x_0) + \{\mathbf{W}_1[\nu_1]\},$$

$$\lim_{\bar{\Omega} \not\ni x \to x_0 \in \partial\Omega} \mathbf{W}_2[\nu_2] = -\frac{1}{2}\nu_2(x_0) + \{\mathbf{W}_2[\nu_2]\},$$

$$\lim_{\Omega \ni x \to x_0 \in \partial\Omega} \sigma^{(n)}V[\mu] = -\frac{1}{2}\mu(x_0) + \{\sigma^{(n)}V[\mu]\}, \tag{6.4.9}$$

$$\lim_{\bar{\Omega} \not\ni x \to x_0 \in \partial\Omega} \sigma^{(n)}V[\mu] = \frac{1}{2}\mu(x_0) + \{\sigma^{(n)}V[\mu]\},$$

$$\lim_{\Omega \ni x \to x_0 \in \partial\Omega} \mathbf{S}^{(n)}V[\mu] = -\frac{1}{2}\mu(x_0) + \{\mathbf{S}(n)V[\mu]\},$$

$$\lim_{\bar{\Omega} \not\ni x \to x_0 \in \partial\Omega} \mathbf{S}^{(n)}V[\mu] = \frac{1}{2}\mu(x_0) + \{\mathbf{S}(n)V[\mu]\},$$

where the terms in braces are direct values of the potential. For definiteness, assume that the region Ω is finite and simply connected. The potential of a double layer with be employed for the solution of the first boundary value problem of the theory of elasticity. By (6.4.9), one obtains for the internal problem the singular integral equation

$$\frac{1}{2}\nu_1(x_0) + \{\mathbf{W}_1[\nu_1]\} = \mathbf{u}(x_0), \tag{6.4.10}$$

or

$$\frac{1}{2}\nu_2(x_0) + \{\mathbf{W}_2[\nu_2]\} = \mathbf{u}(x_0), \tag{6.4.11}$$

for the external problem the equation

$$-\frac{1}{2}\nu_1(x_0) + \{\mathbf{W}_1[\nu_1]\} = \mathbf{u}(x_0), \tag{6.4.12}$$

or

$$-\frac{1}{2}\nu_2(x_0) + \{\mathbf{W}_2[\nu_2]\} = \mathbf{u}(x_0), \tag{6.4.13}$$

For the solution of the second boundary value problem of the theory of elasticity, the potential of a simple layer will be employed. One finds, by (6.4.6), for the internal problem

$$-\frac{1}{2}\mu(x_0) + \{\sigma^{(n)}V[\mu]\} = f(x_0) \tag{6.4.14}$$

and for the external problem

$$\frac{1}{2}\mu(x_0) + \{\sigma^{(n)}V[\mu]\} = f(x_0). \tag{6.4.15}$$

Note 6.4. Using in an analogous manner the pseudo-stress operator $S^{(n)}$, one finds for the internal and external problems the singular integral equations

$$-\frac{1}{2}\mu(x_0) + \{\mathbf{S}^{(n)}V[\mu]\} = \mathbf{q}(x_0) \tag{6.4.16}$$

and

$$\frac{1}{2}\mu(x_0) + \{\mathbf{S}^{(n)}V[\mu]\} = \mathbf{g}(x_0), \tag{6.4.17}$$

respectively. The right hand sides of these systems have no physical significance; nevertheless, equations (6.4.16) and (6.4.17) will be employed in the sequel.

Next, the integral equations (6.4.10) - (6.4.15) will be classified and their solubility studied. Consider the equation

$$\frac{1}{2}\nu(x_0) + \lambda \int_{\partial\Omega} S(x_0, y, n_y)\,\nu(y)\,d_y\partial\Omega = \mathbf{g}(x_0), \tag{6.4.18}$$

the left hand side of which for $\lambda = 1$ coincides with the left hand side of equation (6.4.10), for $\lambda = -1$, with the left hand side of equation (6.4.12).

For a special choice of the parameter α, the components of the matrix S will only contain normal derivatives of r^{-1} and not tangential ones, hence (cf. [28]) for the surface $\partial\Omega$, satisfying the Lyapunov conditions, the kernel of (6.4.18) has for $x_0 \in \partial Q$ and $y \in \partial Q$ a singularity $r^{\gamma-2}$, where γ is the Lyapunov index; consequently, equation (6.4.16) is a Fredholm equation of the second kind. The matrix S will be referred to as Basheleivili matrix. Following [98], the solubility of equation (6.4.16) will now be studied.

Theorem 6.1. Let $\mu \neq 0$, $-\infty < \nu < 1/2$, $\nu = \lambda/2(\lambda + \mu)$, $\mathbf{u} \in C^2(\Omega)$

$$L\mathbf{u} = 0,\ x \in \Omega,$$

$$\mathbf{u}\Big|_{\partial\Omega} = 0, \ x \in \partial\Omega,$$

then $\mathbf{u} \equiv 0$ in Ω.

In fact, by Betti's generalized formulas (6.1.15) and (6.1.16),

$$\int_{\Omega} E(\mathbf{u}, \mathbf{u}) \, d\partial\Omega = 0,$$

hence $\mathbf{u} = const$, and, by the boundary conditions, $\mathbf{u} = 0$.

Consider the simple layer potential

$$\mathbf{V} = 2 \int_{\partial\Omega} \Gamma(x, y) \, f(y) \, d_y \, \partial\Omega. \tag{6.4.19}$$

For the displacement vector $\mathbf{V}$, compute the pseudo-stress vector

$$S^{(n_x)}\mathbf{V} = \int_{\partial\Omega} S(x, y, n_x) \, f(y) \, d_y \, \partial\Omega. \tag{6.4.20}$$

By (6.4.9),

$$[\mathbf{S}^{(n_{x_0})}\mathbf{V}]_i = \lim_{x \to x_0 \in \partial\Omega, x \in \Omega} \mathbf{S}^{(n_x)}\mathbf{V} = -f(x_0) +$$
$$+ \int_{\partial\Omega} S(y, x_0, n_y) \, f(y) \, d_y \, \partial\Omega, \tag{6.4.21}$$

$$[\mathbf{S}^{(n_{x_0})}\mathbf{V}]_e = \lim_{x \to x_0 \in \partial\Omega, x \notin \bar{\Omega}} \mathbf{S}^{(n_x)}\mathbf{V} = f(x_0) +$$
$$+ \int_{\partial\Omega} S(y, x_0, n_y) \, f(y) \, d_y \, \partial\Omega. \tag{6.4.22}$$

Thus

$$2f(x_0) = [\mathbf{S}^{(n_{x_0})}\mathbf{V}]_e - [\mathbf{S}^{(n_{x_0})}\mathbf{V}]_i, \tag{6.4.23}$$

$$2 \int_{\partial\Omega} S f(y) \, d\partial\Omega_y = [\mathbf{S}^{(n_{x_0})}\mathbf{V}]_e + [\mathbf{S}^{(n_{x_0})}\mathbf{V}]_i. \tag{6.4.24}$$

As has already been noted, the solution of the first, internal boundary value problem of the theory of elasticity reduces, with the aid of the double layer potential, by (6.4.9), reduces to the boundary integral equation

$$\nu(x_0) + \int_{\partial\Omega} 2S(x_0, y, n_y) \, \nu(y) \, d_y \, \partial\Omega = 2\mathbf{u}_0(x_0); \tag{6.4.25}$$

for the external problem, to the equation

$$\nu(x_0) - \int_{\partial\Omega} 2S(x_0, y, n_y)\,\nu(y)\,d_y\,\partial\Omega = -2u_0(x_0)\,. \qquad (6.4.26)$$

Equations (6.4.16) and (6.4.17) become for the internal problem

$$\mu(x_0) - \int_{\partial\Omega} 2S(x_0, y, n_{x_0})\,\mu(y)\,d_y\,\partial\Omega = -2g(x_0)\,; \qquad (6.4.27)$$

and for the external problem

$$\mu(x_0) + \int_{\partial\Omega} 2S(x_0, y, n_{x_0})\,\mu(y)\,d_y\,\partial\Omega = -2g(x_0) \qquad (6.4.28)$$

hence it follows that equations (6.4.25) and (6.4.28), and likewise, equations (6.4.26) and (6.4.27) are adjoint.

Theorem 6.2. Let $\mu \neq 0$ and $-\infty < \nu < 1/2$, also $[S^{(n_{x_0})}V(x_0)]_e = 0$, then $f(x_0) = 0$ on $\partial\Omega$.

Proof. By Betti's generalized formula

$$\int_{\partial\Omega} V(x_0)[S^{(n_{x_0})}V(x_0)]_i\,d\partial\Omega = \int_{\Omega} \tilde{E}(V, V)\,d\Omega \qquad (6.4.29)$$

and

$$\int_{\partial\Omega} V(x_0)[S^{(n_{x_0})}V(x_0)]_e\,d\partial\Omega = -\int_{\Re^3/\Omega} \tilde{E}(V, V)\,d\Omega\,, \qquad (6.4.30)$$

hence

$$\int_{\Re^3/\Omega} \tilde{E}(V, V)\,d\Omega = 0\,,$$

and, consequently, $v = const$ in R^3/Ω; however, since V is given in the form of a simple layer, $V(y) \to 0 \mid y \mid \to \infty$ and, therefore, $V \equiv 0$ in R^3/Ω and on $\partial\Omega$. Since the solution of the first boundary value problem of the theory of elasticity is unique, $V \equiv 0$ in Ω; and, by (6.4.23), $f = 0$ on $\partial\Omega$ to be shown.

Since they are Fredholmian, equations (6.4.25) - (6.4.28) may be studied simultaneously. Consider the equation

$$f(x_0) + \lambda \int_{\partial\Omega} 2S(y, x_0, n_{x_0})\,f(y)\,d_y\,\partial\Omega = g(x_0)\,.$$

Theorem 6.3. All poles of the resolvent of equation (6.4.28) are real, simple and satisfy the condition

$$|\lambda| \geq 1,$$

where $\lambda = 1$ is not a singular value, but $\lambda = -1$ is a pole.

Proof. By (6.4.23) and (6.4.24),

$$2\mathbf{g}(x_0) = 2f(x_0) + 2\lambda \int_{\partial\Omega} 2S(y, x_0, n_{x_0})\, f(y)\, d_y\, \partial\Omega =$$

$$= [\mathbf{S}^{(n_{x_0})}\mathbf{V}]_e - [\mathbf{S}^{(n_{x_0})}\mathbf{V}]_i + \lambda\{[\mathbf{S}^{(n_{x_0})}\mathbf{V}]_e + [\mathbf{S}^{(n_{x_0})}\mathbf{V}]_i\} =$$

$$= (1 + \lambda)[\mathbf{S}^{(n_{x_0})}\mathbf{V}]_e - (1 - \lambda)[\mathbf{S}^{(n_{x_0})}\mathbf{V}]_i . \qquad (6.4.31)$$

Introduce two simple layer potentials

$$\mathbf{V}_a(x_0) = 2 \int_{\partial\Omega} \Gamma(x_0, y)\, f_a(y)\, d_y\, \partial\Omega , \qquad (6.4.32)$$

$$\mathbf{V}_b(x_0) = 2 \int_{\partial\Omega} \Gamma(x_0, y)\, f_b(y)\, d_y\, \partial\Omega . \qquad (6.4.33)$$

By Betti's generalized formula

$$\int_{\partial\Omega} \{\mathbf{V}_a(x_0)[\mathbf{S}^{(n_{x_0})}\mathbf{V}_b(x_0)]_i - \mathbf{V}_b(x_0)[\mathbf{S}^{(n_{x_0})}\mathbf{V}_a(x_0)]_i\}\, d\partial\Omega = 0 , \qquad (6.4.34)$$

$$\int_{\partial\Omega} \{\mathbf{V}_a(x_0)[\mathbf{S}^{(n_{x_0})}\mathbf{V}_b(x_0)]_e - \mathbf{V}_b(x_0)[\mathbf{S}^{(n_{x_0})}\mathbf{V}_a(x_0)]_e\}\, d\partial\Omega = 0 . \qquad (6.4.35)$$

If the existence of a complex root $\lambda = a + ib$ is admissible, then the homogeneous equation (6.4.28) admits the solution $f_a + if_b$. By (6.4.31), for $\mathbf{g} = 0$,

$$[\mathbf{S}^{(n_{x_0})}\mathbf{V}_a + i\mathbf{S}^{(n_{x_0})}\mathbf{V}_b]_e = \frac{1 - \lambda_0}{1 + \lambda_0}\, [\mathbf{S}^{(n_{x_0})}\mathbf{V}_a + i\mathbf{S}^{(n_{x_0})}\mathbf{V}_b]_i . \qquad (6.4.36)$$

Multiplying this equation by $\mathbf{V}_a - i\mathbf{V}_b$ and integrating along $\partial\Omega$, taking into account (6.4.34) and (6.4.35), one finds

$$
\begin{aligned}
\int\limits_{\partial\Omega} &\{\mathbf{V}_a[S\mathbf{V}_a]_e + \mathbf{V}_b[S\mathbf{V}_b]_e\}\, d\partial\Omega = \\
&= \frac{1-\lambda_0}{1+\lambda_0} \int\limits_{\partial\Omega} \{\mathbf{V}_a[S\mathbf{V}_a]_i + \mathbf{V}_b[S\mathbf{V}_b]_i\}\, d\partial\Omega ,
\end{aligned}
\tag{6.4.37}
$$

hence $(1 - \lambda_0)/(1 + \lambda_0)$ is real and $b = 0$.

Let λ be a multiple pole. Then there must exist two non-zero functions which satisfy the relations

$$
f_a(x_0) = \lambda \int\limits_{\partial\Omega} 2S(y, x_0, n_{x_0})\, f_a(y)\, d_y\partial\Omega ,
\tag{6.4.38}
$$

$$
f_a(x_0) + f_b(x_0) = \lambda \int\limits_{\partial\Omega} 2S(y, x_0, n_{x_0})\, f_b(y)\, d_y\partial\Omega ,
$$

hence one has for the corresponding $\mathbf{V}_a$ and $\mathbf{V}_b$ the equations

$$
[\mathbf{S}^{(n_{x_0})}\mathbf{V}_a]_e - [\mathbf{S}^{(n_{x_0})}\mathbf{V}_a]_i = \lambda\{[\mathbf{S}^{(n_{x_0})}\mathbf{V}_a]_i + [\mathbf{S}^{(n_{x_0})}\mathbf{V}_a]_e\} ,
\tag{6.4.39}
$$

$$
\begin{aligned}
[\mathbf{S}^{(n_{x_0})}\mathbf{V}_a]_e - [\mathbf{S}^{(n_{x_0})}\mathbf{V}_a]_i &+ [\mathbf{S}^{(n_{x_0})}\mathbf{V}_b]_e - [\mathbf{S}^{(n_{x_0})}\mathbf{V}_b]_i = \\
&= \lambda\{[\mathbf{S}^{(n_{x_0})}\mathbf{V}_b]_i + [\mathbf{S}^{(n_{x_0})}\mathbf{V}_b]_e\} .
\end{aligned}
\tag{6.4.40}
$$

Multiplying (6.4.39) by $\mathbf{V}_b$ and (6.4.40) by $\mathbf{V}_a$, adding the results, integrating the sum over $\partial\Omega$ and using (6.4.34) and (6.4.35), one obtains

$$
\int\limits_{\partial\Omega} \mathbf{V}_a[\mathbf{S}^{(n_{x_0})}\mathbf{V}_a]_e\, d\partial\Omega = \int\limits_{\partial\Omega} \mathbf{V}_a[\mathbf{S}^{(n_{x_0})}\mathbf{V}_a]_i\, d\partial\Omega .
\tag{6.4.41}
$$

However, by (6.4.29) and (6.4.30), the left and right hand sides of this equation can only equal if they vanish, hence, by theorem 6.2, $\mathbf{V}_a = 0$ and, consequently, $f_a = 0$, which is impossible, and the assertion is proved.

Now, construct from (6.4.3). for the solution f of the homogeneous equation (6.4.28) $\mathbf{V}[f]$ and find

$$
\int\limits_{\partial\Omega} \mathbf{V}[\mathbf{S}^{(n_{x_0})}\mathbf{V}]_e\, d\partial\Omega = \frac{1-\lambda}{1+\lambda} \int\limits_{\partial\Omega} \mathbf{V}[\mathbf{S}^{(n_{x_0})}\mathbf{V}]_i\, d\partial\Omega ,
\tag{6.4.42}
$$

hence $(1 - \lambda)/(1 + \lambda) \leq 0$ and, consequently,

$$|\lambda| \geq 1.$$

If one assumes that $\lambda = 1$ is a singular value, then, by (6.4.42)

$$\int_{\partial\Omega} \mathbf{V}_a[\mathbf{S}^{(n_{x_0})}\mathbf{V}]_e \, d\partial\Omega = 0,$$

hence, by theorem 6.2, $f \equiv 0$, which contradicts the assumption.

Finally, for $\lambda = -1$, the existence of an eigenfunction is directly verified.

Next, another group of basic boundary integral equations, namely, equations (6.4.11) and (6.4.13) - (6.4.15) will be investigated. Consider equation (6.4.43), viz.

$$\frac{1}{2}\nu(x_0) + \lambda \int_{\partial\Omega} \sigma(x_0, y, n_y)\,\nu(y)\,d_y\,\partial\Omega = \mathbf{g}(x_0), \qquad (6.4.43)$$

the left hand side of which for $\lambda = 1$ coincides with the left hand side of equation (6.4.11), and, for $\lambda = -1$, with the left hand side of equation (6.4.13).

Considering, as before, the simple layer potential

$$\mathbf{V}(x) = 2 \int_{\partial\Omega} \Gamma(x, y)\, f(y)\, d_y\,\partial\Omega$$

in the capacity of a displacement vector, compute the stress vector

$$\sigma^{(n_x)}\mathbf{V}(x) = \int_{\partial\Omega} 2\sigma(y, x, n_x)\, f(y)\, d_y\,\partial\Omega. \qquad (6.4.44)$$

The solution, with the aid of a simple layer potential, of problems of the theory of elasticity when the stress vector is specified. on the boundary, for the internal problem, is

$$\mu(x_0) - \int_{\partial\Omega} 2\sigma(y, x_0, n_{x_0})\, \mu(y)\, d_y\,\partial\Omega = -2f(x_0), \qquad (6.4.45)$$

for the external problem,

$$\mu(x_0) + \int_{\partial\Omega} 2\sigma(y, x_0, n_{x_0})\, \mu(y)\, d_y\,\partial\Omega = -2f(x_0). \qquad (6.4.46)$$

The investigation of the thus derived boundary integral equation will follow the work of S.G. Mikhlin [45].

For example, equation (6.4.43) will be analysed. Its kernel is a matrix the elements of which have the form

$$
\begin{aligned}
-\frac{1-2\nu}{8\pi(1-\nu)}\frac{\xi_i\delta_{jk}-\xi_j\delta_{ik}}{r^3}\,\alpha_k + O(r^{\gamma-2}) &= \\
= -\frac{1-2\nu}{8\pi(1-\nu)}\frac{\xi_i\alpha_j-\xi_j\alpha_i}{r^3} + O(r^{\gamma-2}),
\end{aligned}
\tag{6.4.47}
$$

where $r = |\,x-y\,|$, $\xi_i = y_i - x_i$, $\cos nx_k = \alpha_k$, n is the normal to the surface $\partial\Omega$ and γ is the Lyapunov index of the surface $\partial\Omega$.

On the surface $\partial\Omega$ introduce the local coordinate system with origin at the point x and the x_3-axis directed along the outward normal, when $\alpha_1 = \alpha_2 = 0$, $\alpha_3 = 1$. For such a change of variable, every element of the symbolic determinant experiences a linear transformation, and, consequently, the entire determinant undergoes such a transformation. In the local coordinate system, by (6.4.47), the principal part of the matrix σ has the form

$$
\sigma_0 \equiv \frac{1-2\nu}{8\pi(1-\nu)}
\begin{pmatrix}
0 & 0 & -\xi_1 \\
0 & 0 & -\xi_2 \\
\xi_1 & \xi_2 & 0
\end{pmatrix}
\frac{1}{r^3}
\tag{6.4.48}
$$

and the system under study becomes

$$
\frac{1}{2}\nu_1(x_0) - \frac{\lambda(1-2\nu)}{8\pi(1-\nu)}\int_{\partial\Omega}\frac{\xi_1}{r^3}\nu_3\,d_y\partial\Omega + T_1(\nu) = u_1(x_0),
$$

$$
\frac{1}{2}\nu_2(x_0) - \frac{\lambda(1-2\nu)}{8\pi(1-\nu)}\int_{\partial\Omega}\frac{\xi_2}{r^3}\nu_3\,d_y\partial\Omega + T_2(\nu) = u_2(x_0),
\tag{6.4.49}
$$

$$
\frac{1}{2}\nu_3(x_0) + \frac{\lambda(1-2\nu)}{8\pi(1-\nu)}\int_{\partial\Omega}\frac{1}{r^3}(\xi_1\nu_1 + \xi_2\nu_2)\,d_y\partial\Omega + T_3(\nu) = u_3(x_0),
$$

where T_k is an integral operator with a weak singularity.

Letting $(1-2\nu)/[(1-\nu)] = \delta$, the symbolic determinant of the system, apart from a numerical factor, is

$$
\begin{pmatrix}
1 & 0 & i\delta\frac{\xi_1}{r} \\
0 & 1 & i\delta\frac{\xi_2}{r} \\
-i\delta\frac{\xi_1}{r} & -i\delta\frac{\xi_2}{r} & 1
\end{pmatrix}
= \frac{3-4\nu}{4(1-\nu)^2}
$$

and, consequently, it differs from zero for admissible values of ν. Then, by the general results of chapter 4, the index of system (6.4.49) and, consequently, of system (6.4.43) is zero. Analogous results are obtained for system (6.4.46).

Note that equations (6.4.43) and (6.4.46) have been derived in the work of N. Kinoshita and T. Mura [93]. However, the authors classified these equations wrongly as Fredholm equations with discontinuous kernels and recommended for their solution an iterative method.

S.G. Mikhlin demonstrated the error in this analysis in his paper [45].

The analysis of the system of equations (6.4.43) will now be continued. In contrast to equations (6.4.28), equations (6.4.43) are singular; the question of their solubility may be investigated by the following scheme. As pointed out by S.G. Mikhlin [45], the singular operators (6.4.43) admit local, left regularization. Under the condition that the surface $\partial\Omega$ is smooth, there also exists a global, left regulator. Furthermore, the operators on the left hand side of (6.4.43) are closed in $L_2[\partial\Omega]$; however, then, by what has been said above, equation (6.4.43) is normally soluble, i.e., for the existence of a solution of the non-homogeneous equation (6.4.43) it is necessary and sufficient that the right hand side will be orthogonal to all the solutions of the adjoint homogeneous operator. Let there exist for some value of λ a right hand side for which there is not single-valued solubility, i.e., the solution does not exist or it is not unique. However, then there exists, by strength of its normal solubility, at least one non-trivial zero in the first case in the given operator, and, in the second case, in its adjoint. Since the index is zero, the number of non-trivial zeroes of the given and its adjoint operator coincide.

It is clear from the above observations that the scheme under consideration for the clarification of the properties of the spectrum of an operator may also be repeated for the operator (6.4.28) by utilization of Betti's formula and the uniqueness of the solution of the Dirichlet problem for the Lame equations.

One finds that the spectrum is real, the special points satisfy the condition $|\lambda| < 1$, $\lambda = 1$ is not, but $\lambda = -1$ is a point of the spectrum.

§5. Direct methods of construction of the boundary integral equations

An alternative to the methods of reduction of boundary value problems of the theory of elasticity, discussed in §3 and §4 is the so-called direct method which is based on an application of Betti's equality and a knowledge of the fundamental solutions of the equations. The advantages of the direct method are linked to the derivation of integral equations in which the unknowns are displacements or stresses, and not auxiliary functions as in the potential method. More distinct advantages of this method tune up when the stress vector is specified on the boundary of a region.

Consider the Lame equations

$$Lu(x) \equiv \mu\Delta u \mid (\lambda + \mu)\,grad\,div\,u = -F\,,\ x \in \Omega \qquad (6.5.1)$$

with the boundary conditions

$$\sigma^{(n)} \equiv 2\mu\frac{\partial u}{\partial n} + \lambda n\,div\,u + \mu n \times rot\,u = f(x)\,,\ x \in \Omega\,. \qquad (6.5.2)$$

Employing the fundamental vectors $\Gamma^{(1)}$, $\Gamma^{(2)}$, $\Gamma^{(3)}$ (cf. (6.2.16)) and Betti's formulas (6.1.12), one finds for the solution of problem (6.5.1) and (6.5.2)

$$\int_{\Omega} [\Gamma^{(i)}Lu - uL\Gamma^{(i)}]\,d\Omega = \int_{\partial\Omega} \{\Gamma^{(i)}\sigma^{(n)}[u] - u\sigma^{(n)}[\Gamma^{(i)}]\}\,d\partial\Omega\,. \qquad (6.5.3)$$

On the basis of equations (6.5.1) and the definition of the vector $\Gamma^{(i)}$, one obtains the integral representation of $u_i(x)$, $x \in \Omega$:

$$-u_i(x) + \int_{\partial\Omega} u(y)\,\sigma^{(n)}[\Gamma^{(i)}]\,d_y\partial\Omega = \int_{\partial\Omega} \Gamma^{(i)}\,f(y)\,d_y\partial\Omega + \int_{\Omega} \Gamma^{(i)}F\,d_y\Omega\,. \qquad (6.5.4)$$

By (6.1.8) and (6.4.15), the vector $\sigma^{(n)}[\Gamma^{(i)}]$ is given by

$$\sigma^{(n)}[\Gamma^{(i)}] = \frac{\mu}{\lambda + 2\mu}\left(\nabla\frac{1}{2\pi r} \times n\right) \times k^i -$$
$$-\left\{\frac{\mu}{\lambda + 2\mu}k^i + \frac{3(\lambda + \mu)}{\lambda + 2\mu}e_i r_0\right\}\frac{\partial}{\partial n}\frac{1}{4\pi r}\,, \qquad (6.5.5)$$

where the k^i are the unit vectors of the coordinate axes, $r_0 = \mid x - y \mid$, $e_i = (y_i - x_i)/\mid x - y \mid$ and $r_0 = e_i k_i$.

Taking in (6.5.4) the limit $x \to x_0 \in \partial\Omega$, using, as in §4, the known relations of potential theory

$$\lim_{x \to x0 \in \partial\Omega} \int_{\partial\Omega} \varphi(y) \frac{\partial}{\partial n_y} \left(-\frac{1}{4\pi r} \right) d_y \partial\Omega =$$
$$= -\frac{1}{2} \varphi(x_0) + \int_{\partial\Omega} \varphi(y) \frac{\partial}{\partial n_y} \left(-\frac{1}{4\pi r} \right) d_y \partial\Omega , \tag{6.5.6}$$

one arrives at the system of integral equations

$$-\frac{1}{2} u_i(x_0) + \int_{\partial\Omega} u(y) \sigma^{(n)} [\Gamma^{(i)}] \, d\partial\Omega =$$
$$= \lim_{x \to x_0 \in \partial\Omega} \left\{ \int_{\partial\Omega} \Gamma^{(i)} f(y) \, d_y \partial\Omega + \int_{\Omega} \Gamma^{(i)} F \, d\Omega \right\} . \tag{6.5.7}$$

In the sequel, for the sake of simple analysis, as usually, the body force $F \equiv 0$; also introduce the notation

$$\lim_{x \to x_0 \in \partial\Omega} \int_{\partial\Omega} \Gamma^{(i)} f(y) \, d_y \partial\Omega = A_i(x_0) .$$

Equation (6.5.7) now becomes

$$-\frac{1}{2} u_i(x_0) + \int_{\partial\Omega} u_i(y) \left\{ \left(\frac{\mu}{\lambda + 2\mu} \left(\nabla \frac{1}{2\pi r} \times n_y \right) \times k^i \right)_j - \right.$$
$$\left. - \left(\frac{\mu}{\lambda + 2\mu} \delta_{ij} + \frac{3(\lambda + \mu)}{\lambda + 2\mu} \right) \frac{\partial}{\partial n_y} \frac{1}{4\pi r} \right\} d_y \partial\Omega = A_i(x_0) , \tag{6.5.8}$$

or, finally, using the identity

$$u_i(A \times k^i)_j = [u \times A]_j , \tag{6.5.9}$$

one obtains

$$-\frac{1}{2} u_i(x_0) + \int_{\partial\Omega} \left\{ \frac{\mu}{\lambda + 2\mu} \left(u \times \left(\nabla \frac{1}{2\pi r} \times n_y \right) \right)_i - \right.$$
$$\left. - u_j(y) \left(\frac{\mu}{\lambda + 2\mu} \delta_{ij} + \frac{3(\lambda + \mu)}{\lambda + 2\mu} e_i e_j \right) \frac{\partial}{\partial n_y} \frac{1}{4\pi r} \right\} d_y \partial\Omega = A_i(x_0) . \tag{6.5.10}$$

The kernel of system (6.5.10) coincides with that of (6.4.43) and, following Mikhlin's scheme, one is led to the assertion: Equation (6.5.10) is singular, its symbol does not degenerate, and its index is zero. Consequently, Fredholm's theorems apply to (6.5.10) as well as to (6.5.4).

The potential method of §4 suggests the possibility of obtaining regular equations for employment of the pseudo-stress vector and the matrices of Basheleishvili. An analogous scheme may be applied in the direct method.

If one uses Betti's generalized formula, one find instead of (6.5.7)

$$-\frac{1}{2}u_i(x_0) + \int_{\partial\Omega} u(y)P^{(n_y)} + [\Gamma^{(i)}]\, d_y\partial\Omega = B_i(x_0), \tag{6.5.11}$$

where

$$B_i(x_0) = \lim_{x\to x_0\in\partial\Omega} \int_{\partial\Omega} \Gamma^{(i)}(x,y)\, P^{(n_y)}[u]\, d_y\partial\Omega .$$

Recall that $P^{(n)}[\Gamma^{(i)}]$ is defined by (6.1.17) and (6.2.15):

$$P^{(n_y)}[\Gamma^{(i)}] - \left[\frac{\alpha}{\mu} \quad \frac{(\alpha+\mu)(\lambda+\mu)}{2(\lambda+\mu)}\right]\left[\nabla\frac{1}{4\pi r}\times n_y\right]\times k^i -$$
$$-\left\{\left[\frac{(\alpha+\mu)(\lambda+3\mu)}{2\mu(\lambda+2\mu)} - \frac{\alpha}{\mu}\right]k^i + \frac{1(\alpha+\mu)(\lambda+\mu)}{2\mu(\lambda+\mu)}\,e_i r_0\right\}\frac{\partial}{\partial n_y}\frac{1}{4\pi r} . \tag{6.5.12}$$

For $\alpha = \frac{\mu(\lambda+\mu)}{\lambda+3\mu}$, the operator P is transformed into a pseudo-stress operator. Hence

$$S^{(n)}[\Gamma^{(i)}] = -\left\{\frac{2\mu}{\lambda+3\mu}\,k^i + 3\frac{\lambda+\mu}{\lambda+3\mu}\,e_i r_0\right\}\frac{\partial}{\partial n_y}\frac{1}{4\pi r} . \tag{6.5.13}$$

Now system (6.5.11) becomes

$$-\frac{1}{2}u_i(x_0) + \int_{\partial\Omega} u(y)S^{(n_y)}[\Gamma^{(i)}]\, d_y\partial\Omega = C_i(x_0), \tag{6.5.14}$$

where

$$C_i(x_0) = \lim_{x\to x_0\in\partial\Omega} \int_{\partial\Omega} \Gamma^{(i)}(x,y)\, S^{(n_y)}[u]\, d_y\partial\Omega . \tag{6.5.15}$$

It has been shown in §4 that system (6.5.14) is Fredholmian; it has been studied in great detail in [28] and [98]. However, the functions C_i have no clear mechanical meaning.

The direct method will now be applied to the system of Beltrami-Michell equations with boundary conditions in terms of stresses and the supplementary boundary conditions of B.E. Pobedrya. Using the fundamental matrix (6.2.16), one obtains the integral representation

$$
\begin{aligned}
\sigma_{ij}(x) = \int_{\partial\Omega} \Big\{ &\sigma_{ij}\frac{\partial F}{\partial n_y} - F\frac{\partial \sigma_{ij}}{\partial n_y} + \frac{\sigma}{2(1+\nu)}(F_{,i}n_j + F_{,j}n_i) - \\
&- \frac{F}{2(1+\nu)}(\sigma_{,i}n_j + \sigma_{,j}n_i) - \frac{\sigma}{2(1+\nu)}\frac{\partial}{\partial n_y}\big[(\delta_{ij} - e_i e_j)F\big] + \\
&+ \frac{\delta_{ij} - e_i e_j}{2(1+\nu)}\, F\frac{\partial \sigma}{\partial n_y} \Big\} \, d_y\partial\Omega\,, \\
F = &-\frac{1}{4\pi r}\,.
\end{aligned}
\tag{6.5.16}
$$

Going to a limit along a non-tangential direction to the boundary, one arrives at the boundary integral equation

$$
\begin{aligned}
\frac{1}{2}\sigma_{ij}(x) = \int_{\partial\Omega} \Big\{ &\sigma_{ij}\frac{\partial F}{\partial n_y} - F\frac{\partial \sigma_{ij}}{\partial n_y} + \frac{\sigma}{2(1+\nu)}(F_{,i}n_j + F_{,j}n_i) - \\
&- \frac{F}{2(1+\nu)}(\sigma_{,i}n_j + \sigma_{,j}n_i) - \frac{\sigma}{2(1+\nu)}\frac{\partial}{\partial n_y}\big[(\delta_{ij} - e_i e_j)F\big] + \\
&+ \frac{\delta_{ij} - e_i e_j}{2(1+\nu)}\, F\frac{\partial \sigma}{\partial n_y} \Big\} \, d_y\partial\Omega\,.
\end{aligned}
\tag{6.5.17}
$$

This equation has so far not yet been studied exhaustively.

§6. New schemes for the classical space problems

It is known that for the solution of boundary value problems of the theory of elasticity "displacements" and "stress" methods may be employed each of which has its advantages and disadvantages. In particular, the use of the displacements method is more natural for the formulation of a problem and writing down of the boundary conditions, while for the definition of the basic unknowns stresses require differentiation which in numerical methods leads to a loss of accuracy.

In turn, for the stress method, the formulation and study of boundary value problems becomes more involved.

An analogous situation applies also in the method of boundary integral equations, where there exist likewise "displacements" and "stresses" approaches. Results which were studied in [36] and which intermediate between the displacements and stresses methods will now be discussed. In the capacity of basic unknowns, displacement gradients will be employed. This step makes possible a determination of the stresses with the aid of algebraic operations.

To start with, the singular integral equations of the basic problems for Laplace's equation

$$\Delta v = 0,$$

will be discussed briefly. Let $u = grad\,v$, when one considers instead of this equation the redefined system of equation

$$\begin{aligned}
u_{1,1} &+ u_{2,2} &+ u_{3,3} &= 0, \\
u_{1,2} &- u_{2,1} & &= 0, \\
u_{1,3} & &- u_{3,1} &= 0, \\
&u_{2,3} &- u_{3,2} &= 0.
\end{aligned} \qquad (6.6.1)$$

Realizing the standard scheme (cf. §5) for the transition to integral equations, using the first three equations, one may write down the representation for the component u_1 of the gradient v

$$u_1(x) = v_{,1}(x) = \int_{\partial\Omega} \left[\frac{\partial v}{\partial n} F_{,1} + \left(v_{,1}n_2 - v_{,2}n_2\right) F_{,2} - \right.$$
$$\left. - \left(v_{,3}n_1 - v_{,1}n_3\right) F_{,3} \right] d\partial\Omega,$$

$$F = -\frac{1}{4\pi r}.$$

In order to obtain u_2 and u_3 exclude from (6.6.1) the third and second equation, respectively. Collecting the results, one finds

$$\nabla v(x) = \int_{\partial\Omega} \left\{ \frac{\partial v}{\partial n_y} \nabla F + \nabla F \times (\nabla v \times n) \right\} d_y \partial\Omega. \qquad (6.6.2)$$

The vector $\nabla v \times n_y$ lies in the tangent plane to the boundary surface. Let k^i be the unit vectors of the coordinate system. Then

$$(\nabla v \times n_y) \cdot k^i = \nabla v \cdot (n_y \times k^i), \quad (\nabla v \times n_y)_i = \frac{\partial v_j}{\partial S},$$

where $S^i = n_y \times k^i$ are vectors tangential to the surface. It is seen that $\nabla v \times n_y$ can be interpreted as Dirichlet data. Proceed now in (6.6.2) to the limit $x \to x_0 \in \partial\Omega$ and form the scalar product of the equation obtained with n_{x_0}. As a result, one arrives at the Dirichlet problem:

$$\frac{1}{2}\frac{\partial v}{\partial n}(x_0) + \int_{\partial\Omega} \frac{\partial v}{\partial n_y}\frac{\partial F}{\partial n_{x_0}}\, d_y\partial\Omega = -\int_{\partial\Omega} (\nabla v \times n_y)(\nabla F \times n_{x_0})\, d_y\partial\Omega. \quad (6.6.3)$$

In an analogous manner, forming the vector product with n_{x_0}, one obtains the system for the Neumann problem:

$$\frac{1}{2}(\nabla v \times x_0) - \int_{\partial\Omega} [\nabla F \times (\nabla v \times n_y)] \times n_{x_0}\, d_y\partial\Omega =$$

$$= \int_{\partial\Omega} \frac{\partial v}{\partial n_y}(\nabla F \times n_{x_0})\, d_y\partial\Omega. \qquad (6.6.4)$$

It is readily shown that the kernel of (6.6.4) has a weak singularity and that in a local coordinate system, linked to the surface, one has two equations with two unknowns [36, 37, 47]. Note that (6.6.3) which solves a Dirichlet problem, is a Fredholm equation of the second kind with a kernel with a weak singularity.

Next, the solution of space problems of the theory of elasticity will be studied. Supplement the Lame system, written in the new coordinates $a_{ij} = u_{i,j}$, by the nine equations of permutation of differentiation:

$$\begin{aligned}
\mu(a_{11,1} + a_{12,2} + a_{13,3}) + (\lambda + \mu)(a_{11,1} + a_{22,1} + a_{33,1}) &= 0,\\
\mu(a_{21,1} + a_{22,2} + a_{23,3}) + (\lambda + \mu)(a_{11,2} + a_{22,2} + a_{33,3}) &= 0,\\
\mu(a_{31,1} + a_{32,2} + a_{33,3}) + (\lambda + \mu)(a_{11,3} + a_{22,3} + a_{33,3}) &= 0,\\
a_{11,2} - a_{12,1} = 0,\ a_{22,1} - a_{21,2} = 0,\ a_{33,1} - a_{31,3} &= 0,\\
a_{11,3} - a_{13,1} = 0,\ a_{22,3} - a_{23,2} = 0,\ a_{33,2} - a_{32,3} &= 0,\\
a_{12,3} - a_{13,2} = 0,\ a_{21,3} - a_{23,1} = 0,\ a_{31,2} - a_{32,1} &= 0.
\end{aligned} \qquad (6.6.5)$$

Construct with introduction of the additional function e a representation for $u_{1,i}$ by separating for this purpose from (6.6.5) the system (6.6.6) and setting $v_i = u_{1,i}$:

$$\begin{aligned}
v_{1,1} + v_{2,2} + v_{3,3} + e_{,1} &= 0,\\
v_{1,2} - v_{2,1} &= 0,\\
v_{1,3} - v_{3,1} &= 0,\\
v_{2,3} - v_{3,2} &= 0,
\end{aligned} \qquad \Delta e = 0. \qquad (6.6.6)$$

The next stages are completely analogous to the case of Laplace's equation. Obtaining a representation for $u_{i,j}$, recall that $e = div\, u$. The final result is:

$$\mu u_{i,j} = \frac{\lambda + \mu}{2(\lambda + 2\mu)} \int_{\partial\Omega} \left[(3e_i e_j - \delta_{ij})\mathbf{q}\nabla F - q_i F_{,i} \right] d_y \partial\Omega +$$
$$+ \int_{\partial\Omega} \left\{ \left[\mu \frac{\partial u_i}{\partial n} + (\lambda + \mu)e n_i \right] F_{,j} + \mu [\nabla F \times (\nabla u_i \times \mathbf{n})]_{,j} \right\} d_y \partial\Omega ;$$

(6.6.7)

where
$$\mathbf{q} = (\lambda + 2\mu)e\mathbf{n} + 2\mu(\omega \times \mathbf{n}), \quad \omega = \frac{1}{2} rot\, \mathbf{u}, \quad F = \frac{1}{4\pi r}.$$

In order to achieve a further simplification, it is expedient to introduce additional integral identities for the variable $\mathbf{q}$. For this purpose, write the Lame system in the form
$$(\lambda + 2\mu)\nabla e - 2\mu\, rot\, \omega = 0, \tag{6.6.8}$$

add to this equation the identity $div\, \omega = 0$ and execute for the resulting system of four equations with four unknowns the construction scheme of integral representations. As a result, one obtains

$$\mathbf{q} + \int_{\partial\Omega} \left[\mathbf{q}\frac{\partial F}{\partial n_x} + \mathbf{q} \times (\nabla F \times \mathbf{n}_x) \right] d_y \partial\Omega = 2\mu \int_{\partial\Omega} \omega \cdot \mathbf{n}[\nabla F \times \mathbf{n}_x] d_y \partial\Omega,$$

(6.6.9)

which will prove useful in the sequel for the regularization of the equations. Form the scalar product of the representation for the gradient by the vector of the normal for $x \in \partial\Omega$. After transformations employing (6.6.9), one finds the system

$$\frac{1}{2}q_i(x_0) + \frac{2\mu}{\lambda + 3\mu} \int_{\partial\Omega} \left[\delta_{ij} + 3\frac{\lambda + \mu}{2\mu} e_i e_j \right] q_j \frac{\partial F}{\partial n_{x_0}} d_y \partial\Omega =$$

$$= -2\mu\frac{\lambda + 2\mu}{\lambda + 3\mu}\left\{ \frac{1}{2}R_i(x_0) + \int_{\partial\Omega} R_i \frac{\partial F}{\partial n_{x_0}} d_y \partial\Omega \right\} + \tag{6.6.10}$$

$$+\frac{\lambda + \mu}{\lambda + 2\mu} \int_{\partial\Omega} \omega \cdot \mathbf{n}[\nabla F \times \mathbf{n}_{x_0}]_i\, d_y \partial\Omega + \int_{\partial\Omega} [\nabla u_i \times \mathbf{n}_y][\nabla F \times \mathbf{n}_{x_0}]\, d_y \partial\Omega .$$

Since $\mathbf{R} \equiv \mathbf{k}^i \times (\nabla u_i \times \mathbf{n})$ and $-2\omega \cdot \mathbf{n} = \partial u_i / \partial s^i$ then the right hand side of this equation is known, provided the boundary conditions are stated in terms of displacements. The integral operator on the left hand side of this system is adjoint to the operator of system (2.5.4). Consequently, by Fredholm's theorems, system (6.6.10) has no more than one solution for an arbitrary right hand side, moreover, the method of successive approximations of §4 may be applied to it.

As a result of the solution of system (6.6.10), the quantities q_j, $j = 1, 2, 3$ are determined which are linked to the components of the stress vector $p_i = \sigma_{ij} n_j$ by the algebraic relations

$$p_i = q_i + 2\mu R_i \,, \quad i = 1, 2, 3 \,. \tag{6.6.11}$$

In an analogous manner, the system of integral equations for boundary value problems in terms of specified stresses may be obtained:

$$2\mu \left\{ \frac{1}{2} (\nabla u_j \times \mathbf{n}_{x_0}) + \int_{\partial \Omega} \left[3 \frac{\lambda + \mu}{\lambda + 2\mu} e_i e_j + \frac{\mu}{\lambda + 2\mu} \sigma_{ij} \right] \right.$$

$$\left[(\nabla u_i \times \mathbf{n}_y) \times \nabla F \right] \times \mathbf{n}_{x_0} \, d_y \partial \Omega + \frac{\mu}{\lambda + 2\mu}$$

$$\left. \int_{\partial \Omega} R_j \nabla F \times \mathbf{n}_{x_0} \, d_y \partial \Omega \right\} = \frac{\lambda + \mu}{\lambda + 2\mu} \tag{6.6.12}$$

$$\int_{\partial \Omega} \left[p_i \nabla r_{,ij} \times \mathbf{n}_{x_0} + 2 p_j \nabla F \times \mathbf{n}_{x_0} \right] d_y \partial \Omega \,.$$

It has been shown in [36] that for $-\infty < \nu < 3/4$, the symbol of system (6.6.12) is non-degenerate and its index is zero.

<h3 align="center">§7. On non–classical integral equations
of the theory of elasticity</h3>

This section deals with method [92] of reduction of problems of the theory of elasticity for unbounded regions with non-zero conditions at infinity to integral equations.

Let Γ be a sufficiently smooth, simple, closed curve in R^2 or in R^3. Denote the outside region by Ω and by n the outward normal to $\Gamma = \partial \Omega$.

The uniqueness of solutions of problems of the theory of elasticity in Ω, when on the boundary $\partial \Omega$ the displacement or the stress vector is specified, is ensured by the conditions (cf. §4).

$$u(x) = o(1) \,, \quad \nabla u(x) = o\left(\frac{1}{|x|}\right) \,, \quad |x| \to \infty \,. \tag{6.7.1}$$

Consider the more general conditions at infinity for problems of the Dirichlet type (i.e., for displacements given on the contour Γ). Introduce the matrix

$$M(x) = \begin{pmatrix} 1 & 0 & -x_2 & 0 & 0 & x_3 \\ 0 & 1 & x_1 & 0 & -x_3 & 0 \\ 0 & 0 & 0 & 1 & x_2 & -x_1 \end{pmatrix},$$

and denote its columns by $m_j(x)$. The vector of a rigid displacement $r(x)$ may be written in the form

$$r(x) = M(x)\omega = m_j(x)\omega_j\,,\ \omega_j \in R^1\,,\ (N = 2,3)\,,\ j = 1,...,3(N-1)\,.$$

Following reference [92], call a displacement field $u(x)$ generalized regular, if the difference $u(x) - r(x)$ satisfies condition (6.7.1).

In the case $N = 2$, it is sufficient for a correct formulation of the first boundary value problem of the theory of elasticity to demand that the solution $u(x)$ be generalized regular, when either the magnitude of ω_3, or the moment of the forces

$$\int_\Gamma m_3(y)\sigma^{(n_y)}[u]\,d\Gamma = b_3\,, \tag{6.7.2}$$

is known. In the case $N = 3$, there arise more possibilities. Let J be some set of the indices of the set $I_6 = \{1,2,3,4,5,6\}$, then conditions "at infinity" are formulated in the following manner: Find a generalized regular solution $u(x)$ for given ω_j, $j \in I_6$ and the conditions of normalization

$$\int_\Gamma m_k(y)\sigma^{(n_y)}[u]\,d\Gamma = b_k\,,\ k \in I_6/J\,, \tag{6.7.3}$$

where b_k are given constants.

For the second boundary value problem of the theory of elasticity under consideration below (for a stress vector specified on Γ) conditions (6.7.1) must be fulfilled. For $N = 2$, it is assumed, in addition, that the loading on the contour Γ is in equilibrium

$$\int_\Gamma \sigma^{(n_y)}[u]\,d\Gamma = 0\,. \tag{6.7.4}$$

If $E(u, u)$ denotes the specific strain energy, then the truth of Betti's generalized formulas

$$\int_\Omega E(u, u)\, dx = -\int_\Gamma u\sigma^{(n)}[u]\, d\Gamma + \int_\Gamma M(y)\omega\sigma^{(n)}[u]\, d\Gamma\,, \ N = 3\,,$$

$$\int_\Omega E(u, u)\, dx = -\int_\Gamma u\sigma^{(n)}[n]\, d\Gamma + \omega_3 \int_\Gamma m_3(y)\sigma^{(n)}[n]\, d\Gamma\,, \ N = 2\,, \tag{6.7.5}$$

is readily verified. Condition (6.7.4) has been used in the deduction of the last identity. The uniqueness theorem for the solutions of the external Dirichlet problems formulated above follows easily from (6.7.5).

For displacement fields holds the representation (generalized Somigliani formula)

$$u(x) = -\int_\Gamma \left\{\Gamma(y, x)\,\sigma^{(n_y)}[u] - \sigma(y, x)\,u(y)\right\}\, d\Gamma + M(x)\omega\,, \ x \notin \Gamma\,, \tag{6.7.6}$$

where $\Gamma(y, x)$ is the fundamental Kelvin-Somigliani matrix ($\Gamma(y, x) = \Gamma(x, y)$) and

$$\sigma(y, x) = (\sigma^{(n_y)}[u]\Gamma(y, x))'.$$

There are several ways of obtaining integral equations on Γ starting from formula (6.7.6). The classical path of deriving singular integral equations is described in §4. The other means will be discussed next.

Proceeding to the limit for $x \to t \in \Gamma$ (along a non-tangential path) in the representation (6.7.6), one finds the integral equation

$$\int_\Gamma \Gamma(y, t)\,\sigma^{(n_y)}[u]\, d\Gamma_y - M(t)\omega = g(t)\,, \ t \in \Gamma\,, \tag{6.7.7}$$

relatively to the surface forces $\sigma^{(n_y)}[u]$, where $g(t)$ is a known function for displacements given on the boundary Γ. The kernel of equations (6.7.7) has a weak singularity and thus the equation is a Fredholm equation of the first kind. The integral equation (6.7.7) of the first boundary value problem of the theory of elasticity has the unique solution $\sigma^{(n_y)}[u]$ for which conditions (6.7.3) are fulfilled. It is sufficient to prove the uniqueness of this solution, to prove that the equation

$$\int_\Gamma \Gamma(y, t)\,\sigma^{(n_y)}[u]\, d\Gamma_y - \sum_{j \notin J} m_j(t)\omega_j = 0\,, \ t \in \Gamma\,, \tag{6.7.8}$$

for the conditions

$$\int_\Gamma m_k(y)\,\sigma^{(n_y)}[u]\,d\Gamma_y = 0\,,\ \ k \in I_6/j\,,\tag{6.7.9}$$

has only the zero solution. Let, in contrast, $q(t)$ be a solution of the system of equations (6.7.8) and (6.7.9), then the function

$$u(x) = -\int_\Gamma \Gamma(y,x)\,q(y)\,d\Gamma_y + \sum_{j\notin J} m_j(t)\omega_j\tag{6.7.10}$$

is the generalized regular solution of a Dirichlet problem, where $\omega_j = 0$, $j \in J$ and $b_k = 0$, $k \in \chi_6/J$, $u|_\Gamma = 0$. Consequently, $u(x) \equiv 0$ for $x \in \Omega$, but, applying to (6.7.10) the stress operator and going to the limit on the boundary, one finds that $\sigma^{(n_y)}[u]|_\Gamma \neq 0$ since $q(t) \neq 0$. This contradiction proves the uniqueness theorem for equation (6.7.7) for the conditions (6.7.3).

In the case of the second boundary value problem of the theory of elasticity, apply for the derivation of the integral equation the stress operator to (6.7.6) and go to the limit on the boundary to obtain

$$-\int_\Gamma \sigma^{(n_x)}\big(\sigma^{(n_y)}\Gamma(x,y)\big)'u(y)\,d\Gamma_y = f(x)\,,\ \ x \in \Gamma\,,\tag{6.7.11}$$

where $f(x)$ is a known function when the stresses are given on the boundary. The kernel of equation (6.7.11) has a singularity of order $|x - y|^{-3}$ and therefore the integral equation is said to be hyper-singular. The uniqueness of its solution is proved by the classical scheme of the potential method. More detailed information on equations of the form (6.7.8) and (6.7.9) is given in §5 of chapter 8.

CHAPTER 7
THE CONTACT PROBLEMS OF THE THEORY OF ELASTICITY

An overwhelming majority of practically important engineering problems formulated in terms of the theory of elasticity, are contact problems, and only mathematical difficulties force investigators to replace the action of the second media by transitional boundary conditions. However, at the present time, the requirement of accuracy of computations all too often creates the necessity to give up simplifications and fulfill completely the conditions on the surface of contact.

It must be noted that the zone of contact requires increased attention: more than 80 mechanisms (cf. [89]) are due to processes at the contact surface and in their direct vicinity. The mathematical study of contact problems began with the fundamental work of Hertz [90] in which were laid the foundations of a new science. Subsequently, there occurred a split into two scientific directions.

The first was concerned with studies, linked to problems of determining the deformed state under stress in contacting bodies with an apriori unknown area of contact. As will be shown below, this was a sufficiently difficult problem which gave rise to a range of various mathematical problems [82,13,1,14,66,97]. In the sequel, these problems will be referred to as proper contact problems or, more simply, as contact problems.

The second direction is connected with studies of failure in the zone of contact. It comprises the study of local contact failure, the creation and spreading of cracks and the formation of fragments in the vicinity of contact, and likewise of problems of abrasion and erosion [26,67]. This direction is referred to in the scientific literature as "mechanics of contact failure".

The timeliness of the problems of contact failure is beyond doubt. However, the theoretical basis of the problems has been developed insufficiently, a fact which demonstrates its difficulties. Any progress is obviously liked to success in the solution of proper contact problems and of problems of crack theory with their eventual unification. One of the means is the application of the methods of boundary integral equations. This chapter is devoted to certain types of contact problems and methods of their solution.

§1. Mathematical formulation of problems

For the sake of definiteness, consider two elastic bodies Q_1 and Q_2 which are in contact on a surface S_{12}. In the regions Q_1 and Q_2, the Lame equations are fulfilled for, generally speaking, different elastic media; on the boundaries

$\partial Q_1/S_{12}$ and $\partial Q_2/S_{12}$ one has the boundary conditions which are traditional for problems of the theory of elasticity

$$u = u_0 , \ \sigma^{(n)} = p. \tag{7.1.1}$$

Naturally, attention will be given to the analysis of particular cases and an explanation of the conditions prevailing in the zone of contact.

The simplest variant of the contact problem is the so-called sectionally homogeneous body, when one assumes that on the boundary S_{12} continuity of the stress vector $\sigma^{(n)}$ and the displacement vector $\mathbf{u}$. In this case, one may also include, when the stress vector $\sigma^{(n)}$ is not continuous on the boundary of the contact, and the displacement vector $\mathbf{u}$ has a given discontinuity. In the sequel, this group of problems will by said to be of type 0. Note that for problems of type 0 the contact is given and spreading is excluded.

Other variants of the contact problem are, following [14], those of type A and B. For type A, the zone of contact is bounded; during the process of deformation, the contact zone may change, but it cannot extend beyond some region, determined by the geometry in the neighbourhood of the zone of contact. For type B, the zone of contact increases * during the deformation.

The condition that the bodies do not penetrate each other, which in contact problems always occurs and indicates the true of the formulation, differs for problems A and B. Following the reasoning of [14], these conditions are:

Type A problem (plane version)

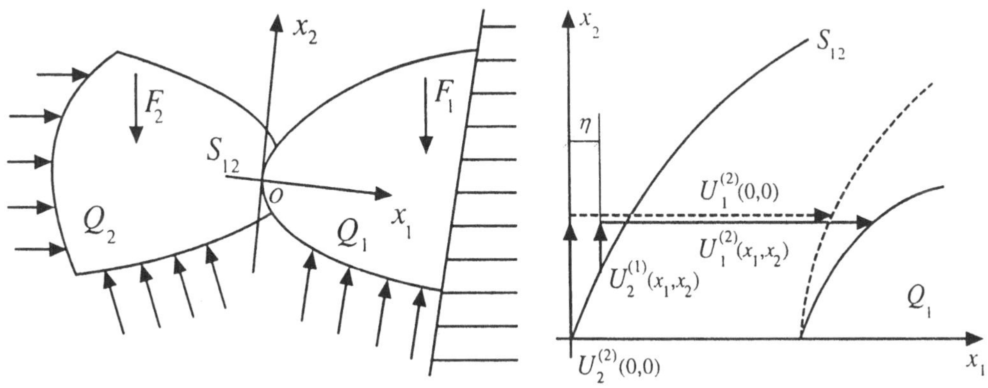

Fig. 5   Fig. 6

---

* Generally speaking, it is unbounded.

Let prior to deformation the bodies Q_1 and Q_2 be in contact along the arc S_{12}. Select a point O on the arc S_{12} as origin of an auxiliary coordinate system, direct the x_1-axis along the normal $\mathbf{n}_2$ to ∂Q_2, and the x_2-axis along the tangent to S_{12}.

During the process of deformation, the points $O_1 \in \partial Q_1$ and $O_2 \in \partial Q_2$ may go away, but they must satisfy the condition of non-penetration

$$u_1^{(2)}(0,0) \leq u_1^{(1)}(x_1, x_2) + \eta, \tag{7.1.2}$$

where (x_1, x_2) is a point on S_{12} such that

$$u_2^{(1)}(x_1, x_2) + x_2 = u_2^{(2)}(0,0).$$

As a consequence of the linearity of the problem, one may assume that $\eta \approx 0$ and

$$u_1^{(2)}(x_1, x_2) \approx u_1^{(1)}(0,0). \tag{7.1.3}$$

(The approximate relation becomes exact if S_{12} is a segment of a straight line).

From (7.1.1) and (7.1.3) follows the simpler condition of non-penetration

$$u_n^{(1)} + u_n^{(2)} \leq 0, \; S_{12}, \tag{7.1.4}$$

where $u_n^{(1)} = u_i^{(1)} n_i^{(1)}$, $u_n^{(2)} = u_i^{(2)} n_i^{(2)}$, $n^{(1)} = -n^{(2)}$.

Note 7.1. Condition (7.1.2) or its simplified version (7.1.4), as can be seen from the reasoning above, are necessary conditions of non-penetration.

Type B problem

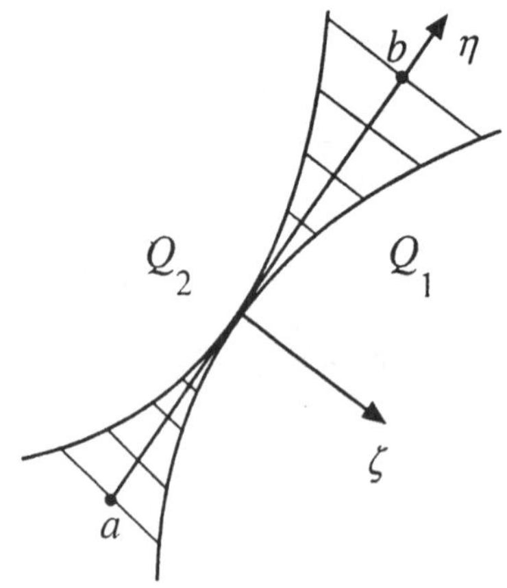

Fig. 7

If during the process of deformation the contact zone may widen, then the condition of non-penetration, developed above, must be replaced. Consider, as an example, the contact between two bulging, elastic surfaces with smooth boundaries and a small zone of contact.

If $\xi = f^{(1)}(\eta)$ and $\xi = -f^{(2)}(\eta)$ are the equations of the boundaries ∂Q_1 and ∂Q_2 in the local coordinate system (ξ, η), then $\varepsilon(\eta) = f^{(1)}(\eta) + f^{(2)}(\eta)$ is the distance between the boundaries of the bodies during deformation.

Reasoning and simplifying in an analogous manner as for type A problem, one arrives at the non-penetration condition

$$u_\xi^{(2)} - u_\xi^{(1)} \leq \varepsilon(\eta). \tag{7.1.5}$$

Note that the minus sign with $u_\xi^{(1)}$ is defined as projection not into $\mathbf{n}^{(2)} = -\mathbf{n}^{(1)}$, but into the ξ-axis.

Consider now the condition in terms of contact stresses.

For type A problem, Newton's law yields

$$\left[\sigma^{(n)}\right]_{S_{12}} = 0, \ \mathbf{n} = \mathbf{n}^{(2)}. \tag{7.1.6}$$

hence

$$\sigma_{nn}^{(2)} = \sigma_{nn}^{(1)}, \ \sigma_{n\tau}^{(2)} = \sigma_{n\tau}^{(1)} \tag{7.1.7}$$

and, in the absence of friction,

$$\sigma_{n\tau}^{(2)} = \sigma_{n\tau}^{(1)} = 0, \tag{7.1.8}$$

while in the presence of dry friction

$$\sigma_{n\tau}^{(2)} = \sigma_{n\tau}^{(1)} = -\kappa \, sign\big(u_\tau^{(2)} - u_\tau^{(1)}\big) \, | \, \sigma_{nn} \, | \, . \tag{7.1.9}$$

For type B problem, the conditions one has

$$\sigma_{n\xi}^{(2)} / \cos \alpha^{(2)} = -\sigma_{n\xi}^{(1)} / \cos \alpha^{(1)} \leq 0, \tag{7.1.10}$$

and likewise in the absence of friction

$$\sigma_{n\eta}^{(2)} = \sigma_{n\eta}^{(1)} = 0, \tag{7.1.11}$$

and, in the presence of dry friction,

$$\sigma_{n\eta}^{(2)} = \sigma_{n\eta}^{(1)} = -\kappa \, sign\big(u_\eta^{(2)} - u_\eta^{(1)}\big) \, | \, \sigma_{n\xi}^{(2)} \, | \tag{7.1.12}$$

where $\alpha^{(1)}$ and $\alpha^{(2)}$ are the angles between the ξ-axis and the tangents to the boundaries ∂Q_1 and ∂Q_2, respectively.

The boundary conditions on the lines of contact are for type A problem

$$
\begin{aligned}
&u_n^{(1)} + u_n^{(2)} \leq 0, \ \sigma_{nn}^{(1)} = \sigma_{nn}^{(2)} \leq 0, \\
&\left(u_n^{(1)} + u_n^{(2)}\right)\sigma_{nn}^{(2)} = 0, \\
&\sigma_{n\tau}^{(2)} = \sigma_{n\tau}^{(1)} = -\kappa \, sign\left(u_\tau^{(2)} - u_\tau^{(1)}\right) \mid \sigma_{nn}^{(2)} \mid .
\end{aligned}
\tag{7.1.13}
$$

where in the absence of friction $\kappa = 0$.

For type B problem

$$
\begin{aligned}
&u_\xi^{(2)} - u_\xi^{(1)} - \varepsilon \leq 0. \\
&\sigma_{n\xi}^{(2)} / \cos\alpha^{(2)} = -\sigma_{n\xi}^{(1)} / \cos\alpha^{(1)} \leq 0, \\
&\left(u_\xi^{(2)} - u_\xi^{(1)} - \varepsilon\right)\sigma_{n\xi}^{(2)} = 0, \\
&\sigma_{n\eta}^{(2)} = \sigma_{n\eta}^{(1)} = -\kappa \, sign\left(u_\eta^{(2)} - u_\eta^{(1)}\right) \mid \sigma_{n\xi}^{(2)} \mid .
\end{aligned}
\tag{7.1.14}
$$

where in the absence of friction $\kappa = 0$.

Consider now as an example one of the simplest sectionally homogeneous problems, solved by C.G. Mikhlin in 1934 [40]: A circular ring

$$
1 \leq x^2 + y^2 \leq \rho^2
$$

with a circular disk of unit radius soldered into it. This problem will be studied by the methods of the theory of functions of a complex variable by finding functions $\varphi_1(z)$ and $\psi_1(z)$, analytic in the region $1 \leq \mid z \mid < \rho$, and functions $\varphi_2(z)$, $\psi_2(z)$, analytic for $\mid z \mid < 1$ which satisfy (for $\mid z \mid = \rho$) the boundary condition

$$
\varphi_1(z) + \overline{z\varphi_1'(z)} + \psi_1(z) = f(\theta)
\tag{7.1.15}
$$

and the contact conditions (for $\mid z \mid = 1$)

$$
\varphi_1(z) + \overline{z\varphi_1'(z)} + \overline{\psi_1(z)} = \varphi_2(z) + \overline{z\varphi_2'(z)} + \overline{\psi_2(z)},
\tag{7.1.16}
$$

$$
\frac{1}{2\mu_1}\left\{\kappa_1\varphi_1(z) - \overline{z\varphi_1'(z)} + \overline{\psi_1(z)}\right\} = \frac{1}{2\mu_1}\left\{\kappa_2\varphi_2(z) - \overline{z\varphi_2'(z)} + \overline{\psi_2(z)}\right\}. \tag{7.1.17}
$$

These conditions ensure the continuity of the stress vector $\sigma^{(n)}$ and of the displacement vector $\mathbf{u}$ during the transition through the contact boundary. Expand the boundary loading $f(\theta)$ in the series

$$f(\theta) = \sum_{n=-\infty}^{+\infty} A_n e^{in\theta}, \tag{7.1.18}$$

and find the solution in the form of power series

$$\varphi_1(z) = \sum_{n=-\infty}^{+\infty} a_n^{(1)} z^n, \ \psi_1(z) = \sum_{n=-\infty}^{+\infty} b_n^{(1)} z^n, \tag{7.1.19}$$

$$\varphi_2(z) = \sum_{n=0}^{+\infty} a_n^{(2)} z^n, \ \psi_2(z) = \sum_{n=0}^{+\infty} b_n^{(2)} z^n, \tag{7.1.20}$$

Comparing coefficients of equal powers of (7.1.15) – (7.1.17) one finds the algebraic system

$$
\begin{aligned}
& a_n^{(1)} \rho^n + (2-n)\overline{a_{2-n}^{(1)}} \rho^{2-n} + \overline{b_{-n}^{(1)}} \rho^{-n} = A_n, \\
& a_n^{(1)} + (2-n)\overline{a_{2-n}^{(1)}} + \overline{b_{-n}^{(1)}} = a_n^{(2)}, \ n \geq 2, \\
& a_{-1}^{(1)} + (2-n)\overline{a_{2+n}^{(1)}} + \overline{b_n^{(1)}} = (2-n)\overline{a_{2+n}^{(2)}} + \overline{b_n^{(2)}}, \ n \geq 1, \\
& a_0^{(1)} + 2\overline{a_2^{(1)}} + \overline{b_0^{(1)}} = a_0^{(2)} + 2\overline{a_2^{(2)}} + \overline{b_0^{(2)}}, \\
& a_1^{(1)} + \overline{a_{-1}^{(1)}} + \overline{b_{-1}^{(1)}} = a_1^{(2)} + \overline{a_1^{(2)}}, \\
& \frac{\kappa_1}{\mu_1} a_n^{(1)} - \frac{2-n}{\mu_1} \overline{a_{-n+2}^{(1)}} + \frac{1}{\mu_1} \overline{b_{-n}^{(1)}} = \frac{\kappa_2}{\mu_2} a_n^{(2)}, \ n \geq 2, \\
& \frac{\kappa_1}{\mu_1} a_n^{(1)} - \frac{2-n}{\mu_1} \overline{a_{n+2}^{(1)}} - \frac{1}{\mu_1} \overline{b_n^{(1)}} = -\frac{2+n}{\mu_1} \overline{a_{n+2}^{(2)}} - \frac{1}{\mu_2} \overline{b_n^{(2)}}, \ n \geq 1, \\
& \frac{\kappa_1}{\mu_1} a_0^{(1)} - \frac{2}{\mu_1} \overline{a_2^{(1)}} - \frac{\overline{b_0^{(1)}}}{\mu_1} = \frac{\kappa_2}{\mu_2} a_0^{(2)} - \frac{2}{\mu_2} \overline{a_2^{(2)}} - \frac{\overline{b_0^{(2)}}}{\mu_2}, \\
& \frac{\kappa_1}{\mu_1} a_1^{(1)} - \frac{\overline{a_1^{(1)}}}{\mu_1} - \frac{\overline{b_1^{(1)}}}{\mu_1} = \frac{\kappa_1}{\mu_1} a_1^{(2)} - \frac{\overline{a_1^{(2)}}}{\mu_2}.
\end{aligned}
\tag{7.1.21}
$$

System (7.1.21) reduces to six algebraic equations in the six unknowns

$$a_n^{(1)}, \ \overline{a_{-n+2}^{(1)}}, \ \overline{b_{-n}^{(1)}}, \ b_{n-2}^{(1)}, \ a_n^{(2)}, \ \overline{b_{n-2}^{(2)}}, \ n \geq 3;$$

$$a_n^{(1)}\rho^n + (2-n)\overline{a_{2-n}^{(1)}}\,\rho^{2-n} + \overline{b_{-n}^{(1)}}\,\rho^{-n} = A_n\,,$$

$$na_n^{(1)}\rho^n + \overline{a_{2-n}^{(1)}}\,\rho^{2-n} + \overline{b_{n-2}^{(1)}}\,\rho^{n-2} = \overline{A}_{-n+2}\,,$$

$$a_n^{(1)} + (2-n)\overline{a_{2-n}^{(1)}} + \overline{b_{-n}^{(1)}} - a_n^{(2)} = 0\,,$$

$$na_n^{(1)} + \overline{a_{-n+2}^{(1)}} - \overline{b_{n-2}^{(1)}} - na_n^{(2)} - \overline{b_{n-2}^{(2)}} = 0\,, \qquad (7.1.22)$$

$$\frac{\kappa_1}{\mu_1}a_n^{(1)} - \frac{2-n}{\mu_1}\overline{a_{2-n}^{(1)}} - \frac{\overline{b_{-n}^{(1)}}}{\mu_1} - \frac{\kappa_2}{\mu_2}a_n^{(2)} = 0\,,$$

$$-\frac{n}{\mu_1}a_n^{(1)} - \frac{\kappa_1}{\mu_1}\overline{a_{2-n}^{(1)}} - \frac{\overline{b_{n-2}^{(1)}}}{\mu_1} + \frac{n}{\mu_2}a_n^{(2)} + \frac{\overline{b_{n-2}^{(2)}}}{\mu_2} = 0\,.$$

After solving system (7.1.22), determine the remaining unknowns $a_0^{(1)}$, $a_1^{(1)}$, $a_2^{(1)}$, $b_0^{(1)}$, $b_{-1}^{(1)}$, $b_{-2}^{(1)}$, $a_0^{(2)}$, $a_1^{(2)}$, $a_2^{(2)}$, $b_0^{(2)}$. The necessary and sufficient condition for solubility of the system is

$$Im\,A_1 = 0\,, \qquad (7.1.23)$$

which is equivalent to the mechanically clear requirement that the resultant moment of all forces, acting on the given region, must vanish.

Determining the unknowns in system (7.1.22) by Cramer's rule, one obtains a majorant of the form

$$\left|a_n^{(1)}\right|\rho^n \leq \frac{B}{A}\left|A_n\right| + \frac{C}{A}n\rho^{-n+2}\left|A_{-n+2}\right|. \qquad (7.1.24)$$

This inequality ensures absolute convergence of the series (7.1.19) and (7.1.20), if the loading on the boundary $f(\theta)$ is sectionally continuous.

Note 7.2. Note that for the sectionally homogeneous problem, in contrast to the problem with cuts, one need not specially verify the absence of a nonzero increment in the displacement vector for exit from the contact area. The single-valued displacement vector ensures fulfillment of the compatibility relations and the conditions of continuity at the boundary of contact.

§2. Formulation of variational problems *

In §1 of this chapter the differential formulation of static contact problems was given one seeks a sufficiently smooth vector function $\mathbf{u}$ defined in $Q_1 \cup Q_2$, that satisfies the Lame equations in Q_1 and Q_2, the boundary conditions (7.1.1) and the contact conditions (7.1.13) or (7.1.14).

Now, following [14], consider the variational formulation of the contact problems. Begin with the case A. Here and further the contact is assumed to be frictionless.

Let, as before,

$$E[\mathbf{u}, \mathbf{v}] = \frac{1}{2} \int\limits_{Q} \sigma_{ij}[\mathbf{u}]e_{ij}[\mathbf{v}]\, dQ \qquad (7.2.1)$$

and the "energy" solution mean a solution $\mathbf{u}$ with finite energy

$$E[\mathbf{u}, \mathbf{u}] = \frac{1}{2} \int\limits_{Q} \sigma_{ij}[\mathbf{u}]e_{ij}[\mathbf{u}]\, dQ. \qquad (7.2.2)$$

Consider the space of virtual displacements

$$V = \{\mathbf{u} \in W_2^1(Q),\ \mathbf{u} = 0,\ \text{on}\ \partial Q/S_{12}\},\ Q = Q_1 \cup Q_2 \qquad (7.2.3)$$

and for the case A the set of admissible displacements

$$K = \{\mathbf{v} \in V : V_n^1 + V_n^2 \leq 0,\ \text{on}\ S_{12}\} \qquad (7.2.4)$$

where $\mathbf{V}^{(1)}$, $\mathbf{V}^{(2)}$ are restrictions of V on Q_1, Q_2, correspondingly.

Definition. The function $\mathbf{u} \in K$ is called the variational solution of the problem A if

$$L[\mathbf{u}] = E[\mathbf{u}, \mathbf{u}] - \int\limits_{Q} F_i u_i\, dQ - \int\limits_{\partial Q} p_i u_i\, d\partial Q \leq$$
$$\leq E[\mathbf{v}, \mathbf{v}] - \int\limits_{Q} F_i v_i\, dQ - \int\limits_{\partial Q} p_i v_i\, d\partial Q \equiv L[\mathbf{v}] \qquad (7.2.5)$$

---

* This paragraph stands rather apart from the main subject of the book, but the notion about contact problems would be essencially incomplete without any information on their variational formulation. So the authors retain this paragraph in the book, but it may be omitted by the reader without any prejudice for understanding of the following paragraph.

for any $\mathbf{v} \in K$. Here F are the volume forces and p the surface forces.

The following statement holds true.

Theorem 7.1. Every classical solution of the problem A is its variational solution. If a variational solution of the problem A is sufficiently smooth, then it is a classical solution of the problem.

The proof of this statement can be supplied by traditional variational inequalities methods.

Analogus results can be obtained for the problems of B type. One should only change the set of admissible displacements

$$K_2 = \left\{ \mathbf{v} \in V : v_\xi^{(1)} - \mathbf{v}_\xi^{(2)} \leq \varepsilon \text{ for allmost all } \eta \in (a, b) \right\}. \qquad (7.2.6)$$

In [14] one can find the existence and uniqueness theorems for the problems of A and B types.

The problem of A type. Formulation of the dual variational problem.

Introduce, following [14], the space H of elements $\hat{e}_{ij}$, $\hat{\sigma}_{ij}$, $\hat{e}_{ij} = \hat{e}_{ji}$, $\hat{e}_{ij} \in L_2(Q)$, $\hat{\sigma}_{ij}$ are found from Hooke's law. Wright down the condition (*)

$$\hat{e}_{ij} = e_{ij}(\mathbf{v}). \qquad (*)$$

It is obvious that the main variational problem of A type is equivalent to the problem of minimizing the functional

$$L_1(\hat{e}, \mathbf{v}) = \frac{1}{2} \int_Q \hat{e}_{ij}\hat{\sigma}_{ij}\, dQ - \int_Q F_i v_i\, dQ - \int_{\partial Q} p_i v_i\, d\partial Q \qquad (7.2.7)$$

under the restriction (*).

This problem can be solved by the Lagrange multipliers method. Introduce the Lagrangian

$$\mathcal{L}(\hat{e}, \mathbf{v}, \lambda) = L_1(\hat{e}, \mathbf{v}) + \int_Q \lambda_{ij}(e_{ij}(\mathbf{v}) - \hat{e}_{ij})\, dQ. \qquad (7.2.8)$$

One can see that

$$\sup_{\lambda \in H} \int_Q \lambda_{ij}(e_{ij}(\mathbf{v}) - \hat{e}_{ij})\, dQ = \begin{cases} 0\,, & \hat{e} = e(v), \\ -\infty\,, & \hat{e} \neq e(v). \end{cases} \qquad (7.2.9)$$

Hence, every solution **u** of the main variational problem of A type satisfies the condition

$$L(\mathbf{u}) = \inf_{v \in K} L(v) = \inf_{v \in K} \sup_{\lambda \in H, \hat{e} \in H} \mathcal{L}(\hat{e}, v, \lambda). \tag{7.2.10}$$

The problem

$$\sup_{\lambda \in H} \inf_{(\hat{e}, V) \in H \times K} \mathcal{L}(\hat{e}, v, \lambda) \tag{7.2.11}$$

is called the dual variational problem [82]. After simple calculation one gets a sequence of equalities

$$\sup_{\lambda \in H} \inf_{(\hat{e}, v) \in H \times K} \mathcal{L}(\hat{e}, v, \lambda) = \sup_{\lambda \in K_{F,p}^+} [-\Pi(\lambda)] = - \inf_{\lambda \in K_{F,p}^+} \Pi(\lambda),$$

$$\Pi(\lambda) = \frac{1}{2} \int_Q \lambda_{ij} \gamma_{ij} \, dQ, \tag{7.2.12}$$

where γ_{ij} are found in a unique fashion from λ_{ij} by inverted Hooke's law, and $K_{F,p}^+$

$$K_{F,p}^+ = \left\{ \lambda \in H : \int_Q \lambda_{ij}(e_{ij}(\mathbf{v})) \, dQ - \int_Q F_i v_i \, dQ - \int_{\partial Q_\tau} p_i v_i \, d\partial Q \geq 0 \right\} \tag{7.2.13}$$

for all $v \in K$. Then the dual variational problem assumes a form that is more usual in mechanics: to find the stress fields $\lambda \in K_{F,p}^+$, such that $\Pi(\lambda) \leq \Pi(\mu)$ for all $\mu \in K_{F,p}^+$.

The inequality of the Prager-Singhe type for contact problems.
The Prager-Singhe equality ([52], [12], [51])

$$\|\sigma^{\mathrm{st}} - \sigma^*\|_H^2 + \|\sigma^{\mathrm{kin}} - \sigma^*\|_H^2 = \|\sigma^{\mathrm{st}} - \sigma^{\mathrm{kin}}\|_H^2, \tag{7.2.14}$$

is widely applied in linear elasticity to obtain estimates for the true solution of the problem $\{\sigma_{ij}^*\}$ using statically admissible $\{\sigma_{ij}^{\mathrm{st}}\}$ and kinematically admissible $\{\sigma_{ij}^{\mathrm{kin}}\}$ stress tensors. The properties of these tensors and of the functional space H are studied in [46]. Here we remined that the norm is introduced by

$$\|\sigma\|_H^2 = \frac{1}{2} \int_Q \sigma_{ij} e_{ij} \, dQ. \tag{7.2.15}$$

Now we cite a generalization of the identity (7.2.14) for the case of contact problems made by B.A. Shoihet. Consider, for example, a domain Ω with a crack Γ. Let F denote the volume forces. The boundary conditions have the form ($\partial\Omega = S_1 \cup S_2$)

$$\mathbf{u}\big|_{S_1} = 0, \ S_1 \,, \tag{7.2.16}$$

$$\sigma^{(n)}\big|_{S_2} = \mathbf{g}, \ S_2 \,. \tag{7.2.17}$$

Taking into account the possibility of contact between the edges of the crack, we define the kinematically admissible stress tensor using the vector $\mathbf{u}$, satisfying the boundary condition on S_1 and an additional condition on the surface of the crack

$$[u_n] \equiv (u_n^+ - u_n^-)\big|_\Gamma \le 0\,. \tag{7.2.18}$$

Knowing $\mathbf{u}$, we can obtain the strains and by Hooke's law the stresses. Here $\mathbf{n}^+$ is the outer normal to one of the edges of the crack, and $u_n^+ = \mathbf{u} \cdot \mathbf{n}^+$ is the projection of the displacement vector on $\mathbf{n}^+$, calculated on this edge. For the other edge we get

$$u^- = \mathbf{u} \cdot \mathbf{n}^+ , \ (\mathbf{n}^+ = -\mathbf{n}^-)\,.$$

In this problem the statically admissible tensor is defined as a tensor satisfying the balance equations, the boundary conditions on S_2 and an additional condition on the surface of the crack *

$$\sigma_{n\tau}^+ = \sigma_{n\tau}^- = 0\,, \ \sigma_{nn}^+ = \sigma_{nn}^- \le 0\,. \tag{7.2.19}$$

A generalization of the Prager-Singhe identity for the case of contact problems takes the form

$$\|\sigma_{ij}^{\mathrm{st}} - \sigma_{ij}^*\|_H^2 + \|\sigma_{ij}^{\mathrm{kin}} - \sigma_{ij}^*\|_H^2 + 2\int_\Gamma \sigma_{nn}^{+*}[u_n^{\mathrm{kin}}]\,d\Gamma +$$

$$+ 2\int_\Gamma \sigma_{nn}^{+\mathrm{st}}[u_n^*]\,d\Gamma = \|\sigma^{\mathrm{st}} - \sigma^{\mathrm{kin}}\|_H^2 + 2\int_\Gamma \sigma_{nn}^{+\mathrm{st}}[u_n^{\mathrm{kin}}]\,d\Gamma\,. \tag{7.2.20}$$

Note, that if there is no contact between the edges of the crack, then the integrals over Γ equal zero and identity (7.2.20) comes to the classical Prager-Singhe identity.

---

* Note that for statically and kinematically admissible tensors the "energy" $E[\sigma, e]$ is finite.

By possible contact of the crack edges the following condition is satisfied

$$[u_n^{\mathrm{kin}}] \le 0, \; [u_n^*] \le 0, \; \sigma_{nn}^*\big|_\Gamma \le 0, \; \sigma_{nn}^{\mathrm{st}}\big|_\Gamma \le 0. \qquad (7.2.21)$$

Note, that the integrals on Γ in the left part of the equation may be omitted due to (7.2.21) and the Prager-Singhe identity can be substituted by a more simple inequality:

$$\|\sigma_{ij}^{\mathrm{st}} - \sigma_{ij}^*\|_H^2 + \|\sigma_{ij}^{\mathrm{kin}} - \sigma_{ij}^*\|_H^2 \le \|\sigma_{ij}^{\mathrm{st}} - \sigma_{ij}^{\mathrm{kin}}\|_H^2 + 2 \int_\Gamma \sigma_{nn}^{+\mathrm{st}} [u_n^{\mathrm{kin}}] \, d\Gamma. \qquad (7.2.22)$$

§3. The integral equations of the simplest contact problems

Consider beforehand the problem of the action of as normal, concentrated force on the boundary of an elastic half-plane, known to be one of the basic problems (cf., for example, [83]) of the theory of elasticity.

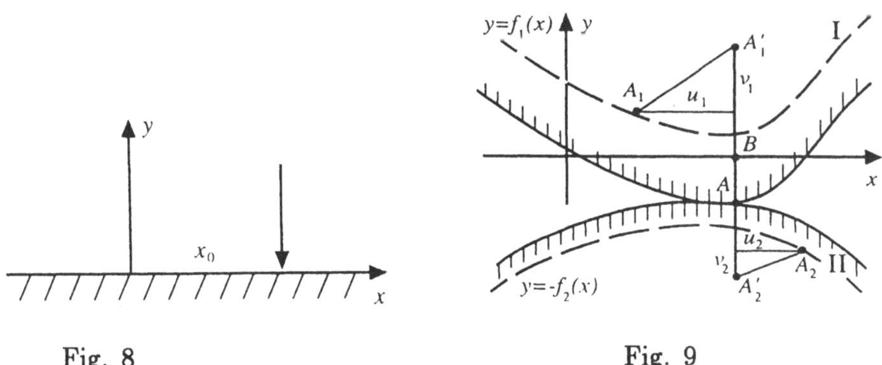

Fig. 8   Fig. 9

The solution of this problem shows that the point of the boundary with abcsissa x undergoes the vertical displacement

$$v = -\frac{2}{\pi E}(1 - \nu^2) p \ln \frac{1}{|x - x_0|} + C_0. \qquad (7.3.1)$$

hence one finds for a load, acting on the boundary over the segment $[a, b]$, the formula

$$v(x) = -\frac{2(1 - \nu^2)}{\pi E} \int_a^b p(y) \ln \frac{1}{|x - y|} \, dy + const. \qquad (7.3.2)$$

Consider now, following I.Ya. Shtaerman [83], the simplest contact problem. Let two plane elastic bodies (cf. Fig. 9) be bounded prior to deformation by the curves $y = f_1(x)$ and $y = f_2(x)$, respectively.

Assume, firstly, that Bodies I and II move to meet each other (without rotation) in the direction of the OY-axis by an amount α ($\alpha = \alpha_1 + \alpha_2$) and, secondly, that the deformations in a small neighbourhood of the zone of contact do not affect the quantity α. Then one has for the points of contact the relation

$$v_1 + v_2 = \alpha - f_1(x) - f_2(x). \tag{7.3.3}$$

In fact, represent the length of the segment $|A_1' A_2'|$ in the form

$$|A_1' A_2'| = |A_1' A| + |AA_2'| = \alpha_1 + \alpha_2 \,,$$

and, on the other hand, by

$$|A_1' A_2'| = |A_1' B| + |BA_2'| = v_1 + f_1(x - u_1) + f_2(x + u_2) \,.$$

Consequently,

$$\alpha_1 + \alpha_2 = v_1 + v_2 + f_1(x - u_1) + f_2(x + u_2) \,.$$

Assuming that the elastic displacements are small and letting

$$f_1(x - u_1) = f_1(x), \quad f_2(x + u_2) = f_2(x),$$

one finds, finally,

$$\alpha_1 + \alpha_2 = v_1 + v_2 + f_1(x) + f_2(x) \,.$$

as was to be proven.

Computing the link between the displacements v_1 and v_2 and the contact stresses from (7.3.2) and substituting into (7.3.3), one arrives at the integral of the first kind

$$\frac{2(1 - \nu^2)}{\pi E} \int\limits_{S} p(y) \ln \frac{1}{|x - y|}\, dy = f_1(x) + f_2(x) + const \equiv f(x). \tag{7.3.4}$$

The solution of this integral equation for $S = (-a, a)$, found in [1] and investigated in [83], is given by

$$P(x) = \frac{1}{\pi\sqrt{a^2 - x^2}} \left[\frac{1}{\pi \ln \frac{2}{a}} \int\limits_{-a}^{a} \frac{f(t)\, dt}{\sqrt{a^2 - t^2}} - \frac{1}{\pi} \int\limits_{-a}^{a} \frac{f(t)\sqrt{a^2 - t^2}}{t - x}\, dt \right],$$

$$f(t) = \frac{\pi E}{2(1 - \nu^2)} \left(f_1(t) + f_2(t) \right).$$

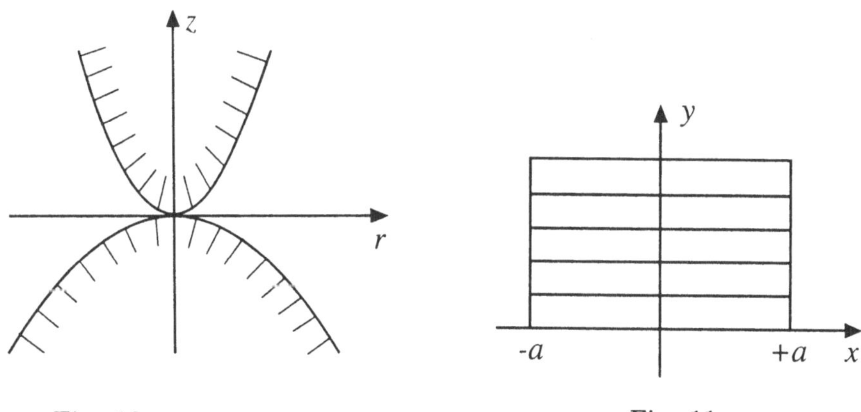

Fig. 10   Fig. 11

In an analogous manner, for the compression of two elastic bodies, bounded by surfaces of revolution, one obtains for the contact stresses the integral equation:

$$\int_{\Sigma} \frac{P(r')\,d\sigma}{\sqrt{(x - x')^2 + (y - y')^2}} = f(r), \quad r^2 = x^2 + y^2, \quad r'^2 = x'^2 + y'^2. \quad (7.3.5)$$

Note 7.3. Note that the problem of the pressure exerted by a plane rigid stamp on an elastic half-plane leads to the following particular case of equation (7.3.4):

$$\int_{-a}^{a} p(y) \ln \frac{1}{|x - y|}\, dy = C, \quad -a < x < a. \quad (7.3.6)$$

In an analogous manner, the problem of pressure exerted by a rigid, cylindrical stamp of circular cross-section on an elastic half-plane reduces to the particular case of equation (7.3.5):

$$\int_{\Sigma} \frac{P(r')\,d\sigma}{\sqrt{(x - x')^2 + (y - y')^2}} = C. \quad (7.3.7)$$

The problem of Hertz type

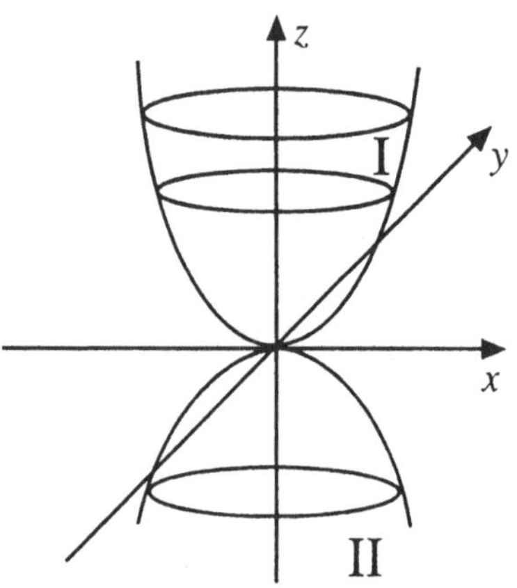

Fig. 12

Consider the closing in of two elastic bodies $E_1; \nu_1$ and $E_2; \nu_2$ with surface equations $z_1 = f_1(x, y)$ and $z_2 = -f_2(x, y)$, respectively (Fig. 12). Assume that the radii of curvature of both bodies, entering into contact, are large compared with the dimensions of the contact region. This assumption allows to apply the formulas for the elastic half-plane.

If it is assumed that the closeness of Bodies I and II equals δ, then one finds, as before, the equation

$$\left[\frac{(1 - \nu_1^2)}{E_1^2} + \frac{(1 - \nu_2^2)}{E_2^2}\right] \int_\Omega \frac{P(x', y')\, dx'\, dy'}{\sqrt{(x - x')^2 + (y - y')^2}} = \delta - f_1(x, y) - f_2(x, y).$$

$$(7.3.8)$$

Setting $z_1 = A_1 x^2 + B_1 y^2 + 2C_1 xy$ and $z_2 = A_2 x^2 + B_2 y^2 + 2C_2 xy$, equation (7.3.8) reduces to algebraic formulas for obtaining the area of contact and the closing in of the bodies (cf., for example, [64]).

§4. Reduction of sectionally homogeneous problems to integral equations

S.G. Mikhlin [42] realized the first reduction of plane contact (sectionally homogeneous) problems of the theory of elasticity to integral equation. With the aid of the Green function of the Dirichlet problem for Laplace's equation, he constructed the function $T(z, \zeta)$, referred to as the Schwartz kernel.

With the aid of this kernel, he presented the integral representation of the contour values of the complex function $\varphi(z)$ for "displacements" $h(z)$, specified on the contour Γ:

$$\varphi(z) = \frac{1}{2\kappa}\, h(z) + \frac{1}{2\pi i \kappa} \int_{\Gamma} \frac{h(\zeta)}{\zeta - z}\, d\zeta + \int_{\Gamma} \left\{ K_1(z,\zeta)\, h(\zeta) + K_2(z,\zeta)\, \overline{h(\zeta)} \right\} d\zeta \,.$$

$$(7.4.0)$$

In the case of the sectionally homogeneous problem and the contour $\Gamma = \Gamma_0 \cup \Gamma_1 \cup ... \cup \Gamma_n$ (Fig. 13), performing the limit $z \to z_0 \in \Gamma_{ik}$, $z \in \Omega_k$, he obtained

$$\varphi_k(z_0) = \frac{1}{2\kappa}\, h(z_0) + \frac{1}{2\pi i \kappa} \int_{\Gamma_{ik}} \frac{h(\zeta)}{\zeta - z}\, d\zeta + \int_{\Gamma} \Big\{ K_1^{(k)}(z_0,\zeta)\, h(\zeta) +$$
$$+ K_2^{(k)}(z,\zeta)\, \overline{h(\zeta)} \Big\}\, d\zeta \,.$$

$$(7.4.1)$$

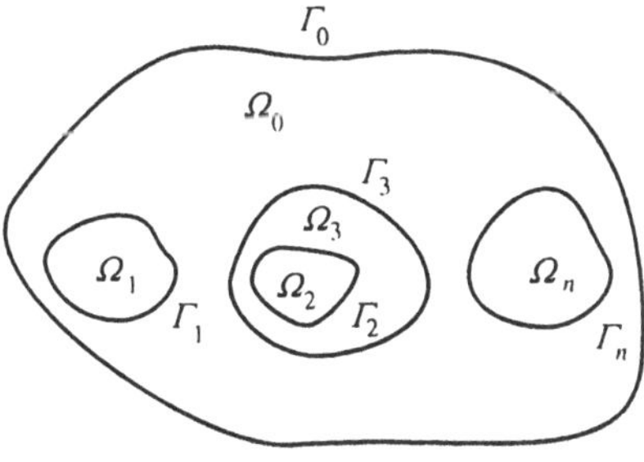

Fig. 13

Following S.G. Mikhlin, the boundary integral equation will now be derived. The continuity of the stress vector $\sigma^{(n)}$ and the displacement vector $\mathbf{u}$ for passage through the boundary Γ_{ik} yields the relations

$$\varphi_i(t) + t\overline{\varphi_i'(t)} + \overline{\psi_i(t)} = \varphi_k(t) + t\overline{\varphi_k'(t)} + \overline{\psi_k(t)} \,,$$
$$\frac{1}{2\mu_i}\left[\kappa_i \varphi_i(t) - t\overline{\varphi_i'(t)} - \overline{\psi_i(t)} \right] = \frac{1}{2\mu_k}\left[\kappa_k \varphi_k(t) - t\overline{\varphi_k'(t)} - \overline{\psi_k(t)} \right] \equiv g_{ik}$$

hence

$$2(\mu_k - \mu_i) g_{ik} = (1 + \kappa_k)\varphi_k - (1 + \kappa_i)\varphi_i \,. \tag{7.4.2}$$

(where repeated subscripts do not imply summation!!). Using (7.4.0), one arrives at the system of integral equations

$$\left[\frac{\mu_k(\kappa_k-1)}{\kappa_k} - \frac{\mu_i(\kappa_i-1)}{\kappa_i}\right]g_{ik}(z_0) = \frac{1}{\pi i}\left[\frac{\mu_k(\kappa_k+1)}{\kappa_k} + \frac{\mu_i(\kappa_i+1)}{\kappa_i}\right]$$

$$\int_{\Gamma_{ik}}\frac{g_{ik}(\zeta)}{\zeta-z_0}\,d\zeta + \sum_{r\neq s}\int_{\Gamma_{rs}}\left\{A_{rs}^{(ik)}(z_0,\zeta)g_{rs}(\zeta) + B_{rs}^{(ik)}(z_0,\zeta)\overline{g_{rs}}\right\}d\zeta + \Phi_{ik}(z_0).$$

$$(7.4.3)$$

(where repeated subscripts do not imply summation!!). This system can
be regularized by F. Noether's method and its solubility established on the
basis of results in [42] which prove the solubility of sectionally homogeneous
problems of the plane theory of elasticity.

In his papers [77,81] and, finally, in [99], D.I. Sherman executed his
progressive analysis of sectionally homogeneous problems of the theory of
elasticity. His work in the last paper will be presented here with insignificant
cuts.

For the sake of simplicity, consider a region with one "stamp". The
doubly connected region Ω is bounded by the curves l_0 and l_1: $l = l_0 \cup l_1$
(Fig. 14). As in [99], the methods of the theory of functions of a complex
variable will be employed. The, for stresses specified on the boundary l_0, the
boundary condition assumes the form

$$\varphi(t) + t\overline{\varphi(t)} + \overline{\psi(t)} = f(t).\tag{7.4.4}$$

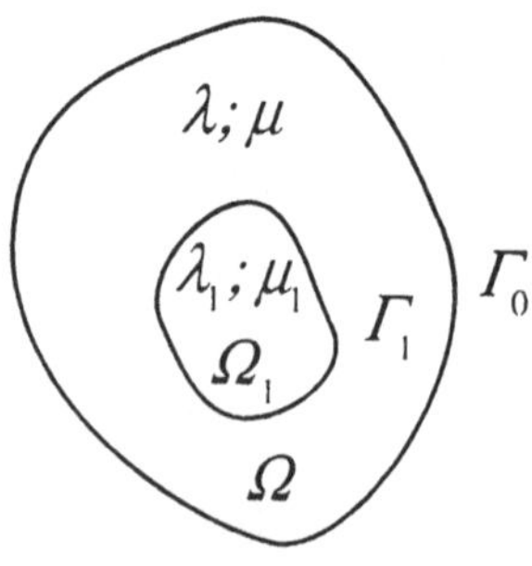

Fig. 14

In turn, one finds on the contact region l_1 the relations

$$\varphi(t) + t\overline{\varphi'(t)} + \overline{\psi(t)} = \varphi_1(t) + t\overline{\varphi_1'(t)} + \overline{\psi_1(t)},\tag{7.4.5}$$

$$\frac{\kappa}{\mu}\,\varphi(t) - \frac{1}{\mu}\left[\overline{t\varphi(t)} + \overline{\psi(t)}\right] = \frac{\kappa_1}{\mu_1}\,\varphi_1(t) - \frac{1}{\mu_1}\left[\overline{t\varphi_1(t)} + \overline{\psi_1(t)}\right]. \qquad (7.4.6)$$

Recall that vanishing of the resultant vector and moment of the external forces leads to the relation

$$\mathrm{Re}\int_{l_0} f(t)\,\overline{dt} = 0. \qquad (7.4.7)$$

Following D.I. Sherman, introduce the auxiliary function $w(t)$ defined by

$$\varphi(t) + \overline{t\varphi'(t)} + \overline{\psi(t)} = 2w(t),\ t \in l_0. \qquad (7.4.8)$$

which replaces the more involved condition

$$\varphi(t) + \overline{t\varphi'(t)} + \overline{\psi(t)} = 2w(t) + \frac{B}{t} + C, \qquad (7.4.9)$$

$$B = i\,\mathrm{Re}\int_l w(t)\,\overline{dt}\,,\ C = \int_l w(t)\,ds\,.$$

It follows from (7.4.5) and (7.4.8) that on l_1

$$\varphi(t) - \varphi_1(t) = w(t)\,,\ \psi(t) - \psi_1(t) = -\overline{w(t)} - \bar{t}w'(t)\,. \qquad (7.4.10)$$

Next introduce the Cauchy type integral

$$\frac{1}{2\pi i}\int_{l_1}\frac{w(t)}{t - z}\,dt\,. \qquad (7.4.11)$$

which allows to rewrite (7.4.10) in the form

$$\varphi(t_0) - \lim_{z\to t_0,z\in\Omega}\frac{1}{2\pi i}\int_{l_1}\frac{w(t)}{t-z}\,dt = \varphi_1(t_0) - \lim_{z\to t_0,z\in\Omega}\frac{1}{2\pi i}\int_{l_1}\frac{w(t)}{t-z}\,dt, \qquad (7.4.12)$$

$$\psi(t_0) + \lim_{z\to t_0,z\in\Omega}\frac{1}{2\pi i}\int_{l_1}\frac{\overline{w(t)} + \bar{t}w(t)}{t-z}\,dt = \psi_1(t_0) +$$

$$+ \lim_{z\to t_0,z\in\Omega}\frac{1}{2\pi i}\int_{l_1}\frac{\overline{w(t)} + \bar{t}w(t)}{l-z}\,dt\,. \qquad (7.4.13)$$

In an analogous manner, introducing the integrals

$$
\frac{1}{2\pi i}\int_{l_0}\frac{\omega(t)}{t-z}\,dt,\ \ F(z)=\frac{1}{4\pi i}\int_{l_0}\frac{f(t)}{t-z}\,dt,\ \ G(z)=
$$

$$
=\frac{1}{4\pi i}\int_{l_0}\frac{\overline{f(t)}-\bar{t}f'(t)}{t-z}\,dt,
\tag{7.4.14}
$$

one obtains on the contour l_0 from (7.4.4) and (7.4.8) the formulas

$$
\varphi(t_0)=\omega(t_0)+\frac{1}{2}f(t_0),
\tag{7.4.15}
$$

$$
\psi(t_0)=-\overline{\omega(t_0)}-\bar{t_0}\omega'(t_0)+\frac{1}{2}\,\overline{f(t_0)}-\frac{1}{2}\,\bar{t_0}f'(t_0),
\tag{7.4.16}
$$

and hence

$$
\varphi(t_0)-\lim_{z\to t_0,\,z\in\Omega}\frac{1}{2\pi i}\int_{l_0}\frac{\omega(t)}{t-z}\,dt-F_i(t_0)=
$$

$$
=-\lim_{z\to t_0,\,z\notin\overline{\Omega}\cup\overline{\Omega}_1}\frac{1}{2\pi i}\int_{l_0}\frac{\omega(t)}{t-z}\,dt-F_e(t_0),
\tag{7.4.17}
$$

$$
\psi(t_0)+\lim_{z\to t_0,\,z\in\Omega}\frac{1}{2\pi i}\int_{l_0}\frac{\overline{\omega(t)}+\bar{t}\omega(t)}{t-z}\,dt-G_i(t_0)=
$$

$$
\lim_{z\to t_0,\,z\notin\overline{\Omega}\cup\overline{\Omega}_1}\frac{1}{2\pi i}\int_{l_0}\frac{\overline{\omega(t)}+\bar{t}\omega'(t)}{t-z}\,dt-G_e(t_0).
\tag{7.4.18}
$$

Now seek the solution of the problem in the form

$$
\varphi(z)=\frac{1}{2\pi i}\int_{l}\frac{\omega(t)}{t-z}\,dt+F(z),
\tag{7.4.19}
$$

$$
\Psi(z)=-\frac{1}{2\pi i}\int_{l}\frac{\overline{\omega(t)}+\bar{t}\omega'(t)}{t-z}\,dt+G(z),
\tag{7.4.20}
$$

$$
\varphi_1(z)=\frac{1}{2\pi i}\int_{l}\frac{\omega(t)}{t-z}\,dt+F_1(z),
$$

$$\Psi_1(z) = -\frac{1}{2\pi i} \int_l \frac{\overline{\omega(t)} + \bar{t}\overline{\omega'(t)}}{t - z} \, dt + G_1(z).$$

Since the functions $F(z)$ and $G(z)$ are continuous, one has, by the formulas of Sokhotskii-Plemelj,

$$\varphi(t_0) + t\overline{\varphi'(t_0)} + \overline{\psi(t_0)} = \varphi_1(t_0) + t\overline{\varphi_1'(t_0)} + \overline{\psi_1(t_0)}, \tag{7.4.21}$$

i.e., condition (7.4.5) is satisfied.

Using (7.4.19), condition (7.4.6) and (7.4.20) leads to the integral equations

$$\frac{1}{2}\left(\frac{\kappa+1}{\mu} + \frac{\kappa_1+1}{\mu_1}\right)\omega(t_0) + \frac{1}{2\pi i}\left(\frac{\kappa+1}{\mu} - \frac{\kappa_1+1}{\mu_1}\right)\int_{l_1}\frac{\omega(t)}{t - t_0}\,dt =$$

$$= -\left(\frac{1}{\mu} - \frac{1}{\mu_1}\right)M[\omega(t), t_0] - \frac{1}{2\pi i}\left(\frac{\kappa-1}{\mu} - \frac{\kappa_1-1}{\mu_1}\right)\int_{l_1}\frac{\omega(t)}{t - t_0}\,dt + Q(t_0),$$

$$Q(t_0) = -\left(\frac{\kappa-1}{\mu} - \frac{\kappa_1-1}{\mu_1}\right)F'(t_0) - \left(\frac{1}{\mu} - \frac{1}{\mu_1}\right)P(t_0) + g(t_0), \tag{7.4.22}$$

$$M[\omega(t), t_0] = \frac{1}{2\pi i}\int_l\left\{\omega(t)\,d\ln\frac{t - t_0}{t - t_0} + \omega(t)\,d\frac{t - t_0}{t - t_0}\right\}. \tag{7.4.23}$$

It is useful to note that the expression

$$P(z, \bar{z}) \equiv F(\bar{z}) - z\overline{F'(z)} - \overline{G(z)}, \tag{7.4.24}$$

when crossing l_0 from inside and outside at the point t_0, has as limit one and the same expression

$$P(t_0) = \frac{1}{4\pi i}\int_{l_0}\left\{f(t)\left[\frac{dt}{t - t_0} - \frac{\overline{dt}}{\bar{t} - \bar{t}_0}\right] + \overline{f(t)}\,d\frac{t - t_0}{t - t_0}\right\}. \tag{7.4.25}$$

The boundary conditions on the curve will be completed next. By (7.4.19) and the Sokhotskii-Plemelj formulas, one has

$$\varphi_i - \varphi_e = \omega(t_0) + \frac{1}{2}f(t_0), \quad \varphi_i' - \varphi_e' = \omega'(t_0) + \frac{1}{2}f'(t_0),$$

$$t_0\left(\overline{\varphi_i'} - \overline{\varphi_e'}\right) - t_0\overline{\omega'(t_0)} + t_0\frac{1}{2}\overline{f'(t_0)},$$

$$\overline{\varphi'_i} - \overline{\varphi'_e} = -\omega(t_0) - t_0\overline{\omega'(t_0)} + \frac{1}{2}\left(f(t_0) - t_0\overline{f'(t_0)}\right),$$

or

$$\varphi'_i - \varphi'_e - t_0\left(\overline{\varphi'_i} - \overline{\varphi'_e}\right) - \left(\overline{\psi_i} - \overline{\psi_e}\right) = 2\omega(t_0)$$

or

$$\varphi_i - t_0\overline{\varphi'_i} - \overline{\psi_i} - 2\omega(t_0) = \varphi_e - t_0\overline{\varphi'_e} - \overline{\psi_e}. \tag{7.4.26}$$

By (7.4.9), one has on l_0

$$\frac{B}{t} + C = \varphi_e - t_0\overline{\varphi'_e} - \overline{\psi_e} \tag{7.4.27}$$

or, for

$$\varphi^* = i\varphi_e, \ \psi^* = i\varphi^*_e, \ \varphi^*_e + t_0\overline{\varphi^*_e} + \overline{\psi^*_e} = \frac{iB}{t} - iC. \tag{7.4.28}$$

Now, from the uniqueness we have

$$\varphi^* = C = 0, \ \psi^* - \frac{B}{z} = 0. \tag{7.4.29}$$

Expanding $\psi^* - B/z$ into a series in negative powers of z, one finds for z^{-1} the coefficient

$$\int_l \left[\overline{\omega(t)}\,dt - \omega(t)\,\overline{dt}\right] - \frac{1}{2}\int_{l_0}\left[\overline{f}\,dt + f\,\overline{dt}\right] - 2\pi iB = 0, \tag{7.4.30}$$

which, by (7.4.29), must be equal to zero. Hence

$$B = 0, \ \int_l \overline{\omega(t)}\,dt - \omega(t)\overline{dt} = 0 \text{ and, consequently}, \ \int_l \omega(t)\,\overline{dt} = \int_l \overline{\omega}\,dt = 0.$$
$$\tag{$*$}$$

Thus, if (7.4.9) is satisfied, then (7.4.8) is fulfilled. Furthermore, if (7.4.8) is fulfilled, then, by the condition $\varphi_e = \psi_e$ and (7.4.19), one obtains

$$\varphi(t_0) - \omega(t_0) - \frac{1}{2}f(t_0) = 0,$$

$$\psi(t_0) + \overline{\omega(t_0)} + \overline{t_0}\omega'(t_0) - \frac{1}{2}\overline{f(t_0)} + \frac{1}{2}\overline{t_0}f'(t_0) = 0,$$

hence

$$\varphi(t_0) + t_0\overline{\varphi'(t_0)} + \overline{\psi(t_0)} = f(t_0).$$

In turn, condition (7.4.9) leads to the equation

$$\omega(t_0) = M[\omega(t_0), t_0] - \int_l \left\{ i(\overline{t_0})^{-1} \, Re\left[\omega(t)\,\overline{dt}\right] + \omega(t)\,ds \right\} + P(t_0). \quad (7.4.31)$$

Thus, the sectionally homogeneous problem under consideration is reduced to the system of equations (7.4.21) and (7.4.31). Its solubility will be proved next, following D.I. Sherman.

Let $P(t_0) = 0$, $Q(t_0) = 0$, then, using (7.4.24), replacing $F(z)$ and $G(z)$ by $iF(z)$ and $iG(z)$, and using the uniqueness theorem of solutions of problems of the theory of elasticity, one finds

$$F(z) = kz + C_1, \; G(z) = \overline{C}_1, \quad (7.4.32)$$

when (7.4.22) yields

$$g(t) = (kt + C_1)\left(\frac{\kappa - 1}{\mu} - \frac{\kappa_1 - 1}{\mu_1}\right). \quad (7.4.33)$$

Finally, from the uniqueness theorem of problems of the theory of elasticity and from the form of the functions $F(z)$ and $G(z)$, one obtains $F_e = G_e = 0$, and, consequently,

$$f(t) = 2(kt + C_1). \quad (7.4.34)$$

Replacing $\varphi^{(0)}$, $\psi^{(0)}$, $\varphi_1^{(0)}$, $\psi_1^{(0)}$ by

$$\varphi^{(0)} - kz - C_1, \; \psi^{(0)} - \overline{C}_1, \; \varphi_1^{(0)} - kz - C_1, \; \psi_1^{(0)} - \overline{C}_1$$

and use in the new expressions the notation $\varphi^{(0)}$, $\psi^{(0)}$, $\varphi_1^{(0)}$, $\psi_1^{(0)}$, one obtains from equations (7.4.19) and (7.4.20)

$$\varphi^{(0)} = \frac{1}{2\pi i} \int_l \frac{\omega_1(t)}{t - z}\, dt, \; \psi^{(0)} = -\frac{1}{2\pi i} \int_l \frac{\overline{\omega_0(t) + \overline{t}\omega_0'(t)}}{t - z}\, dt,$$

$$\varphi_1^{(0)} = \frac{1}{2\pi i} \int_l \frac{\omega_0(t)}{t - z}\, dt, \; \psi_1^{(0)} = -\frac{1}{2\pi i} \int_l \frac{\overline{\omega_0(t) + \overline{t}\omega_0'(t)}}{t - z}\, dt. \quad (7.4.35)$$

By these expressions and since ω_0 satisfies equations (7.4.21) and (7.4.31) as well as condition (*), one finds that $\varphi^{(0)}$, $\psi^{(0)}$, $\varphi_1^{(0)}$, $\psi_1^{(0)}$ satisfy the homogeneous conditions (7.4.4) - (7.4.6), and hence

$$\varphi^{(0)} = iC^{(0)}z + E^{(0)}, \; \psi^{(0)} = -\overline{E}^{(0)},$$

$$\varphi_1^{(0)} = iC_1^{(0)}z + E_1^{(0)}, \; \psi_1^{(0)} = -\overline{E}_1^{(0)}, \quad (7.4.36)$$

where

$$\frac{\kappa+1}{\mu} C^{(0)} = \frac{\kappa_1+1}{\mu_1} C_1^{(0)}, \quad \frac{\kappa+1}{\mu} E^{(0)} = \frac{\kappa_1+1}{\mu_1} E_1^{(0)}. \tag{7.4.37}$$

Moreover, equations (7.4.8) and (7.4.9) yield

$$\omega_0(t) = i\big[C^{(0)} - C_1^{(0)}\big]\,t + \big(E^{(0)} - E_1^{(0)}\big), \quad \text{on } l_1 \tag{7.4.38}$$

and

$$\omega_0(t) = iC^{(0)}t + E^{(0)}, \quad \text{on } l_0.$$

Finally, by condition (*), add to (7.4.37)

$$C^{(0)}\Omega + C_1^{(0)}\Omega_1 = 0, \tag{7.4.39}$$

$$(L_0 + L_1)E^{(0)} - L_1 E_1^{(0)} = 0. \tag{7.4.40}$$

Note that in deriving condition (7.4.40), the arbitrariness in the choice of the coordinate origin has been used to ensure the conditions

$$\int_l x\,ds = \int_l y\,ds = 0. \tag{7.4.41}$$

hence, for

$$\frac{\kappa_1+1}{\mu_1} \neq \frac{L_1}{L_1 + L_0} \frac{\kappa+1}{\mu} \tag{7.4.42}$$

$C^{(0)} = C_1^{(0)} = E^{(0)} = E_1^{(0)} = 0$ and, consequently, $\omega_0(t) \equiv 0$, as was to be proved.

§5. The new singular integral equations of contact problems of the theory of elasticity

For certain classes of contact problems, the method of reduction to singular integral equations is effectively applied by utilization of the contact fundamental solutions.

Let there be given the plane region Ω (cf. Fig. 15), comprising the region Ω_0 with elastic Lame constants λ_0 and μ_0 an inclusion Ω_1 with elastic constants λ_1 and μ_1, $\Omega_0 = \Omega/\Omega_1$. On the outer boundary $\partial\Omega$, there are given boundary conditions ch are traditional for the theory of elasticity, but on the boundary $\partial\Omega_1$ a variant of the contact conditions, for example,*

$$\mathbf{u}_1\big|_{\partial\Omega_1} = \mathbf{u}_0\big|_{\partial\Omega_1}, \quad \sigma^{(n)}(u_1)\big|_{\partial\Omega_1} = \sigma^{(n)}(u_0)\big|_{\partial\Omega_1}. \tag{7.5.1}$$

---

* Note that the gap in the contact zone is excluded.

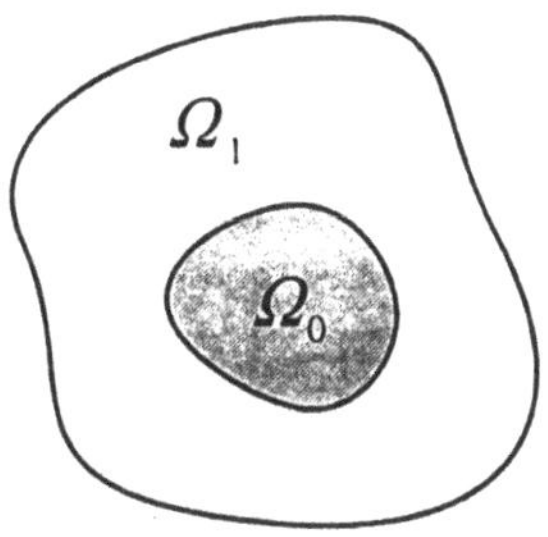

Fig. 15

Let the contact fundamental solution (cf. [89,97]) be determined by the rule

$$J(x,y) = \begin{cases} \Gamma^{(i)}(x-y) + P(x,y), & x \in \Omega,\ y \in \Omega_i,\ i = 0,1; \\ P(x,y), & x \in \Omega_i,\ y \in \Omega_j,\ i,j = 0,1; \end{cases} \qquad (7.5.2)$$

where $\Gamma^{(i)}$ is one of the columns of the Kelvin-Somigliani matrix with the Lame parameters λ_i, and μ_i, and the vector $P(x,y)$ is the solution (with respect to x) of the Lame equations in the compound region Ω such that $J(x,y)$ satisfies the homogeneous contact conditions on the boundary $\partial\Omega_1$.

In order to construct the Green contact tensor, the vector $P(x,y)$ is chosen such that the constructed sum satisfies not only the contact, but also the boundary conditions.

Existence theorems for the unknown vector $P(x,y)$ have been proved in [97]. The construction of the tensor $J(x,y)$ for the simplest case of contact of two elastic half-planes with different elastic constants has been developed in [89c].

Knowledge of the contact stress tensor permits introduction of the potential

$$\frac{1}{\pi} \int_S J(x,y)\,\varphi(y)\,dS ,$$

which is analogous to the simple layer potential, and to reduce the contact problem to a singular equation on a non-contact surface.

§6. On a method of successive approximations for the solution of 3D–problems

V.D. Kupradze [28] has studied by methods of integral equations problems of a finite inclusion Ω_1 (with elastic constants μ_1 and λ_1) in an infinite, isotropic medium $\Omega_0 \equiv R^3/\Omega_1$ (with elastic constants μ_0 and λ_0). The problem reduces to the determination of a displacement vector field $u = u(x)$ which satisfies the system of equations of the theory of elasticity

$$\mu(x)\Delta u + (\lambda(x) + \mu(x))\,\mathrm{grad}\,\mathrm{div}\,u = 0\,;\ \lambda(x),\mu(x) \begin{cases} \lambda_1,\mu_1\,,\ x \in \Omega_1\,, \\ \lambda_0,\mu_0\,,\ x \in R^3/\Omega_1 \end{cases},$$

the boundary conditions on the interface S of the elastic media

$$u^+ - u^- = u_0\,,\ (\sigma_n^1)^+ - (\sigma_n^0)^- = f_0 \tag{7.6.1}$$

and the conditions at infinity

$$u(x) = o(1)\,,\ \nabla u(x) = o\big(|x|^{-1}\big)\,,\ |x| \to \infty\,,$$

where σ_n^i in is the stress vector in the i-th medium, $i = 0,1$; $\sigma_n^i \equiv \sigma^{(n)}[u_{(x)}]$, $x \in \Omega_i$. Employing an integral representations for the displacement vectors in homogeneous media (cf. chapter 6), one finds

$$2u(x) = \int_S \Gamma^1(x,y)(\sigma_n^1)^+\, dS - \int_S \big(\sigma_n^1 \Gamma^1(x,y)\big)' u^+(y)\, dS\,,\ x \in \Omega_1\,,* \tag{7.6.2}$$

---

* Recall that Γ is the Kelvin-Somagliani matrix (cf. (6.4.2)), $g^{(p)}$ is its columns, and $\sigma_n\Gamma$ is introduced according to the rule

$$\sigma_n\Gamma \overset{\mathrm{def}}{=} \begin{pmatrix} \sigma_n(g^{(1)})k_1 & \sigma_n(g^{(2)})k_1 & \sigma_n(g^{(3)})k_1 \\ \sigma_n(g^{(1)})k_2 & \sigma_n(g^{(2)})k_2 & \sigma_n(g^{(3)})k_2 \\ \sigma_n(g^{(1)})k_3 & \sigma_n(g^{(2)})k_3 & \sigma_n(g^{(3)})k_3 \end{pmatrix},$$

where k_i, $i = 1,2,3$ are the coordinate unit vectors, A' is the transposed matrix A:

$$A \equiv \begin{pmatrix} a_{11} & a_{12} & a_{13} \\ a_{21} & a_{22} & a_{23} \\ a_{31} & a_{32} & a_{33} \end{pmatrix}, A' = \begin{pmatrix} a_{11} & a_{21} & a_{31} \\ a_{12} & a_{22} & a_{32} \\ a_{13} & a_{23} & a_{33} \end{pmatrix}.$$

$$0 = \int_S \Gamma^1(x,y)(\sigma_n^1)^+ \, dS + \int_S \left(\sigma_n^1 \Gamma^1(x,y)\right)' u^+(y) \, dS \, , \quad x \in R^3/\overline{\Omega}_1 \, , \quad (7.6.3)$$

$$2u = \int_S \Gamma^0(x,y)(\sigma_n^0)^+ \, dS + \int_S \left(\sigma_n^0 \Gamma^0(x,y)\right)' u^-(y) \, dS \, , \quad x \notin R^3/\overline{\Omega}_1 \, , \quad (7.6.4)$$

$$0 = -\int_S \Gamma^0(x,y)(\sigma_n^0)^+ \, dS + \int_S \left(\sigma_n^0 \Gamma^0(x,y)\right)' u^-(y) \, dS \, , \quad x \in \Omega_1 \, . \quad (7.6.5)$$

The integral representations (7.6.2) - (7.6.5) and equation (7.6.1) yield the representations

$$2u(x) = -\int_S \Gamma^0(x,y)(\sigma_n^1)^+ \, dS + \int_S \left(\sigma_n^0 \Gamma^0(x,y)\right)' u^+(y) \, dS + \qquad (7.6.6)$$
$$+ F(x) \, , \quad x \notin R^3/\overline{\Omega}_1 \, ,$$

$$0 = -\int_S \Gamma^0(x,y)(\sigma_n^1)^+ \, dS + \int_S \left(\sigma_n^0 \Gamma^0(x,y)\right)' u^+(y) \, dS + F(x) \, , \quad x \in \Omega_1 \, , \qquad (7.6.7)$$

$$F(x) = \int_S \Gamma^0(x,y) f_0(y) \, dS + \int_S \left(\sigma_n^0 \Gamma^0 1(x,y)\right)' u_0(y) \, dS \, .$$

Taking in (7.6.3) and (7.6.7) the limit $x \to t \in S$ (from Ω and from $R^3/\overline{\Omega}$, respectively), one finds

$$\int_S \Gamma^1(t,y)(\sigma_n^1)^+ \, dS - u^+(t) - \int_S \left(\sigma_n^1 \Gamma^1(x,y)\right)' u^+(y) \, dS = 0 \, , \qquad (7.6.8)$$

$$\int_S \Gamma^0(t,y)(\sigma_n^1)^+ \, dS + u^+(t) - \int_S \left(\sigma_n^0 \Gamma^0(t,y)\right)' u^+(y) \, dS = F(t) \, . \qquad (7.6.9)$$

The system (7.6.8), (7.6.9) comprises six equations in the six unknowns $(\sigma_n^1)^+$, u^+ serves for their values on the boundary. For $x \notin S$, the solution is determined with the aid of the representations (7.6.2) and (7.6.6).

In the simple particular case when the Poisson coefficients for the medium and the inclusion Ω_1 coincide

$$\frac{\lambda_0}{2(\lambda_0 + \mu_0)} = \frac{\lambda_1}{2(\lambda_1 + \mu_1)} \, . \qquad (7.6.10)$$

The system of equations becomes two consecutive, soluble equations. In fact, it follows from (7.6.10) that

$$\lambda_0 \mu_1 - \lambda_1 \mu_0 = 0 \, , \quad \Gamma^0 = \frac{\mu_1}{\mu_0} \Gamma^1 \, , \quad \sigma_n^0 = \frac{\mu_0}{\mu_1} \upsilon_n^1 \, ;$$

and the system of equations (7.6.8) and (7.6.9) assumes the form

$$\int_S \Gamma^1(t,y)(\sigma_n^1)^+ \, dS - u^+(t) - \int_S \left(\sigma_n^1 \Gamma^1(x,y)\right)' u^+(y) \, dS = 0 \,,$$

$$\frac{\mu_1}{\mu_0} \int_S \Gamma^1(t,y)(\sigma_n^1)^+ \, dS + u^+(t) - \int_S \left(\sigma_n^1 \Gamma^1(x,y)\right)' u^+(y) \, dS = F(t) \,.$$

Multiplying the first equation by μ_1/μ_0 and subtracting the result from the second equation, one finds

$$u^+(t) - \frac{\mu_0 - \mu_1}{\mu_0 + \mu_1} \int_S \left(\sigma_n^1 \Gamma^1(x,y)\right)' u^+(y) \, dS = \frac{\mu_0}{\mu_1 + \mu_0} F(t) \,. \tag{7.6.11}$$

This equation has a unique solution for any values of μ_0, $\mu_1 > 0$ (cf. chapter 6.). Thus, one has for the unknown vector $(\sigma_n^1)^+$ the equation

$$\int_S \Gamma^1(t,y)(\sigma_n^1)^+ \, dS = u^+(t) + \int_S \left(\sigma_n^1 \Gamma^1(x,y)\right)' u^+(y) \, dS \,, \tag{7.6.12}$$

which has been studied in chapter 6, §7.

For many elastic bodies, condition (7.6.10) is not fulfilled, but the Poisson coefficients differ by a small amount, i.e., $\nu_0 - \nu_1 = \varepsilon$, where ε is a small parameter.

The components of the stress vector acting at the point $y \in S$ on the plane with normal $n = (n1, n2, n3)$ can be given the form

$$\left(\sigma_n^i\right)_j^+ = \mu_i \left[\frac{2\nu_i}{1 - 2\nu_i} n_j \operatorname{div} u + \sum_{k=1}^{3} \left(\frac{\partial u_j}{\partial y_k} + \frac{\partial u_k}{\partial y_j} \right) n_k \right], \quad i = 0, 1;$$

and the components of the Kelvin-Somiliagni tensor the form

$$\Gamma^i_{jk}(x,y) = \frac{1}{\mu_i 8\pi} \left[\frac{3 - 4\nu_i}{2(1 - \nu_i)} \frac{\delta_{jk}}{|x - y|} - \frac{1}{2(1 - \nu_i)} \frac{(y_j - x_j)(y_k - x_k)}{|x - y|^3} \right],$$
$$i = 0, 1 \,.$$

Thus, the tensor $T^i(\nu_i, x, y) \equiv \left(\sigma_n^i \Gamma^i(x,y)\right)'$ depends only on the given arguments ν_i, x, y and is an algebraic function of ν_i ($\nu_i \neq 1/2; \ \nu_i \neq 1$).

Assume that the functions $(\sigma_n^1)^+$ and u^+ depends analytically on ε in some neighbourhood of the point $\varepsilon = 0$, i.e.,

$$\left(\sigma_n^1\right)^+(t) = \sum_{k=0}^{\infty} \varepsilon^k p_k(t) \,, \quad u^+(t) = \sum_{k=0}^{\infty} \varepsilon^k q_k(t) \,, \tag{7.6.13}$$

Formal substitution of (7.6.13) into (7.6.8) and (7.6.9) and comparison of the coefficients of equal powers of ε yields a recurrent sequence of integral equations of the form (7.6.11) and (7.6.12).

In fact, one finds

$$T^i(\nu_0, x, y) = \sum_{k=0}^{\infty} \varepsilon^k \frac{D^k T^i(\nu_1, x, y)}{k!}\, , \quad \Gamma^0 = \frac{\mu_1}{\mu_0} \sum_{k=0}^{\infty} \varepsilon^k \frac{D^k \Gamma^1(x, y)}{k!}\, ,$$

where D^k denotes the k-th derivative with respect to the variable ν_1. One finds from (7.6.8) and (7.6.9)

$$\int_S \Gamma^1(t, y)\, p_k(y)\, dS - q_k(t) - \int_S T^1(\nu_1, t, y)\, q_k(y)\, dS - 0\,, \qquad (7.6.14)$$

$$\sum_{k,j=0}^{\infty} \frac{\varepsilon^{k+j}}{k!} \left[\frac{\mu_1}{\mu_0} \int_S D^k \Gamma^1(t, y)\, p_j(y)\, dS - \int_S D^k T^1(\nu_1, t, y)\, q_j(y)\, dS \right] +$$

$$+ \sum_{k=0}^{\infty} \varepsilon^k q_k(t) = F(t)\,.$$

Comparing coefficients of the same powers of ε in the second equation, one obtains

$$\sum_{j=0}^{k} \frac{1}{(k-j)!} \int_S \left[\frac{\mu_1}{\mu_0} D^{k-j} \Gamma^1(t, y)\, p_j(y) - D^{k-j} T^1(\nu_1, t, y)\, q_j(y) \right] dS +$$

$$+ q_k(t) = \delta_{0k} F(t)\,.$$

Transform this equation, retaining on the left hand side terms which depend on $p_k(y)$ and $q_k(y)$, to obtain

$$\frac{\mu_1}{\mu_0} \int_S \Gamma^1(t, y)\, p_k(y)\, dS - \int_S T^1(\nu_1, t, y)\, q_k(y)\, dS + q_k(t) =$$

$$= \delta_{0k} F(t) - \sum_{j=0}^{k-1} \frac{1}{(k-j)!} \int_S \left[\frac{\mu_1}{\mu_0} D^{k-j} \Gamma^1(t, y)\, p_j(y) - \qquad (7.6.15) \right.$$

$$\left. - D^{k-j} T^1(\nu_1, t, y)\, q_j(y) \right] dS\,.$$

As before, equations (7.6.14) and (7.6.15) yields

$$
q_k(t) - \frac{\mu_0 - \mu_1}{\mu_0 + \mu_1} \int_S \sigma_n^1 \Gamma^1(t,y)' q_k(y)\,dS = \frac{\mu_0}{\mu_1 + \mu_0}\, \delta_{0k} F(t) - \frac{\mu_0}{\mu_1 + \mu_0} \times
$$

$$
\times \sum_{j=0}^{\infty} \frac{1}{(k-j)!} \int_S \left[\frac{\mu_1}{\mu_0}\, D^{k-j}\Gamma^1(t,y)\, p_j(y) - D^{k-j} T^1(\nu_1,t,y)\, q_j(y) \right] dS .
$$

$$(7.6.16)$$

Thus, the determination of the functions $p_0(t)$ and $q_0(t)$ has been reduced
to solution of the from (7.6.11) and (7.6.12). Moreover, once the functions
$p_0(t), p_1(t), ..., p_{k-1}(t)$ and $q_0(t), q_1(t), ..., q_{k-1}(t)$ have been found, the right
hand side of (7.6.11) is known. This means that the determination of $p_k(t)$
again is reduced to solving an equation of the form (7.6.11) in which the right
hand side depends on $p_0(t), ..., p_{k-1}(t)$ and $q_0, ..., q_{k-1}(t)$. The function $q_k(t)$
is determined from (7.6.14) which has the same form as equation (7.6.12).

In order to base the assumption of analyticity of $(\sigma_n^1)^+$ and u^+ as a
function of the parameter $\varepsilon = \nu_0 - \nu_1$ at the point $\varepsilon = 0$, it is sufficient
to obtain estimates $\leq \|p_k\| \leq C_1^k$ and $\|q_k\|_{H^{-1/2}(S)} \leq C_2^k$, $k = 0, 1, ...,$ to
ensure convergence of the series (7.6.13) for small $|\varepsilon|$. Here the norms are in
the spaces $L_2(S)$ and $H^{-1/2}(S)$, respectively, C_1 and C_2 are constants which
do not depend on k. Such estimates have been proved in [61] and are readily
found on the basis of the properties of the operators of equations (7.6.11) and
(7.6.12), studied in chapter 8, §5.

CHAPTER 8
PROBLEMS OF THE THEORY OF CRACKS

A general approach to the study of the integral equations for plane as well as space problems for regions with non smooth boundaries was proposed by V.G. Maz'ja [35] and developed by him for the plane problem in cooperation with S.S.Zargaryan [24,25]. An other approach was employed independently by W. Wendland, E. Stephan and M. Costabel [87, 101, 102].

R.V. Duduchava [20], relying on the classical method of Radon [65], investigated in detail the equations of Muskhelishvili and Sherman–Lauricella for the basic plane problems of the theory of elasticity in the case of bodies with singular points on the boundaries (without points of reversal). The form of these integral equations of Potential theory does not depend on the presence of singular points. The situation is different in the case of Muskhelishvili's integral equations, obtained with the aid of conformal mapping. For simply connected regions with smooth boundaries, it was investigated by V.A. Fok and N.I. Muskhelishivili (cf. §3, chapter 5). In the section 1 of this chapter, an analogue of this equations is studied for the outside of a simply connected, bounded region with singular points [60]. Section 2 treats the concretization of Muskhelishivili's equation for the case of branching cracks [3, 4, 62].

In the section 3 the problem of a crack in a layered medium is considered [48, 49]. Section 4 deals with the problem of inclusions the boundaries of which have points of reversal [22]. The last section 5 contains certain results relating to space problems of the theory of elasticity for bodies with cracks [87, 101, 102].

§1. Muskhelishvili's integral equations

1^0. Let Ω be a simply connected region, bounded by a closed, sectionally smooth contour Γ which is oriented in such a manner that during a positive circuit the region Ω remains on the left hand side; also, the angles α_i between the half tangents at angular points c_i do not equal 0, π, 2π $(i = 1, 2, ..., n)$ (Fig. 16). As it is known (cf. chapter 5), the solutions of the plane problems of the theory of elasticity for a body, occupying the region Ω reduces to determining the two functions $\varphi(z)$ and $\psi(z)$, analytic in Ω and satisfying on the boundary Γ the conditions

$$k\varphi(t) - \overline{t\varphi'(t)} - \overline{\psi(t)} = g(t)\,, \ t \in \Gamma, \qquad (8.1.1)$$

where $k = 3 - 4\nu$ for the first and $k = -1$ for the second boundary value problem, and $g(t)$ is a function given on Γ. It is assumed that $g(t)$ satisfies a Lipshitz condition on the contour Γ and is a smooth curve away from the angular points, and that the resultant moment of the external forces vanishes, i.e.,

$$\mathrm{Re} \int_{\Gamma} \overline{f(t)}\, dt = 0 \,. \tag{8.1.2}$$

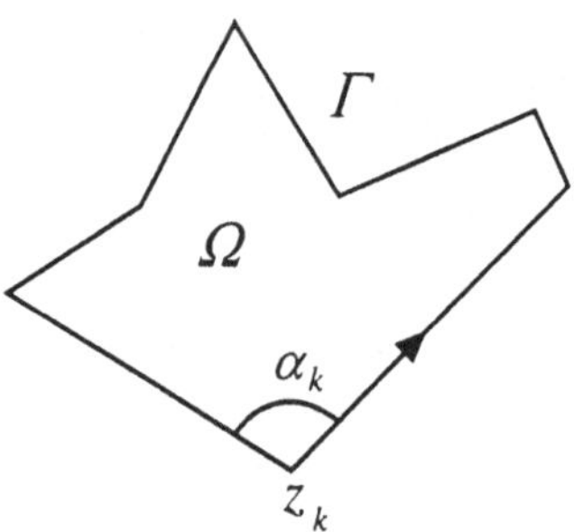

Fig. 16

Let Ω be an infinite region of the complex z-plane and $z = \omega(\xi)$ be the function which realizes a conformal mapping of the unit circle on this region, where

$$\omega(\xi) = c_0 \xi^{-1} + \omega_0(\xi) \,,$$

and $\omega_0(\xi)$ is holomorphic in this circle.

After setting $z = \omega(\xi)$ in (8.1.1), it assumes in terms of the functions $\Phi(\xi) = \varphi(\omega(\xi))$ and $\Psi(\xi) = \psi(\omega(\xi))$ the form

$$k\Phi(t) - \frac{\omega(t)}{\omega'(t)}\, \overline{\Phi'(t)} - \overline{\Psi(t)} = f(t) \,, \quad |t| = 1 \,. \tag{8.1.3}$$

N.I.Muskhelishvili's classical integral equation (cf. chapter 5, §3) for this problem is

$$k\Phi'(t) - \frac{1}{2\pi i} \int_{\gamma} k(t,\tau)\, \overline{\Phi'(\tau)}\, d\tau = A(t) \,, \quad |t| = 1 \,. \tag{8.1.4}$$

where γ is the unit circle, $A(t)$ is a known function and

$$K(t,\tau) = \frac{1}{\omega'(\tau)} \frac{\partial}{\partial t}\left(\frac{\omega_0(\tau) - \omega_0(t)}{\tau - t} \right) = \frac{1}{\omega'(\tau)} \int_{0}^{1} \omega_0''(\tau - \lambda(\tau - t))\lambda\, d\lambda \,.$$

The basic assumption used for the derivation of equation (8.1.4) is that the mapping $z = \omega(\xi)$ is conformal up to the boundary, that is continuously differentiable for $\omega'(\xi)\ |\xi| \leq 1$ and $\omega'(\xi) \neq 0$ for $|\xi| \leq 1$. This condition ensures the continuity of the kernel $K(t,\tau)$, and hence the applicability of the Fredholm theory to equation (8.1.4).

Next, let there be a violation of the conformity of the mapping $z = \omega(\xi)$ at a finite number of points $\{t_k\}_{k=1}^n$ of the unit circle, where

$$\omega(\xi) = z_k + \left(\xi - t_k\right)^{\frac{\alpha_k}{\pi}} g_k(\xi)\,, \quad g_k(t_k) \neq 0\,, \tag{8.1.5}$$

with $g_k(\xi)$ an analytic function at the point t_k, $0 < \alpha_k < 2\pi$. In particular, this condition is fulfilled, if the region Ω coincides with the external region of the n polygon with outside angles α_k, $k - 1, ..., n$.

The presence of angular points on the boundary of Ω requires modifications of the form of equation (8.1.4). In fact, consider the behaviour of the kernel $K(t,\tau)$ when conditions (8.1.5) apply. Fix arbitrarily $t \neq t_k$, $k = 1, 2, ..., n$, when as $\tau \to t_k$,

$$\lim_{\tau \to t_k} \overline{\omega'(\tau)}\, K(t,\tau) = \frac{\alpha_k}{\pi}\left(\frac{\alpha_k}{\pi} - 1\right)\left(t - t_k\right)^{\frac{\alpha_k}{\pi} - 2} G_k(t) + O\left(\left|t - t_k\right|^{\frac{\alpha_k}{\pi} - 1}\right),$$

$$G_k(t) = \int_0^{\varepsilon} \frac{g_k(t_k - \lambda(t_k - t))\, d\lambda}{\lambda^{1 - \alpha_k/\pi}}\,, \quad \varepsilon > 0\,. \tag{8.1.6}$$

Next, an estimate for the kernel $K(t,\tau)$ will be obtained for t and τ in the neighbourhood of t_k. For the sake of definiteness, let $|t - t_k| \leq |\tau - t_k|$, when

$$|K(t,\tau)| \leq \frac{c_k}{|\omega'(\tau)|} \int_0^1 \frac{\lambda\, d\lambda}{|\tau - \lambda(\tau - t) - t_k|^{2 - \alpha_k/\pi}} \leq$$

$$\leq \frac{c_k}{|\omega'(\tau)|} \int_0^1 \frac{\lambda\, d\lambda}{\left[\delta_k^2(t,\tau) + (1 - 2\lambda)^2 y^2\right]^{1 - \alpha_k/2\pi}}\,.$$

where t' is the point on the unit circle which lies at the distance $|t - t_k|$ from the points t_k and $t' \neq t$; δ_k is the distance from the point t to the segment which joins the points t and t', and $2y = |t - t'|$ is the length of this segment (when $|t - t_k| \geq |\tau - t_k|$, one must replace in this notation the point t by the point τ.)

Introducing the new variable $\mu = \delta_k^{-1}(1 - 2\lambda)y$ into the last integral, one finds

$$|K(t,\tau)| \leq \frac{c_k}{4|\omega'(\tau)|}\left(\int_{-\delta_k^{-1}y}^{\delta_k^{-1}y} \frac{(1 - \mu\frac{\delta_k}{y})d\mu}{(1 + \mu^2)^{1-\alpha_k/2\pi}}\right) \frac{\delta_k^{\frac{\alpha_k}{\pi}-1}}{y} =$$

$$= \frac{c_k}{4|\omega'(\tau)|}\left(\int_{-\delta_k^{-1}y}^{\delta_k^{-1}y} \frac{d\mu}{(1 + \mu^2)^{1-\alpha_k/2\pi}}\right) \frac{\delta_k^{\frac{\alpha_k}{\pi}-1}}{y} \leq$$

$$\leq \frac{c_k\delta_k^{\frac{\alpha_k}{\pi}-1}}{4y|\omega'(\tau)|}\left\{2 + 2\int_1^{\delta_k^{-1}y} \frac{d\mu}{(1 + \mu^2)^{1-\alpha_k/2\pi}}\right\} \leq$$

$$\leq \frac{c_k\delta_k^{\frac{\alpha_k}{\pi}-1}}{2y|\omega'(\tau)|}\left\{\frac{\pi}{\alpha_k - \pi}\delta_k^{-\frac{\alpha_k}{\pi}+1}y^{\frac{\alpha_k}{\pi}-1} - \frac{2\pi - \alpha_k}{\alpha_k - \pi}\right\}.$$

Thus, for any points t and τ in some neighbourhood of the point t_k, one has the estimate

$$|K(t,\tau)| \leq \begin{cases} \dfrac{c_k\pi\max\left\{|t - t_k|^{\frac{\alpha_k}{\pi}-2}, |\tau - t_k|^{\frac{\alpha_k}{\pi}-2}\right\}}{2(\alpha_k - \pi)\,|\omega'(\tau)|} \,, & \pi < \alpha_k\,, \\[4mm] \dfrac{c_k(2\pi - \alpha_k)\delta_k^{\frac{\alpha_k}{\pi}-1}}{2(\pi - \alpha_k)\,|\omega'(\tau)|\min\{|t - t_k|, |\tau - t_k|\}}\,, & \alpha_k < \pi\,. \end{cases} \tag{8.1.7}$$

Estimates (8.1.6) and (8.1.7) permit to draw the conclusion that under condition (8.1.5) the kernel of the classical equation (8.1.4) is not summable on $\gamma \times \gamma$. This circumstance also leads to the conclusion that equation (8.1.4) must be modified when the boundary contains angular points.

2^0. To start with, the results of [20] will be formulated which are required in the sequel. Select as the space in which the solution of (8.1.1) will be sought the Sobolev space $W_p^1(\Gamma,\rho)$, $\rho(t) = \prod_{k=1}^{n} |t - z_k|^{\gamma_k}$, $(1 < p < \infty$, $-1 < \gamma_k < p - 1)$.

Theorem 8.1 ([20]). There exists a unique solution of the equations of Muskhelishvili (5.2.5) and of Sherman-Lauricella (5.5.5) in the space $W_p^1(\Gamma,\rho)$, where

$$\delta_j < \frac{1 + \gamma_j}{p} < 1, \ j = 1, 2, ..., n\,, \ 1 < p < \infty\,, \tag{8.1.8}$$

and δ_j is the unique solution of the transcendental equation

$$(1 - \delta_j)^2 \sin^2 \alpha_j' = k^2 \sin^2 \alpha_j'(1 - \delta_j), \ 1 - \frac{\pi}{\alpha_j'} < \delta_j < \frac{1}{2};$$

where $\alpha_j' = \alpha_j$ for $\pi < \alpha_j < 2\pi$ and $\alpha_j' = 2\pi - \alpha_j$ for $0 < \alpha_j < \pi$. The solution $\varphi(t) \in H_{1/2}(\Gamma)$, $\varphi(t)$ is continuous in the sense of Hoelder with index $\mu \in (0,1)$ everywhere on Γ with the exception of angular points, in the vicinity of which $\varphi(t)$ is Hoelder continuous with index $0 < \mu_j < 1 - \delta_j$, where $0 < \delta_j < 1/2$.

Note 8.1. In R.V. Duduchav [20], an analogous assertion is proved for the equation of Sherman-Lauricella for the case of a multiply connected (possibly infinite) region. In the same reference, the mixed problem is treated in detail.

Note 8.2. The existence of a summable derivative of the solution of the integral equation is used to prove the equivalence of these integral equations with the initial basic boundary value problems of the theory of elasticity.

3^0. The remainder of this section is based on the special properties of the coefficient of $\overline{\Phi'(t)}$ in equation (8.1.3). Define on the unit circle the sectionally constant, non-decreasing function $\alpha(\theta)$, $\theta \subset [0, 2\pi]$ with discontinuities at the points t_k with jumps α_k, $k = 1, 2, ..., n$. Moreover, let $\eta(t) = \exp(2i\alpha(\theta))$ for $t = \exp(i\theta)$, when the function $\eta(t)$ is defined single-valuedly exactly apart from a factor.

Lemma 8.1. (On distribution of singularities) For any point $t_0 \in \gamma$ there exists analytic functions $g_{t_0}(s)$ and $h_{t_0}(s)$, defined in some neighbourhood of the point t so that $h_{t_0} = 0$ and there is satisfied the relation

$$\frac{\overline{\omega(t)}}{\omega'(t)} = \lim_{s \to t, |s| < 1} \frac{\overline{\omega(s)}}{\omega'(s)} = \frac{\overline{\omega(t_0)}}{\omega'(t)} + \eta(t)g_{t_0}(t)\overline{h_{t_0}(t)}. \tag{8.1.9}$$

Proof. It follows from (8.1.5) that

$$\omega'(s) = \left(s - t_k\right)^{\frac{\alpha_k}{\pi}} g_k'(s) + \frac{\alpha_k}{\pi}\left(s - t_k\right)^{\frac{\alpha_k}{\pi} - 1} g_k(s)$$

for point s of the unit circle, close to the point t_k. One has

$$\frac{\overline{\omega(s)}}{\omega'(s)} - \frac{\overline{\omega(t_k)}}{\omega'(t_k)} = \frac{\overline{\left(s - t_k\right)^{\frac{\alpha_k}{\pi}} g_k(s)}}{\left(s - t_k\right)^{\frac{\alpha_k}{\pi}} g_k'(s) + \frac{\alpha_k}{\pi}\left(s - t_k\right)^{\frac{\alpha_k}{\pi}} g_k'(s)} =$$

$$= \frac{\overline{\left(s - t_k\right) g_k(s)}}{\left(s - t_k\right)g_k'(s) + \frac{\alpha_k}{\pi} g_k(s)} \exp\left\{\frac{2\alpha_k}{\pi} i \arg(s - t_k)\right\}.$$

It is readily verified that for $t \in \gamma$ there is fulfilled the relation

$$\arg(t - t_k) = \begin{cases} \dfrac{\pi + \arg t + \arg t_k}{2} \,, & \arg t > \arg t_k \,, \\[2mm] \dfrac{-\pi + \arg t + \arg t_k}{2} \,, & \arg t < \arg t_k \,. \end{cases} \tag{8.1.10}$$

i.e.

$$\exp\left\{ \frac{2\alpha_k}{\pi} i \arg(t - t_k) \right\} = \eta_k(t)\, t^{\frac{\alpha_k}{\pi}}\, t_k^{\frac{\alpha_k}{\pi}} \,, \quad t \in \gamma \,,$$

where

$$\eta_k(t) = \begin{cases} \exp(i\alpha_k) \,, & \arg t > \arg t_k \,, \\[1mm] \exp(-i\alpha_k) \,, & \arg t < \arg t_k \,. \end{cases}$$

Thus, there exists always a constant $c_k \neq 0$ such that

$$c_k \eta_k(t)\, t^{\frac{\alpha_k}{\pi}}\, t_k^{\frac{\alpha_k}{\pi}} = \eta(t)\, t^{\frac{\alpha_k}{\pi}} \,, \quad t \in \gamma \,.$$

Let $g_{t_k}(s) = g_k(s)\, s^{\frac{\alpha_k}{\pi}}$, where the function $s^{\frac{\alpha_k}{\pi}}$ is analytic in the vicinity of the point $s = t_k$ and

$$h_{t_k}(s) = \frac{(s - t_k)\bar{c}_k^{-1}}{(s - t_k)\, g_k'(s) + \frac{\alpha_k}{\pi}\, g_k(s)} \,.$$

Then relation (8.1.9) is satisfied for all $t_0 = t_k$, $k = 1, 2, ..., n$. For $t_0 \neq t_k$, $k = 1, 2, ..., n$, condition (8.1.3) is fulfilled with $\alpha_k = \pi$ and the function $\eta(t)$ is continuous, hence relation (8.1.9) is satisfied also in this case. Clearly, the function $h_{t_k}(s)$ is analytic at the point t_k, since $g_k(t_k) \neq 0$.

  Note 8.3. There exists a finite cover $\{U_{x_k}\}$ of the circle, where U_{x_k} is the circle of the points x_k, $|x_k| = 1$, such that the points $\{x_k\}$ comprise all points $\{t_k\}_{k=1}^n$, and the functions $g_{x_k}(s)$ and $h_{x_k}(s)$ are analytic in U_{x_k}. The different functions $G_{x_k}(s) = g_{x_k}(s)\overline{h_{x_k}(s)}$, determined locally for $s \in U_{x_k}$, are interlinked in the following manner:

$$G_{x_l}(t) = G_{x_k}(t) + \frac{\omega(x_k) - \omega(x_l)}{\eta(t)\,\omega'(t)} \,, \quad |t| = 1 \,, \; t \in U_{x_k} \cap U_{x_l} \,.$$

  Consider next the derivation of the modified integral equation for an infinite region Ω under condition (8.1.5). Without limiting generality, it may be assumed that $\omega(t_j) \neq 0$, $j = 1, ..., n$.

It follows from theorem 8.1 that for the solution $\Phi(t)$ of problem (8.1.4) hold the estimates

$$\int\limits_{\gamma'_k} |\Phi(t)|^p \, |t - t_k|^{\frac{2\alpha_k}{\pi}\gamma_k + \frac{2\alpha_k}{\pi} - 1} \, |dt| < \infty\,, \quad k = 1, ..., n\,,$$

$$\int\limits_{\gamma'_k} \left(\frac{\Phi'(t)}{\omega'(t)}\right)^p |t - t_k|^{\frac{2\alpha_k}{\pi}\gamma_k + \frac{2\alpha_k}{\pi} - 1} \, |dt| < \infty\,, \quad k = 1, ..., n\,, \tag{8.1.11}$$

where

$$\delta_k < \frac{1 + \gamma_k}{p} < 1\,, \ 1 < p < \infty\,, \ 0 < \delta_k < \frac{1}{2}\,.$$

Thus, for $p > 1$ chosen such that $p\delta_k\alpha_k/\pi < 1$, the functions $\Phi(t)$ and $\Phi'(t)/\omega'(t)$ are summable with degree p on the unit circle. In particular, if $\alpha_k < \pi$, then these functions are summable with degree $p = 2$; if for certain k_j, $j = 1, ..., m \le n$, for $\alpha_{x_j} > \pi$, polynomial with simple zeroes at (corresponding) points $t_{k_j} \in \gamma$ are denoted by $Q^*_m(t)$, then the function $Q^*_m(t)\Phi'(t)/\omega'(t)$ is quadratically summable. This preliminary information will be use for the derivation of an integral equation.

Multiplication of (8.1.3) by $Q^*_m(t)(2\pi i(t - \xi))^{-1}$ and integrating around the unit circle, one finds

$$\frac{k}{2\pi i}\int\limits_\gamma \frac{\Phi(t)\,\overline{Q^*_m(t)}}{t - \xi}\,dt - \frac{1}{2\pi i}\int\limits_\gamma \frac{\overline{Q^*_m(t)}\,\omega(t)}{\overline{\omega'(t)}\,(t - \xi)}\,\overline{\Phi'(t)}\,dt = F(\xi)\,, \ |\xi| < 1\,. \tag{8.1.12}$$

where $F(\xi)$ is a known function. In the derivation of this equation use has been made of the relation

$$\frac{1}{2\pi i}\int\limits_\gamma \frac{\overline{\Psi(t)\,Q^*_m(t)}}{t - \xi}\,dt = \overline{\Psi(0)\,Q^*_m(0)}$$

and the normalization $\Psi(0) = 0$ which does not affect the state of stress and strain has been taken into consideration.

Represent the first integral on the left hand side of (8.1.12) in the form

$$\frac{k}{2\pi i}\int\limits_\gamma \frac{\Phi(t)Q_m(t)}{t - \xi}\,\frac{dt}{t^m} = k\Phi(\xi)Q_m(\xi)\xi^{-m} - \sum_{k=0}^{m-1} c_k \xi^{k-m}\,,$$

where $Q_m(t) = t^m \overline{Q_m^*(t)}$ and the constants c_k are such that the right hand side of the last equality is analytic for $\xi = 0$. In order to transform the second integral in equation (8.1.12), select a special cover of the unit circle, using lemma 8.1 and note 8.3. Let $\{\gamma_k'\}$ be the union of simple arcs of the unit circle which covers γ, $t_k \in \gamma_k'$ and the functions $g_{t_k}(s)$ and $h_{t_k}(s)$, analytic in the circle γ_k'. The point t_k subdivides the arc γ_k' into two parts, $\gamma_k' = \gamma_{2k-1} \cup \gamma_{2k}'$, within each of which the function $\eta(t)$ is constant.

For the second integral on the left hand side of (8.1.12), one has

$$I(\xi) = \frac{1}{2\pi i} \int_\gamma \frac{Q_m^*(t)\omega(t)}{\omega'(t)\,(t-\xi)}\, \overline{\Phi'(t)}\, dt = \frac{1}{2\pi i} \sum_{k=1}^n \omega(t_k) \times$$

$$\times \int_{\gamma_k'} \frac{Q_m^*(t)\,\Phi'(t)}{\omega'(t)\,(t-\xi)}\, dt + \frac{1}{2\pi i} \sum_{k=1}^n \Big\{ \eta(t_k - 0) \times$$

$$\times \int_{\gamma_{2k-1}} \frac{G_{t_k}(t)\,Q_m^*(t)}{\omega'(t)\,(t-\xi)}\, \overline{\Phi'(t)}\, dt + \eta(t_k + 0) \int_{\gamma_{2k}} \frac{G_{t_k}(t)\,Q_m^*(t)}{\omega'(t)\,(t-\xi)}\, \Phi'(t)\, dt \Big\}.$$

From (8.1.12) follows now the equation

$$k\Phi(\xi) = Q_m^{-1}(\xi)\Big\{ \sum_{k=0}^{m-1} c_k\xi^k + \xi^m I(\xi) + \xi^m F(\xi) \Big\}, \ |\xi| < 1. \qquad (8.1.13)$$

For the summability of the function $\Phi(t)$ on the unit circle it is necessary that

$$\sum_{l=0}^{m-1} c_l t_{jk}^l + t_{jk}^m I(t_{jk}) + t_{jk}^m F(t_{jk}) = 0, \ k = 1,...,m. \qquad (8.1.14)$$

which permits to determine single-valuedly the constants c_k, $k = 0, 1, ..., m - 1$. In fact, the system of algebraic equations (8.1.14) in terms of the unknowns c_k has a non-zero determinant.

In order to obtain the integral equation for $\Phi'(t)$ on the unit circle, it is sufficient to differentiate (8.1.13) and execute the limit (radially) $\xi \to t \in \gamma$, $t \neq t_k$, $k = 1,...,n$. Consider the details of the limit process.

From the analyticity of the function $Q_m^*(\xi)\Phi'(\xi)$ follow the equalities

$$\int_\gamma \frac{\overline{Q_m^*(t)\Phi(t)}}{\omega'(t)(t-\xi)}\, dt = 0, \ \int_\gamma \frac{\overline{Q_m^*(t)\Phi'(t)}}{(t-\xi)}\, \frac{dt}{t} = 0, \ |\xi| < 1.$$

by means of which the integral $I(\xi)$ can be given the form

$$I(\xi) = \frac{1}{2\pi i} \sum_{l=1}^{n} [\omega(t_l) - \omega(t_k)] \int_{\gamma_l'} \frac{\overline{Q_m^*(t)\Phi'(t)}}{\omega'(t)(t-\xi)}\, dt +$$

$$+ \frac{1}{2\pi i} \sum_{l=1}^{n} \int_{\gamma_l'} \frac{[tG_{t_l}(t) - \xi G_{t_l}(\xi)]}{t-\xi}\, \overline{Q_m^*(t)\Phi'(t)}\, \frac{\eta(t)\, dt}{t} +$$

$$+ \frac{\xi}{2\pi i} \left\{ \sum_{l=1}^{n} \int_{\gamma_l'} [G_{t_l}(\xi)\eta(t) - G_{t_k}(\xi)\eta(t_k \pm 0)] \frac{\overline{Q_m^*(t)\Phi'(t)}}{t-\xi}\, \frac{dt}{t} \right\}, \qquad (8.1.15)$$

where $k = 1, 2, ..., n$ is arbitrary and the choice of the sign in the expression $\eta(t_k \pm 0))$ leads to the exclusion from the last sum of the integral along γ_{2k-1} or γ_{2k}, respectively.

Thus, the function $I(\xi)$ is infinitely differentiable in some neighbourhood of an arbitrary point $t \in \gamma$, $t \neq t_k$, $k = 1, 2, ..., n$.

Now let $t \in \gamma_k'$, $t \neq t_k$: in order to perform the limit $\xi \to t$ in the equation

$$k\Phi'(\xi) = \frac{d}{dt} \left\{ Q_m^{-1}(\xi) \left[\sum_{k=0}^{m-1} c_k \xi^k + \xi^m I(\xi) + \xi^m F(\xi) \right] \right\}, \qquad (8.1.16)$$

use the representation (8.1.15). We obtain the integral equation

$$k\Phi'(\xi) = \frac{d}{dt} \left\{ Q_m^{-1}(t) \left[\sum_{k=0}^{m-1} c_k t^k + t^m I(t) \right] \right\} + F_1(t), \ t \in \gamma, \qquad (8.1.17)$$

where $F(t)$ is a known function, $t \neq t_k$, $k = 1, ..., n$ and the function $I(t)$ is defined by (8.1.15) for $t \in \gamma_k'$, the term $\eta(t_k \pm 0)$ denoting the plus sign for $t \in \gamma_{2k}'$ and the minus sign for $t \in \gamma_{2k-1}'$. Thus, the integral equation (8.1.17) does not contain singular integrals and has unremovable singularities at the points t_k, $k = 1, ..., n$.

Note 8.4. The derivative $\Phi'(t)$ of problem (8.1.4) is a summable function on the unit circle under condition (8.1.5).

In fact, select $p_k > 1$ so that $\alpha_k/2\pi < p_k^{-1} < \alpha_k/\pi$ for $k = 1, ..., n$. Hoelder's inequality yields the estimate

$$\int_{\gamma_k'} |\Phi'(t)|\, |dt| \leq \left(\int_{\gamma_k'} \left| \frac{\Phi'(t)}{\omega'(t)} \right|^{p_k} |dt| \right)^{\frac{1}{p_k}} \left(\int_{\gamma_k'} |\omega'(t)|^{\frac{p_k}{p_k-1}}\, |dt| \right)^{1-\frac{1}{p_k}}.$$

where the first integral on the right hand side is finite through the choice of the parameter p_k: $\delta_k \alpha_k / \pi \leq \alpha_k / 2\pi < p_k^{-1}$ and Properties (8.1.11). The second integral is also finite by (8.1.5) and the inequality

$$\left(\frac{\alpha_k}{\pi} - 1 \right) \frac{p_k}{p_k - 1} > -1 \, ,$$

which is equivalent to the inequality $p_k^{-1} < \alpha_k / \pi$ and ensures the summability of $\omega'(t)$ with degree $p_k / (p_k - 1)$

Note 8.5. For $\alpha_k \in (0, 2\pi)$, the functions $\Phi'(t)$ and $\Phi'(t)/\omega'(t)$ are summable, hence it is unnecessary for the derivation of the integral equation to multiply equation (8.1.3) by the polynomial $Q_m^*(t)$. (Such a necessity arises in the presence of cracks, $\alpha_k = 2\pi$, for the integral entering into (8.1.17) to make sense). For $Q_m^*(t) \equiv 1$, equation (8.1.17) assumes the form

$$
\begin{aligned}
k\Phi'(\tau) = \; & \frac{1}{2\pi i} \sum_{l=1}^{n} [\omega(t_l) - \omega(t_k)] \int_{\gamma_l'} \frac{\Phi'(t)\, dt}{\omega'(t)(t - \tau)^2} + \\
& + \frac{1}{2\pi i} \sum_{l=1}^{n} \int_{\gamma_l'} \frac{d}{d\tau} \left[\frac{t G_{t_l}(t) - \tau G_{t_l}(\tau)}{t - \tau} \right] \frac{\Phi'(t)\eta(t)\, dt}{t} + \\
& + \frac{1}{2\pi i} \sum_{l=1}^{n} \int_{\gamma_l'} \frac{d}{d\tau} \left[\frac{(G_{t_l}(\tau)\eta(t) - G_{t_l}(\tau)\eta(\tau))}{t - \tau} \, \frac{\tau}{t} \right] \overline{\Phi'(t)}\, dt + \\
& + F_1(\tau), \; \tau \in \gamma_k' \, ,
\end{aligned}
\tag{8.1.18}
$$

where the integrals make sense for $\tau \neq t_k$, $k = 1, ..., n$.

Theorem 8.2. Let condition (8.1.5) be fulfilled and $0 < \alpha_k < 2\pi$, $k = 1, 2, ..., n$, then equation (8.1.18) has a unique solution in the class of summable functions $\Phi'(t)$ for which $\Phi'(t)/\omega'(t)$ is summable and

$$\int_{\gamma} \frac{\overline{\Phi'(t)}\, dt}{(t - \xi)t} = 0 \, , \; |\xi| < 1 \, . \tag{8.1.19}$$

Proof. Let $R(t)$ be the solution of the homogeneous integral equation corresponding to (8.1.18). Define an analytic function in the unit circle by

$$\Phi(\xi) = k^{-1} \frac{1}{2\pi i} \int_{\gamma} \frac{\omega(t)\overline{R(t)}}{\omega'(t)(t - \xi)}\, dt \, , \; |\xi| < 1 \, . \tag{8.1.20}$$

Using lemma 8.1 and the properties

$$\int_\gamma \frac{\overline{R(t)}}{(t-\xi)}\,\frac{dt}{t} = 0, \quad \int_\gamma \frac{\overline{R(t)}}{\omega'(t)(t-\xi)}\,dt = 0, \ |\xi| < 1,$$

one find, as during the derivation of equation (8.1.18) that the function $\Phi(\xi)$ satisfies the condition $\Phi'(\tau) = R(\tau)$, $\tau \in \gamma$, $\tau \neq t_k$, $k = 1, 2, ..., n$. Consequently, one finds

$$k\Phi(\xi) = \frac{1}{2\pi i}\int_\gamma \frac{\omega(t)\,\overline{\Phi'(t)}}{\omega'(t)\,(t-\xi)}\,dt, \ |\xi| < 1$$

or

$$\frac{1}{2\pi i}\int_\gamma \left[k\Phi(t) - \frac{\omega(t)}{\overline{\omega'(t)}}\,\overline{\Phi'(t)}\right]\frac{dt}{(t-\xi)} = 0, \ |\xi| < 1.$$

i.e., there exists the function $\Psi_1(\xi)$ which is analytic outside the unit circle, vanishes at infinity $\Psi_1(\infty) = 0$ and has the limiting value

$$\Psi_{1e}(t) = k\Phi(t) - \frac{\omega(t)}{\overline{\omega'(t)}}\,\overline{\Phi'(t)}.$$

The analytic function, defined by (8.1.20), and the function $\Psi(\xi) = \overline{\Psi_1(\bar\xi^{-1})}$ are the solution of the boundary value problem of the theory of elasticity

$$k\Phi(t) - \frac{\omega(t)}{\overline{\omega'(t)}}\,\overline{\Phi'(t)} - \Psi(t) = 0,$$

where $\Psi(0) = 0$. Thus, $\Phi'(\tau) = R(\tau) = 0$, $\tau \in \gamma$ which completes the proof of the theorem.

Note 8.6. In the presence of cracks ($\alpha_k = 2\pi$ for certain $k = 1, 2, ..., n$) there may not exist a solution of problem (8.1.3) with a summable function $\Phi'(t)/\omega'(t)$. In that case is convenient to employ equation (8.1.17) which has a unique solution in the class of summable functions $Q_m^*(t)\,\Phi'(t)$ for which $Q_m^*(t)\Phi'(t)/\omega'(t)$ is a summable function and

$$\int_\gamma \frac{\overline{Q_m^*(t)\Phi'(t)}}{t-\xi}\,\frac{dt}{t} = 0, \ |\xi| < 1,$$

where the constants c_k, $k = 1, 2, ..., m$ are determined by system (8.1.14). The proof of this statement is analogous to the proof of theorem 8.2.

An example of the application of equation (8.1.17) is given in the next section.

§2. The equations for branching cracks

An interesting problem of crack theory to which the developed method may be applied is presented the plane with n cuts Γ_n which radiate from the origin (Fig. 17). The conformal mapping of the unit circle onto C/Γ has the form (cf. [15])

$$\omega_n(\xi) = \xi^{-1} \prod_{j=1}^{n} \left(1 - \xi \exp(-i\beta_j)\right)^{\alpha_j/\pi}, \quad \sum_{j=1}^{n} \alpha_j = 2\pi,$$

where α_j are the angles between the joining cuts, and the points $t_j = \exp(i\beta_j)$, $j = 1, ..., n$ are images of zero.

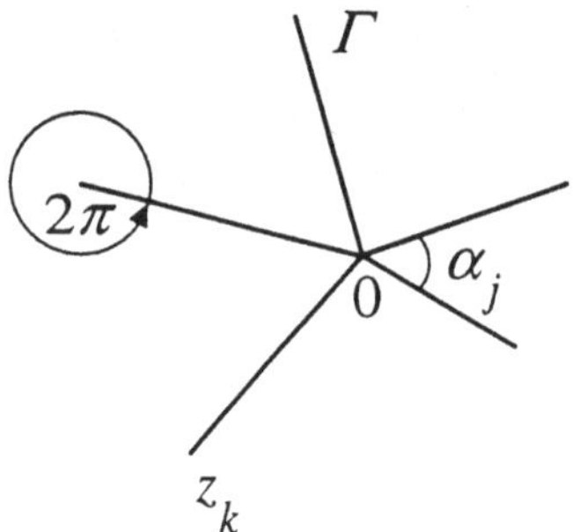

Fig. 17

Verify the relation (specified by lemma 8.1)

$$\lim_{\xi \to t} \frac{\omega_n(\xi)}{\omega_n'(\xi)} = \eta(t)\frac{P_n(t)}{tQ_n(t)}, \quad |t| = 1,$$

where the sectionally constant function $\eta(t)$ is defined in §1, the polynomial $P_n(t)$ has zeroes at the points t_j, $j = 1, 2, ..., n$, corresponding to ends of the cuts. One has

$$\frac{\overline{\omega_n(t)}}{\omega_n'(t)} = t^{-1}\frac{P_n(t)}{Q_n(t)} \exp(2i \arg \omega_n(t)), \quad |t| = 1,$$

where $P_n(t) = (t - \exp(i\beta_1))...(t - \exp(i\beta_n))$, $Q_n(t) = t^n \overline{Q_n^*(t)}$ and $Q_n^*(\xi)$ is some polynomial of degree n. Note that at the ends z_k, $k = 1, , ..., n$, of the cuts,

$$\omega_n(\xi) = z_k + (\xi - s_k)^2 g_k(\xi), \quad g_k(s_k) \neq 0,$$

hence the polynomial $Q_n(t)$ has zeroes at the points s_k, $k = 1, 2, ..., n$. Besides, using (8.1.10), it is easily verified that the function

$$\arg \omega_n(t) = \arg t + \pi^{-1} \sum_{j=1}^{n} \alpha_j (1 - t \exp(-i\beta_j))$$

is sectionally constant for $t \in \gamma$ and has jumps at the points α_j, $j = 1, 2, ..., n$.

Write equation (8.1.8) in the form

$$kt^{-n}Q_n(t)\Phi(t) - \eta(t)\frac{P_n(t)}{t^{n+1}}\overline{\Phi'(t)} - t^{-n}Q_n(t)\Psi(t) = \frac{Q_n(t)f(t)}{t^n}.$$

Multiply both sides of this equation by $[2\pi i(t - \xi)]^{-1}$ and integrate it around the unit circle γ, to obtain

$$\frac{k}{2\pi i} \int_\gamma t^{-n}Q_n(t)\Phi(t)(t - \zeta)^{-1}\, dt -$$

$$- \frac{1}{2\pi i} \int_\gamma \eta(t)\frac{P_n(t)}{t^{n+1}}\overline{\Phi(t)}(t - \zeta)^{-1}\, dt = F(\zeta), \tag{8.2.1}$$

where $F(\zeta)$ is a known function (cf.(8.1.12)) and use has been made of the equation

$$\frac{1}{2\pi i} \int_\gamma t^{-n}Q_n(t)\overline{\Psi(t)}(t - \zeta)^{-1}\, dt = \frac{1}{2\pi i} \int_\gamma \overline{Q_n^*(t)\Psi(t)}(t - \zeta)^{-1}\, dt =$$

$$= \overline{Q_n^*(0)\Psi(0)} = 0,$$

following from the normalization $\Psi(0) = 0$.

Transform equation (8.2.1), using the analytic function $\Phi(\zeta)$:

$$k\zeta^{-n}Q_n(\zeta)\Phi(\zeta) = F(\zeta) + k\sum_{k=0}^{n-1} c_k\zeta^{k-n} + \frac{1}{2\pi i}\int_\gamma \eta(t)\frac{P_n(t)}{t^{n+1}}\overline{\Phi'(t)}(t - \zeta)^{-1}\, dt,$$

$$c_k = \frac{D^k(q_n(t)\Phi(t))\big|_{t=0}}{k!} = \sum_{j=0}^{k} \frac{D^{k-j}Q_n(0)}{(k - j)!}\frac{1}{2\pi i}\int_\gamma \Phi(t)t^{-j-1}\, dt.$$

As a result, one finds the equation in terms of $\Phi(\zeta)$:

$$\Phi(\zeta) = \frac{1}{kQ_n(\zeta)}\left\{\zeta^n F(\zeta) + \frac{\zeta^n}{2\pi i}\int_\gamma \eta(t)\frac{P_n(t)}{t^{n+1}}\overline{\Phi'(t)}(t-\zeta)^{-1}\,dt + \right.$$

$$\left. + k\sum_{k=0}^{n-1}c_k\zeta^k\right\}. \tag{8.2.2}$$

Now determine the sectionally constant function $\eta(\zeta)$, $|\zeta| < 1$ which has discontinuities along the rays $\arg\zeta = \arg t_j$, $j = 1, 2, ..., n$:

$$\eta(\zeta) = \eta(s_k), \quad |s_k| = 1, \quad \arg t_k < \arg s_k < \arg t_{k+1},$$

for ζ: $|\zeta| < 1$, $\arg t_k < \arg\zeta < \arg t_{k+1}$, $k = 1, ..., n$.

Since $\Phi'(\zeta)$ is analytic in the unit circle, one has

$$\frac{1}{2\pi i}\int_\gamma \frac{P_n(t)}{t^{n+1}}\overline{\Phi'(t)}(t-\zeta)^{-1}\,dt = 0, \quad |\zeta| < 1. \tag{8.2.3}$$

This relation permits to rewrite equation (8.2.2) in the form

$$\Phi(\zeta) = \frac{1}{kQ_n(\zeta)}\left\{\zeta^n F(\zeta) + \frac{\zeta^n}{2\pi i}\int_\gamma\left[\eta(t) - \eta(\zeta)\right]\frac{P_n(t)}{t^{n+1}}\overline{\Phi'(t)}(t-\zeta)^{-1}\,dt + \right.$$

$$\left. + k\sum_{k=0}^{n-1}c_k\zeta^k\right\}. \tag{8.2.4}$$

It follows from the summability of the function $\Phi(t)$ on the unit circle that the numerator on the right hand side of (8.2.4) vanishes at the points $\zeta = s_k$, $k = 1, 2, ..., n$ (the zeroes of the polynomial $Q_n(\zeta)$):

$$s_k^n F(s_k) + \frac{s_k^n}{2\pi i}\int_\gamma\left[\eta(t) - \eta(s_k)\right]\frac{P_n(t)}{t^{n+1}}\overline{\Phi(t)}(t-s_k)^{-1}\,dt + $$

$$ + k\sum_{j=0}^{n-1}c_j s_k^j, \quad k = 1, ..., n. \tag{8.2.5}$$

If the function $\Phi(\zeta)$ has already been determined, then the potential $\Psi(\zeta)$ may be found from the boundary condition (8.1.8) which in this case may be written in the form

$$t^{-n}Q_n^*(t)\Psi(t) = kt^{-n}Q_n^*(t)\overline{\Phi(t)} - \overline{\eta(t)P_n(t)}\Phi'(t) - t^{-n}Q_n^*(t)\overline{f(t)}. \tag{8.2.6}$$

Integration of this equation with the kernel $[2\pi i(t - \zeta)^{-1}]$ around the unit circle yields

$$\zeta^{-n}Q_n^*(\zeta)\Psi(\zeta) = kQ_n(0)\overline{\Phi(0)} - \frac{1}{2\pi i}\int_\gamma \overline{\eta(t)P_n(t)}\,t(t - \zeta)^{-1}\Phi'(t)\,dt -$$

$$- \frac{1}{2\pi i}\int_\gamma Q_n^*(t)\overline{f(t)}t^{-n}(t - \zeta)^{-1}\,dt + \sum_{k=0}^{n-1} c_k^*\zeta^{k-n}\,.$$

$$(8.2.7)$$

It follows from the summability of the function $\Psi(t)$ that the right hand side of this equation vanishes at the points s_k, $k = 1, 2, ..., n$. It is readily verified that this condition leads to a single-valued determination of the constant c_k^*, $k = 1, 2, ..., n - 1$.

Thus, the problem of the determination of the two complex potentials $\Phi(\zeta)$ and $\Psi(\zeta)$ has been reduced to solution of the functional equation (8.2.4).

Differentiating (8.2.4) and performing the limit $\zeta \to t \in \gamma_k = \{t \in \gamma : \arg t_k < \arg t < \arg t_{k+1}\}$, one obtains

$$\Phi'(t) = H(t) + \frac{k^{-1}}{2\pi i}\int_\gamma \frac{\partial}{\partial t}\left\{\frac{\eta(\tau) - \eta(s_k)}{\tau - t}\cdot\frac{P_n(\tau)}{Q_n(t)}(t/\tau)^n\right\}\overline{\Phi'(\tau)}\tau^{-1}\,d\tau\,, t \in \gamma_k\,,$$

$$H(t) = \frac{\partial}{\partial t}\left\{\frac{k^{-1}}{Q_n(t)}\left[t^n F(t) + k\sum_{k=0}^{n-1}c_k\zeta^k\right]\right\}.\qquad (8.2.8)$$

This integral equation does not contain singular integrals; the constants c_k, entering into the expression for the function $H(t)$, are determined from the system of algebraic equations (8.2.5).

Theorem 8.3. Equation (8.2.8) has only one solution in the class of summable functions $P_n(t)t^{-n}\overline{\Phi'(t)}$, for which condition (8.2.3) is fulfilled.

The proof is based on note 8.6. In fact, set $m = n$ and $\omega(t) = \omega_n(t) + z_0$, $z_0 \neq 0$. Then the summability of the function $Q_n^*(t)\Phi(t)/\omega'(t)$ is equivalent to summability of the function $P_n(t)t^{-n}\overline{\Phi'(t)}$, and the summability of the function $Q_n^*(t)\Phi'(t)$ follows from that of the function $\Phi'(t)$ in the neighbourhood of the points s_k, $k = 1, 2, ..., n$.

Condition (8.2.3) may be reformulated in the following manner. Expand the function $P_n(t)t^{-n-1}\overline{\Phi'(t)}$ in the Fourier series on the unit circle γ:

$$P_n(t)t^{-n-1}\overline{\Phi'(t)} = \sum_{k=-\infty}^{\infty} a_k t^k\,,\ |t| = 1\,.$$

Now, condition (8.2.3) is equivalent to the relation: $a_k = 0$, $k = 1, 2, ...$, hence

$$t^n \overline{P_n(t)} \Phi'(t) = \sum_{k=1}^{\infty} \overline{a}_{-k} t^{k-1} = \sum_{k=0}^{\infty} b_k t^k , \ t \in \gamma ,$$

where $\{b_k\}_{k=0}^{\infty}$ are arbitrary complex numbers.

In particular, consider problem (8.1.4) for the semi-infinite crack, comprising the ray $(-\infty, 0)$ and the interval $\{z = r\exp(i\varphi_0) : 0 \le r < r_0\}$ along the angle φ_0 to the ray. The conformal mapping of the unit circle onto the corresponding region has the form

$$z = \frac{w}{w-1} , \ w = \zeta^{-1}(1 - \zeta\exp(-i\beta_1))^{\alpha_1/\pi}(1 - \zeta\exp(-i\beta_2))^{\alpha_2/\pi} , \ |\zeta| < 1 .$$

where it is assumed that $w_2(s_1) = 0$ with s_1 and s_2 the points on the unit circle γ which correspond the ends of the cuts in the w-plane.

Write equation (8.1.8) in the form

$$\frac{k(t-s_2)}{t}\Phi(t) + \theta(t)\frac{P_2(t)}{t^3}\left(\overline{\frac{1-w_2(t)}{1-\overline{s}_1 t}}\right)\overline{\Phi'(t)} + \overline{(1-\overline{s}_1 t)\Psi(t)} = g(t) ,$$

$$\theta(t) = \eta(t)\frac{\overline{1-w_2(t)}}{1-w_2(t)} , \ g(t) = \frac{t-s_2}{t}f(t) . \tag{8.2.9}$$

Note certain properties of the functions, appearing in this equation: $|\theta(t)| = 1$ for all $t \in \gamma$ and $\theta(t) = \eta(s_1)$ for $t \in \gamma_1$. Moreover

$$\int_{\gamma} \frac{P_2(t)}{t^3}\left(\overline{\frac{1-w_2(t)}{1-\overline{s}_1 t}}\right)\Phi'(t)\frac{dt}{\zeta-t} = 0 , \ |\zeta| < 1 .$$

Taking this into consideration, one obtains, as during the derivation of equation (8.2.8), from (8.2.9) the integral equation for the function $\Phi'(t)$:

$$\Phi'(t) = \frac{k^{-1}}{2\pi i}\int_{\gamma} \frac{\partial}{\partial\tau}\left(\frac{\tau}{\tau-s_2}\cdot\frac{\theta(\tau)-\theta(t)}{\tau-t}\right)\frac{P_2(t)}{t^3}\overline{\frac{1-w_2(t)}{1-\overline{s}_1 t}}\Phi'(t) +$$

$$+ \frac{\partial}{\partial\tau}\left(\frac{c_0}{\tau-s_2} + \frac{\tau F(\tau)}{k(\tau-s_2)}\right) , \ \tau \in \gamma, \ \tau \ne \exp(i\beta_j), \ j = 1, 2 , \tag{8.2.10}$$

where the constant

$$c_0 = \frac{s_2 F(s_2)}{k} - \frac{s_2}{2\pi i k}\int_{\gamma}\left(\frac{\theta(s_2)-\theta(t)}{s_2-t}\right)\frac{P_2(t)}{t^3}\frac{1-\overline{w_2(t)}}{1-s_1\overline{t}}\overline{\Phi(t)}\, dt .$$

In an analogous manner to theorem 8.3, it may be proved that there exists no more than one solution of the integral equation (8.2.10).

An example of the manipulations involved in the proposed method will now be presented [62]. Consider the state of stress and strain for the external region, shown in Fig. 17, under the action of an internal, uniform, normal unit pressure. The function $\omega(t)$

$$\omega(t) = \frac{2\sqrt{1 - b^2\frac{(t-1)^2}{(t+1)^2}}}{d - \sqrt{1 + b^2\frac{(t-1)^2}{(t+1)^2}}} \, , \quad d = \frac{2+l}{l} \, , \quad b = \sqrt{d^2 - 1} \, , \qquad (8.2.11)$$

maps the unit circle conformally (Fig. 17.b) onto the region shown in Fig. 17a. One obtains the stress concentration factor K_I at the end of the crack (Fig. 17c). The results agree with the known solution for $l \gg 1$ and $l \ll 1$ (cf., for example, [57]).

§3. Problem of a crack in a layered medium

Consider an elastic medium consisting of several layers with different elastic properties, which has a semi-infinite crack, or an angular cut (Fig. 18). On the boundary of the crack, there are given conditions in terms of displacements, stresses or both, so that there occurs in the region plane strain or anti-plane shear (cf. [48,49]).

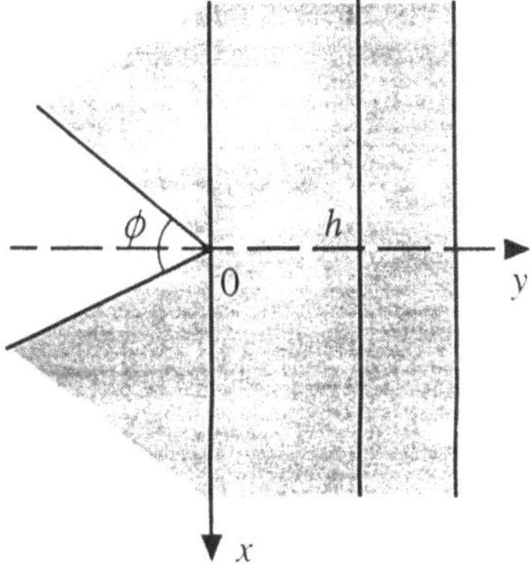

Fig. 18

After application of the Fourier transformation (with respect to x in the region $y > 0$) and the Mellin transformation (for $y < 0$ in each angle) and their subsequent joining at the boundary $y = 0$, the stated problems of the theory of elasticity reduce to a system of integral equations on the semi-axis $(x > 0, y = 0)$ of the form

$$(Cg)(x) \equiv g(x) + \int_0^\infty K(x,\tau)\,\Psi(\xi,\tau)\,L(x,\tau)\,g(\tau)\,d\tau = f(x)\,, \qquad (8.3.1)$$

where K and L are measurable, bounded on R_+^2 matrix functions on order n, $\Psi(x,\tau)$ is a homogeneous, degree -1 matrix function of the same dimensionality, $\Psi(x,\tau) = \lambda^{-1}\Psi(\lambda x, \lambda \tau)$, $\lambda > 0$, satisfying for certain $\alpha, \beta \in R$ and any $i,j = 1, 2, ..., n$ the conditions:

$$\int_0^\infty |\Psi_{ij}(x,1)|\rho_{\alpha,\beta}(x)\,\frac{dx}{x} < \infty\,, \quad \rho_{\alpha,\beta}(x) = \begin{cases} x^\alpha,\ 0 < x < 1,\\ x^\beta,\ 1 < x < \infty. \end{cases} \qquad (8.3.2)$$

It is natural to consider the operator C in the space of vector-functions $L_{p,\rho}(R_+^1)$, $\rho = \rho_{\alpha,\beta}$ with norm

$$\|u\|_{p,\rho}^\rho = \sum_{j=1}^n \int_0^\infty |u_j(x)|^p \rho_{\alpha,\beta}^n(x) x^{-1}\,dx\,, \quad 1 \le p < \infty.$$

The idea of joining of the Fourier and Mellin transforms will be demonstrated by the example of the simplest problem of anti-plane deformation.

According to the formulation of the problem of anti-plane deformation, the non-zero stress tensor components σ_{xz} and σ_{yz} are linked to the displacement $u = u_z$ in the following manner:

$$\sigma_{xz} = \mu\frac{\partial u}{\partial x}\,, \ \sigma_{yz} = \mu\frac{\partial u}{\partial y}\,,$$

when the compatibility conditions are satisfied identically and the equilibrium equation in the absence of body forces assumes the form:

$$\frac{\partial}{\partial x}\sigma_{xz} + \frac{\partial}{\partial y}\sigma_{yz} = 0\,.$$

Thus, the function $u = u_z$ which decreases towards infinity satisfies the equation

$$\Delta u = 0,\ (x,y) \in \Omega_1 \cup \Omega_0^+ \cup \Omega_0^-\,, \qquad (8.3.3)$$

where Ω_1 is the elastic strip with shear modulus μ_1 and $\Omega_0^{\pm}$ is the elastic wedge with shear modulus μ_0, symmetrically located with respect to the y-axis and rigidly joined to the strip Ω_1.

On the boundary elastic strip, glued to the wedge, one has the boundary conditions of a rigid joint

$$u^{(0)} = u^{(1)}, \quad \mu_0 \frac{\partial u^{(0)}}{\partial y} = \mu_0 \frac{\partial u^{(1)}}{\partial y}, \quad (x, y) \in \Gamma_0, \qquad (8.3.4)$$

where

$$u^{(0)} = u\big|_{\Omega_0^{\pm}}, \quad u^{(1)} = u\big|_{\Omega_1}, \quad \Gamma_0 = \{(x, y)\}: y = 0\}.$$

Applying to (8.3.3) in corresponding regions the Fourier transform

$$(Ff)(\lambda) = (2\pi)^{-1} \int_{-\infty}^{\infty} f(x)e^{i\lambda x}\, dx$$

and Mellin transform

$$f(x) = \int_{0}^{\infty} f(x)r^{s-1}\, dr, \quad x = re^{i\theta}$$

one arrives at the following representation (cf, [71]):

$$Fu^{(1)}(\lambda, y) = c_1^{(1)}(\lambda)e^{|\lambda|y} + c_2^{(1)}(\lambda)e^{-|\lambda|y},$$
$$\tilde{u}^{(0)}(s, \theta) = A(s)\cos\theta s + B(s)\sin\theta s. \qquad (8.3.5)$$

If symmetric loading is specified, then this means, under conditions of anti-plane deformation, that $u(-x, y) = -u(x, y)$. Therefore the functions $c_1^{(1)}(\lambda)$ and $c_2^{(1)}(\lambda)$ are purely imaginary and odd in λ, i.e., inverting the Fourier transform, one find from (8.3.5)

$$u^{(1)}(x, y) = \int_{-\infty}^{\infty} \left(c_1^{(1)}(\lambda)e^{|\lambda|y} + c_2^{(1)}(\lambda)e^{-|\lambda|y} \right) e^{-i\lambda x}\, d\lambda =$$

$$= -2i \int_{-\infty}^{\infty} \left(c_1^{(1)}(\lambda)e^{|\lambda|y} + c_2^{(1)}(\lambda)e^{-|\lambda|y} \right) \sin\lambda x\, d\lambda,$$

$$\frac{\partial u^{(1)}}{\partial y} = -2i \int_{-\infty}^{\infty} \left(c_1^{(1)}(\lambda)e^{|\lambda|y} + c_2^{(1)}(\lambda)e^{-|\lambda|y} \right) \lambda \sin\lambda s\, d\lambda.$$

Next, conditions (8.3.4) will be satisfied along the boundary Γ_0:

$$u^{(0)}(x,0) = -2i \int\limits_{-\infty}^{\infty} \left(c_1^{(1)}(\lambda) + c_2^{(1)}(\lambda) \right) \sin \lambda x \, d\lambda \, , \quad x > 0 \, ,$$

$$\mu_0 \frac{\partial u^{(0)}}{\partial y}(x,0) = \mu_0 x^{-1} \frac{\partial u^{(0)}}{\partial \theta}(x,0) = \tag{8.3.6}$$

$$= \mu_1 (-2i) \int\limits_{0}^{\infty} \left(c_1^{(1)}(\lambda) - c_2^{(1)}(\lambda) \right) \lambda \sin \lambda x \, d\lambda \, , \quad x > 0 \, .$$

Applying to both sides of (8.3.6) the Mellin transform, one finds

$$A(s) = -2i \int\limits_{0}^{\infty} \left(\int\limits_{0}^{\infty} \left[c_1^{(1)}(\lambda) + c_2^{(1)}(\lambda) \right] \sin \lambda r \, d\lambda \right) r^{s-1} \, dr \, , \quad r > 0 \, ,$$

$$sB(s) = -2i \frac{\mu_1}{\mu_0} \int\limits_{0}^{\infty} \left(\int\limits_{0}^{\infty} \left[c_1^{(1)}(\lambda) - c_2^{(1)}(\lambda) \right] \sin \lambda r \, d\lambda \right) r^{s} \, dr \, , \quad r > 0 \, . \tag{8.3.7}$$

The interchange of the order of the integrations on the right hand side may be based on (8.3.7). Using from an integral table the result

$$\int\limits_{0}^{\infty} r^{s-1} \sin \lambda r \, dr = \Gamma(s) \lambda^{-s} \sin \frac{\pi s}{2} \, , \quad |\mathrm{Re} s| < 1 \, ,$$

the expressions of (8.3.7) may be transformed to

$$A(s) = -2i \Gamma(s) \sin \frac{\pi s}{2} \int\limits_{0}^{\infty} \left[c_1^{(1)}(\lambda) + c_2^{(1)}(\lambda) \right] \lambda^{-s} \, d\lambda \, ,$$

$$sB(s) = -2i \frac{\mu_1}{\mu_0} \Gamma(s+1) \cos \frac{\pi s}{2} \int\limits_{0}^{\infty} \left[c_1^{(1)}(\lambda) - c_2^{(1)}(\lambda) \right] \lambda^{-s} \, d\lambda \, , \tag{8.3.8}$$

These equations are fulfilled for s, lying in some strip of the complex plane which contains the imaginary axis. The transformation on the right hand side, acting on the functions $c_1^{(1)}(\lambda) \pm c_2^{(1)}(\lambda)$ are easily expressed in terms of

Mellin transforms, hence (8.3.8) is easily inverted, expressing the functions $c_1^{(1)}(\lambda)$ and $c_2^{(1)}(\lambda)$ in terms of $A(s)$ and $B(s)$.

The loading (or displacements) specified on the surfaces of the wedge link linearly the functions $A(s)$ and $B(s)$ in the following manner

$$A(s)\,p(s) + B(s)\,q(s) = g(s) \tag{8.3.9}$$

to the known functions p, q and g. Note that a presence of an arbitrary number of elastic strips, interlinked rigidly and linked to the strip Ω_1, does not principally complicate the problem under consideration. Thus, when the surfaces of the external strips are unloaded, one easily establishes the link:

$$c_2^{(1)}(\lambda) + H(\lambda)\,c_1^{(1)}(\lambda) = 0\,, \tag{8.3.10}$$

where $H(\lambda)$ depends on the elastic constants and the thicknesses of all the strips.

Substituting the expressions for $A(s)$ and $B(s)$ into equations (8.3.9) and taking relations (8.3.10) into account, one obtains the integral equation in terms of the function $c_1^{(1)}(\lambda)$ which is readily given the form (8.3.1). In the case under consideration

$$(Cg)(\lambda) \equiv g(\lambda) + \mu\pi^{-1} \int\limits_0^\infty K(\lambda,\xi)\,\Psi(\lambda,\xi)\,g(\xi)\,d\xi\,, \quad \lambda > 0\,,$$

$$\Psi(\lambda,\xi) = \frac{1}{2\pi i \xi} \int\limits_{-i\infty}^{i\infty} (\xi^{-1}\lambda)^s \frac{ds}{\kappa - \cos\pi s}\,, \quad \lambda,\xi > 0\,, \tag{8.3.11}$$

$$\kappa = \frac{\mu_0 - \mu_1}{\mu_0 + \mu_1}\,, \quad \mu = \frac{(H(0) - 1)(1 - \kappa)\mu_0}{\mu_1 + \mu_0 H(0)}\,, \quad g(\lambda) = \lambda^{1/2} c_1^{(1)}(\lambda)\,,$$

$$K(\lambda,\xi) = \frac{(H(\xi) - 1)(\mu_1 + H(0)\mu_0)}{(H(0) - 1)(\mu_1 + H(\lambda)\mu_0)}\,, \quad \lambda,\xi > 0\,.$$

Note that

$$K(0,0) = 1\,, \quad |K(\lambda,\xi)| \le M \exp(-2\xi h_1)\,, \quad \lambda,\xi > 0\,, \tag{8.3.12}$$

where M is constant and h_1 is the thickness of the strip Ω_1.

Let $H_{\alpha,\beta}(R_+^1)$ denote the space of functions $L_{1,\rho}(R_+^1)$ with weight

$$\rho(x) = \begin{cases} x^{\alpha+1}\,, & 0 < x < 1\,, \\ x^{-\beta+1}\,, & 1 < x < \infty\,. \end{cases}$$

In the case of cracks, one has

$$\Psi(\lambda,\xi) = \frac{1}{\pi \sin \pi\nu^*}\,(\lambda/\xi)^{-\nu^*}\,\frac{(\lambda/\xi)^{2\nu^*-2}-1}{(\lambda/\xi)^{-2}-1}\,,\quad \nu^* = \frac{1}{\pi}\arccos\kappa\,.$$

It is easily verified that for $-\nu^* + 1/2 < \beta < \nu^* + 1/2$, one has

$$\Psi(1,\xi) \in H_{-\beta,\beta}(R_+^1)\,. \tag{8.3.13}$$

This establishes a direct verification of the behaviour of the function Ψ at the origin and at infinity. Correspondingly, the operator

$$(Tg)(\lambda) = \frac{1}{\pi}\int_0^\infty K(\lambda,\xi)\,\Psi(\lambda,\xi)\,g(\xi)\,d\xi\,,$$

may be viewed in the space $H_{-1/2,\beta}(R_+^1)$, $-1/2 \le -\beta < -1/2+\nu^*$. For $r > 0$, the class $H_{\alpha,\beta}(0,r) = H_\alpha(0,r)$ is considered with the norm $\|\cdot\|_{-\alpha,\alpha} = \|\cdot\|_\alpha$.

Lemma 8.2. The operator T is in the space $H_{-1/2,\beta}(R_+^1)$ for $-1/2 \le -\beta < -1/2 + \nu^*$.

Proof. Using Estimate (8.3.13), one obtains

$$\|Tg\|_{-1/2,\beta} \le \frac{M}{\pi}\int_0^\infty |g(\xi)|\exp(-2\xi h_1)\Big(\frac{1+\xi}{\xi}\Big)^{1/2}\frac{1}{(1+\xi)^\beta}$$

$$\int_0^\infty \Big(\frac{\xi}{\lambda}\Big)^{1/2}\Big(\frac{1+\lambda}{1+\xi}\Big)^{1/2-\beta}|\Psi(\lambda,\xi)|\,d\xi\,d\lambda\,.$$

Estimate the expression

$$\int_0^\infty \Big(\frac{\xi}{\lambda}\Big)^{1/2}\Big(\frac{1+\lambda}{1+\xi}\Big)^{1/2-\beta}|\Psi(\lambda,\xi)|\,d\lambda = \int_0^\infty t^{-1/2}\Big(\frac{1+t\xi}{1+\xi}\Big)^{1/2-\beta}|\Psi(1,t)|\,dt \le$$

$$\le \int_0^\infty t^{-1/2}(\max\{1,t^{1/2-\beta}\})\,|\Psi(1,t)|\,dt\,,$$

which is bounded by (8.3.13), hence the lemma is proved.

It is seen from (8.3.13) and the proof of the lemma that the operator

$$T\colon H_{-\alpha,\beta}(R_+^1) \to H_{-\alpha,\beta}(R_+^1), \quad 1/2 - \nu^* < \beta \le \alpha < \nu^* + 1/2,$$

is bounded.

Denote by $P_r\colon H_{-1/2,\beta}(R_+^1) \to H_{-1/2,\beta}(R_+^1)$ the multiplication operator on the characteristic function of the manifold $[o, r]$. The following relation is true:

$$\|T - P_r T P_r\| = o(1), \quad r \to \infty.$$

In fact, estimate the integral $(r > 1)$

$$\int\limits_r^\infty \left(\frac{\lambda+1}{\lambda}\right)^{1/2} \frac{1}{(1+\lambda)^\beta} \left| \int\limits_0^r K(\lambda,\xi)\,\Psi(\lambda,\xi)\,g(\xi)\,d\xi \right| d\lambda \le$$

$$\le \frac{\sqrt{2}\,M}{\sin\pi\nu^*} \int\limits_0^r |g(\xi)| \exp(-2\xi h_1) \int\limits_0^\infty \frac{1}{(1+\lambda)^\beta} \frac{\lambda^{2-2\nu^*} - \xi^{2-2\nu^*}}{(\lambda\xi)^{1/2-\nu^*}(\lambda^2 - \xi^2)}\, d\lambda\, d\xi \le$$

$$\le \frac{\sqrt{2}\,M}{\sin\pi\nu^*} \max_{\xi\to 0}(\xi^{\nu^*} \exp(-2\xi h_1)) \int\limits_r^\infty \frac{1}{(1+\lambda)^\beta} \frac{d\lambda}{\lambda^{1/2-\nu^*}} \|g\|_{-1/2,\beta}\,,$$

where we have been used the inequality

$$\lambda^{2-2\nu^*} - \xi^{2-2\nu^*} \le \lambda^{-2\nu^*}(\lambda^2 - \xi^2), \quad \lambda \ge \xi > 0.$$

Since $\beta > 1/2 - \nu^*$, the required result is obtained.

Note now that the equation

$$g(\lambda) + \mu P_r T P_r g(\lambda) = f(\lambda)$$

in the space $H_{-1/2,\beta}(R_+^1)$ is equivalent to the equation

$$g + \mu P_r T P_r g = P_r f = f_r$$

in the space $H_{1/2}(o, r)$. The latter, after introduction of the variables $\lambda = rt$, $\xi = rs$, is reduced to the form

$$a(t) + \frac{\mu}{\pi}\left\{ \int\limits_0^1 \Psi(t,s)\,a(s)\,ds + (K_r a)(t) \right\} = f_r(t), \quad 0 < t \le 1. \qquad (8.3.14)$$

Verify now that the integral operator

$$(K_r a)(t) = \int_0^1 [K(rt, rs) - 1]\Psi(t, s)a(s)\,ds$$

is completely continuous in the space $H_{1/2}(0, 1)$.

First of all, note that for $\xi, \lambda > 0$, one has the inequality

$$\left|\xi^{2-2\nu^*} - \lambda^{2-2\nu^*}\right| \leq \left|\xi^2 - \lambda^2\right|\min\{\xi^{-2\nu^*}, \lambda^{-2\nu^*}\}. \qquad (8.3.15)$$

In order to prove the complete continuity of the operator K_r it is sufficient to verify that the manifold of functions

$$\frac{1}{\sqrt{t}}\int_0^1 [K(rt, rs) - 1]\,\Psi(t, s)\,a(s)\,ds\,, \quad \|a\|_{1/2} \leq 1\,.$$

is completely continuous in $L_1(0, 1)$.

One has for $h > 0$ that

$$\left|\int_0^1 \left(\frac{1}{\sqrt{t}} - \frac{1}{\sqrt{t+h}}\right)\int_0^1 [K(rt, rs) - 1]\Psi(t, s)\,a(s)\,ds\,dt\right| \leq$$

$$\leq \int_0^1 |a(s)|\int_0^1 \left(\frac{1}{\sqrt{t}} - \frac{1}{\sqrt{t+h}}\right)|[K(rt, rs) - 1]| \times$$

$$\times \min\{s^{-1/2-\nu^*}t^{\nu^*-1/2}, t^{-1/2-\nu^*}s^{\nu^*-1/2}\}\,dt\,ds \leq$$

$$\leq \max_{0 < t, s < 1}\{s^{-1}|[K(rt, rs) - 1]|\}J\,,$$

$$J = \int_0^1 |a(s)|\int_0^1 \left(\frac{1}{\sqrt{t}} - \frac{1}{\sqrt{t+h}}\right)s^{1/2-\nu^*}\,t^{\nu^*-1/2}\,dt\,ds\,.$$

Obviously, the case $\nu^* \geq 1/2$ leads to the zero estimate; consider therefore the case $\nu^* < 1/2$:

$$J \leq \int_0^1 [t^{\nu^*-1} - (t+h)^{\nu^*-1}]\,dt\,.$$

Thus, $J = 0(1 - (1 + h)^{\nu^*} + h^{\nu^*})$ for $h \to o$.

One still has to obtain estimates for the expression

$$\int\limits_0^1 \frac{1}{\sqrt{t}} \left| \int\limits_0^1 |[K(rt, rs) - 1]| \, |\Psi(t + h, s) - \Psi(t, s)| \, |a(s)| \, ds \right| \, dt \le$$

$$\le \max_{0 < t, s \le 1} \{ s^{-1} |[K(rt, rs) - 1]| \} \int\limits_0^1 \frac{|a(s)|}{\sqrt{s}} \, s \times$$

$$\times \int\limits_0^{s^{-1}} |\Psi(\xi + hs^{-1}, 1) - \Psi(\xi, 1)| \frac{d\xi}{\sqrt{\xi}} \, ds \, .$$

The inconvenience of the estimate of the right hand side of this inequality is linked to the necessity attending simultaneously to the compensation of the singularities at $\xi = 1$, at the origin and at infinity. Therefore, deal first of all with the neighbourhood of the point $\xi = 1$. One has

$$\int\limits_0^1 \frac{|a(s)|}{\sqrt{s}} \, s \int\limits_{1-h^\varepsilon}^{1+h^\varepsilon} |\Psi(\xi + hs^{-1}, 1) - \Psi(\xi, 1)| \frac{d\xi}{\sqrt{\xi}} \, ds \le$$

$$\le \int\limits_0^1 \frac{|a(s)|}{\sqrt{s}} \, s \int\limits_{1-h^\varepsilon}^{1+h^\varepsilon} \left(|\Psi(\xi + hs^{-1}, 1)| + |\Psi(\xi, 1)| \right) \frac{d\xi}{\sqrt{\xi}} \, ds \, , \quad \varepsilon = 2^{-m} \, .$$

In order to arrive at the required estimate, use the definition of the function $\Psi(\lambda, \xi)$ and the inequality (8.3.15).

Estimate the remaining integral

$$\int\limits_0^1 \frac{|a(s)|}{\sqrt{s}} \, s \left(\int\limits_0^{1-h^\varepsilon} + \int\limits_{1+h^\varepsilon}^{s^{-1}} \right) |\Psi(\xi + hs^{-1}, 1) - \Psi(\xi, 1)| \frac{d\xi}{\sqrt{\xi}} \, ds = J_1 \, ,$$

where either $\xi \le 1 - h^\varepsilon$, or $\xi \ge 1 + h^\varepsilon$. Multiplying the first integral by $1 - (1 - h^\varepsilon)^{1/2} \le 1 - \sqrt{\xi}$ and dividing the second by $1 - (1 + h^\varepsilon) \le 1 - \sqrt{\xi}$, cancel the numerator and the denominator of the integrand by the factor $|1 - \sqrt{\xi}|$, to obtain

$$
J_1 \le \int_0^1 \frac{|a(s)|}{\sqrt{s}} \int_0^{s^{-1}} \left| \frac{(\xi + hs^{-1})^{2-2\nu^*} - 1}{(\xi + hs^{-1})^{1/2-\nu^*}(\xi + hs^{-1} + 1)(\sqrt{\xi + hs^{-1}} + 1)} \right.
$$
$$
\left. - \frac{\xi^{2-2\nu^*} - 1}{\xi^{1/2-\nu^*}(\xi + 1)(\sqrt{\xi} + 1)} \right| \frac{d\xi}{\sqrt{\xi}} \, ds \, .
$$

After reduction to a common denominator and grouping of the terms of the same type, use the simple relation

$$
(\xi + hs^{-1})^p - \xi^p = p \int_0^{hs^{-1}} (\xi + t)^{p-1} \, dt \, , \ 0 < p < 1 \, .
$$

For $p < 1$, it is sufficient to employ a rougher estimate. For a typical term, one finds

$$
\int_0^1 \frac{|a(s)|}{\sqrt{s}} s \int_0^{s^{-1}} \left|(\xi + hs^{-1})^p - \xi^p\right| \frac{d\xi \, ds}{(\xi + 1)\sqrt{\xi}} \le
$$
$$
\le p \int_0^1 \frac{|a(s)|}{\sqrt{s}} s \int_0^{s^{-1}} \frac{d\xi \, dt \, ds}{(\xi + 1)(\xi + t)^{1-p}\sqrt{\xi}} \le
$$
$$
\le p \int_0^1 \frac{|a(s)|}{\sqrt{s}} s \int_0^{hs^{-1}} \frac{dt}{t^{1-p}} \int_0^\infty \frac{d\xi}{(\xi + 1)\sqrt{\xi}} \, ds = 0(h^p) \, , \ h \to o \, .
$$

The final estimate worsens by order $\varepsilon = 2^{-m}$, however, selecting the integer m sufficiently large, one finds $J_1 = 0(h^\delta)$, $h \to 0$, $\delta > 0$, which completes the proof of the complete continuity of the operator K_r.

Consider in the space $H_{1/2}(0,1)$, the equation

$$
a(t) + \frac{\mu}{\pi} \int_0^1 \Psi(t, s) \, a(s) \, ds = f_r(t), \ 0 < t < 1 \, , \tag{8.3.16}
$$

the left hand side of which coincides with the principal part of equation (8.3.14). After substitution of the variables $t = \exp(-x)$, $s = \exp(-y)$, this equation becomes an equation on the half-axis with a difference kernel to which the results of ([14], ref. part I) may be applied. A basic role in the theory of such equations is played by the function

$$1 + \mu N(t), \quad N(t) = \frac{1}{\operatorname{ch}\pi t - \chi_0}.$$

Obviously, for $\mu > \chi_0 - 1$, the equation $1 + \mu N(t) = 0$ does not have a root on the real axis. Since for sufficiently small $|\mu|$ the equation has a unique solution in the class $H_{1/2}(0,1)$, then, for $\mu > x_0 - 1$, it has a unique solution in the space $H_{1/2}(0,1)$. For $\mu < \chi_0 - 1$, the complex roots of the equation $1 + \mu N(z) = 0$ have the form

$$z_k^{\pm} = \frac{1}{\pi}\ln(\chi_0 + \mu \pm \sqrt{(\chi_0 + \mu)^2 - 1}) + 2ik, \quad k \in Z.$$

Consequently, on the straight lines $\operatorname{Im}z = \pm\nu^*$, the characteristic equation has no roots, and, by theorem 15.4 ([14], ref. part I), the homogeneous equation

$$a(t) + \frac{\mu}{\pi}\int_0^1 \Psi(t,s)\, a(s)\, ds = 0 \tag{8.3.17}$$

has exactly the single non-trivial solution

$$a_0(t) = d_1 t^{-1/2}\sin(z_0^+\ln\frac{1}{t} + d_2) + o(t^{-1/2+\nu^*}), \quad t \to o.$$

Finally, if $\mu = \chi_0 - 1$, then the unique solution of equation (8.3.17) has the form

$$a_0(t) = d_1 t^{-1/2}\ln\frac{1}{t} + d_2 t^{-1/2} + o(t^{-1/2-\nu^*}), \quad t \to o.$$

Theorem 8.4. For $\mu > \chi_0 - 1$, equation (8.3.14) is Noetherian with index zero in the space $H_{-1/2,\beta}\ (R_+^1)$.

In this case $\mu \leq \chi_0 - 1$, the situation changes: Equation (8.3.16) does not, in general, have a solution in the space $H_{1/2}(0,1)$. It is soluble in the space $H_{1/2-\varepsilon}(0,1) \supset H_{1/2}(0,1)$, $0 < \varepsilon < \nu^*$, with index equal to unity, for almost all $\mu \leq \chi_0 - 1$. The compactness of the operator K_r in the space $H_{1/2-\varepsilon}(0,1)$ for $0 \leq \varepsilon < \nu^*$ leads to the assertion of

Theorem 8.5. For almost all $\mu \leq \chi_0 - 1$, the index of equation (8.3.14) in the space $H_{-1/2+\varepsilon,\beta}(R_+^1)$, $0 < \varepsilon < \nu^*$, $1/2-\nu^* < \beta \leq 1/2-\varepsilon$ equals unity.

Consider now the asymptotic behaviour of the stresses $\sigma_{xz}^{(1)}$ and $\sigma_{yz}^{(1)}$ in the neighbourhood of zero:

$$\sigma_{xz}^{(1)} \sim K_{III}\frac{\cos(1-\nu^*)\theta}{r^{1-\nu^*}}\,,\ \ \sigma_{yz}^{(1)} \sim -K_{III}\frac{\sin(1-\nu^*)\theta}{r^{1-\nu^*}}\,,\ \ \theta = \frac{\pi}{2} - \varphi_0\,.$$

These formulae generalize the known relations for the asymptotic of the principal terms of the stresses, obtained for homogeneous media.

Examples of the evaluation of the concentration factor K_{III} in the problem of anti-plane bending of media are shown in Fig. 19. (cf., the table). The computations have been performed by the projection method; at each step, the integral operator (8.3.11) has been discretized on the segment $[\varepsilon, 1/\varepsilon]$ by Simpson's rule with a logarithmic scale. The calculations have been controlled in the norm of the operator C and the relative error of the stress concentration factor.

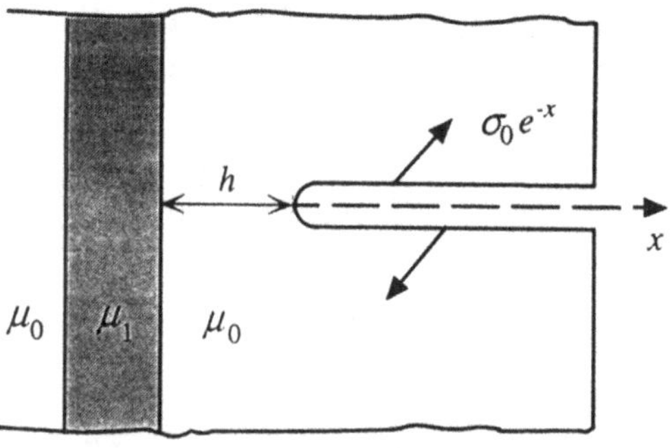

Fig. 19

h	$\|\mathbb{C}\|_{1,0,0}$	$\frac{1}{2\sqrt{\pi}\sigma_0}K_{III}$	$\frac{\Delta K_{III}}{K_{III}}$	$1/\varepsilon$	$t,$ min
		$\mu_1 = 40\mu_0$			
10	0.5	0.159174	0.0015	$3.2 \cdot 10^4$	12
1	0.5	0.160366	0.0032	$3.2 \cdot 10^4$	11.5
0.1	0.5	0.17925	0.022	$3.2 \cdot 10^4$	12
0.01	0.5	0.27387	0.017	$6.4 \cdot 10^4$	14
		$\mu_0 = 40\mu_1$			
10	0.5	0.15913	0.0019	$3.2 \cdot 10^4$	11
1	0.5	0.157964	0.0015	$3.2 \cdot 10^4$	11
0.1	0.5	0.1419677	0.0019	$1.6 \cdot 10^4$	9.5
0.02	0.5	0.095	$5 \cdot 10^{-3}$	10^4	9

Fig. (19)

§4. The problem of elastic inclusions with sharp points

Consider the problem of torsion of an elastic cylinder with shear modulus μ_-, which is reinforced by an elastic rod with shear modulus μ_+ and the cross-section of which is the simply connected region G^+ with the sectionally smooth boundary Γ (Fig. 20). The last property indicates that Γ is a rectified arc of length $|\Gamma|$, where, if $t = t(s)$, $0 \le s \le |\Gamma|$, is its equation in arc coordinates, then $t'(s)$ is sectionally continuous with points of discontinuity of order 1.

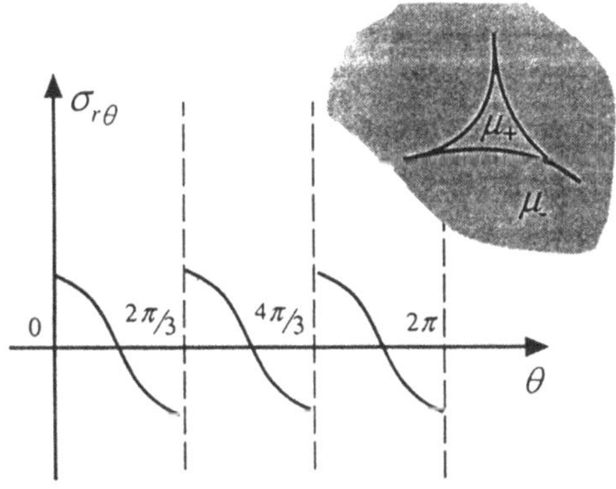

Fig. 20

For the sake of simplicity, assume that the diameter of G^+ is much less than the diameter of the cylinder, so that one will consider the elastic inclusion G^+ in the complex plane.

To begin with, the torsion function F will be found, i.e., the sectionally analytic function with a line of jumps Γ which tends to zero at infinity and satisfies on the boundary Γ the boundary conditions (cf. [58]):

$$F_+(t) + \overline{F_+(t)} = F_-(t) + \overline{F_-(t)}, \tag{8.4.1}$$

$$\mu_+[F_+(t) - \overline{F_+(t)}] - \mu_-[F_-(t) - \overline{F_-(t)}] = (\mu_+ - \mu_-)g(t), \tag{8.4.2}$$

where $g(t) = i|t|^2$. Condition (8.4.1) expresses the continuity of the displacements, condition (8.4.2) the equilibrium of the forces acting on the elements of the side surface of the reinforced bar.

Multiplying (8.4.1) by μ_+ and adding the result to (8.4.2), one finds the condition, equivalent to these two conditions,

$$\frac{2\mu_+}{(\mu_+ + \mu_-)} F_+(t) - F_-(t) - \lambda\overline{F_-(t)} = \lambda g(t) , \ t \in \Gamma , \tag{8.4.3}$$

where $\lambda = (\mu_+ - \mu_-)/(\mu_+ + \mu_-)$.

Multiplying both sides of (8.4.3) by $[2\pi i(t - \zeta)]^{-1}$ and integrating along the contour Γ, one obtains

$$F(\zeta) - \lambda \frac{1}{2\pi i} \int_\Gamma \frac{\overline{F_-(t)}\, dt}{t - \zeta} = \lambda g_1(\zeta)\,, \quad \zeta \notin \overline{G^+}\,, \tag{8.4.4}$$

where $g_1(\zeta)$ is a known function, and the analyticity of $F(\zeta)$ for $\zeta \notin \Gamma$ and the fact that $F(\zeta) \to o$ for $|\zeta| \to \infty$ have been used.

Performing the limit $\zeta \to t \in \Gamma$, one finds, on the basis of the Sokhotskii-Plemelj formulas

$$F_-(t) - \lambda(-\overline{F_-(t)} + (S\overline{F_-})(t))/2 = \lambda g_1(t)\,, \tag{8.4.5}$$

where $S = S_\Gamma$ is a Cauchy singular integral operator (cf. chapter 3).

Prior to a study of the integral equation (8.4.5), the basic properties of the operator S will be investigated which hold for arbitrarily sectionally smooth curves (cf. [17]). The operator S acts in the space $L_p(\Gamma)$, $1 < p < \infty$:

$$\int_\Gamma (S\varphi)f\, dt = -\int_\Gamma \varphi(Sf)\, dt\,,$$

$$S(fS\varphi) = f\varphi + (Sf)(S\varphi) + S(\varphi Sf)\,, \tag{8.4.6}$$

where $\varphi \in L_p(\Gamma)$, $f \in L_p(\Gamma)$, $p^{-1} + q^{-1} = 1$, $p > 1$. In particular, for closed piecewise smooth contour Γ the Poincaré-Bertrand formula $S^2\varphi = \varphi$ is true.

Theorem 8.6. Integral equation (8.4.5) has not more than one solution in $L_p(\Gamma)$, $p > 1$.

Proof. Let $G \in L_p(\Gamma)$ be a solution of the homogeneous equation, corresponding to (8.4.5):

$$G(t) - \lambda(-\overline{G(t)} + (S\overline{G}))/2 = 0\,. \tag{8.4.7}$$

Define the analytic function $F(\zeta)$ in $C/\overline{G^+}$ by the formula

$$F(\zeta) = \frac{\lambda}{2\pi i} \int_\Gamma \frac{\overline{G(t)}}{t - \zeta}\, dt\,, \quad \zeta \notin \overline{G^+}\,,$$

then, due to (8.4.7), we have $F_-(t) = G(t)$ for almost all $t \in \Gamma$. Then for $\zeta \in G^+$ we define the function $F(\zeta)$ by the formula

$$F(\zeta) = \lambda \frac{(\mu_+ + \mu_-)}{2\mu_+} \frac{1}{2\pi i} \int_\Gamma \frac{\overline{G(t)}\, dt}{t - \zeta}, \quad \zeta \in G^+.$$

Thus,

$$F_+(t) = \frac{(\mu_+ + \mu_-)}{2\mu_+}(\lambda\overline{G(t)} + \lambda(S\overline{G})(t))\frac{1}{2} = \frac{(\mu_+ + \mu_-)}{2\mu_+}(G(t) + \lambda\overline{G(t)})$$

for almost all $t \in \Gamma$ and $F(\zeta)$ is a torsion function, satisfying the homogeneous boundary condition

$$\frac{2\mu_+}{(\mu_+ + \mu_-)}\, F_+(t) - F_-(t) - \lambda\overline{F_-(t)} = 0, \ t \in \Gamma.$$

Problems I and II of the theory of elasticity admit the equivalent formulation in terms of harmonic functions $u = \mathrm{Re}F$:

$$\begin{cases} \Delta u = 0, \ \zeta \notin \Gamma, \\ u^+ = u^-, \ \zeta \in \Gamma, \\ (1 + \lambda)\dfrac{\partial u^+}{\partial \nu} - (1 - \lambda)\dfrac{\partial u^-}{\partial \nu} = \lambda g_2, \ \zeta \in \Gamma, \end{cases} \tag{8.4.8}$$

where $g_2(t) = g'(t)$, $t \in \Gamma$ and $|\nabla u| = 0(|\zeta|^{-2})$ for $|\zeta| \to \infty$. The, corresponding to (8.4.8), quadratic form (energy of the deformed elastic cylinder) may be written as

$$\mu_+ \int_{G^+} |\nabla u|^2 dG^+ + \mu_- \int_{G^-} |\nabla u|^2 dG^- = \int_\Gamma \left(\mu_+\frac{\partial u^+}{\partial \nu} - \mu_-\frac{\partial u^-}{\partial \nu}\right) u\, d\Gamma.$$

when, for $g_2(t) = 0$, $t \in \Gamma$, $F(\zeta) = iC$, where C is a constant. However, in the case of the homogeneous problem (8.4.3), the function $g(t) = 0$, hence $F_-(t) = G(t) = 0$, $t \in \Gamma$, and theorem 8.6 has been proved.

Introducing the notation of the identity operator $If = f$, the adjoint complex operator $Vf = \overline{f}$ and setting $K^-f = (-f + Sf)/2$, equation (8.4.5) becomes

$$(I - \lambda K^- V)F_- = \lambda K^- g, \ |\lambda| < 1. \tag{8.4.9}$$

Its solution for $|\lambda|^{-1} > r_{K^-}$, (r_{K^-} is the spectral radius of the operator K^-) is given by the series

$$F_-(t) = \sum_{n=0}^{\infty} \lambda^{n+1}(K^-V)^n(K^-g)(t), \qquad (8.4.10)$$

which converges in Banach function space an Γ, where
1. the operator K^- acts continuously,
2. the operator V acts continuously,
3. contains the function $g(t)$.

For example, the function space $H_\alpha(\Gamma)$, which satisfies on Γ a Hoelder condition with index $\alpha \in (0,1)$ with the ordinary norm

$$\|f\|_{H_\alpha(\Gamma)} = \max_{t \in \Gamma}|f(t)| + \sup_{t_1,t_2 \in \Gamma} |f(t_1) - f(t_2)|/|t_1 - t_2|^\alpha,$$

has the properties 1 - 3, hence follows the assertion of

Theorem 8.7. The boundary value F_- of the torsion function for problem (8.4.1) and (8.4.2) on any sectionally smooth contour Γ satisfies a Hoelder condition with index $\alpha \in (0,1)$ and represents the, uniformly convergent with respect to $t \in \Gamma$, series (8.4.10) for sufficiently small $|\lambda|$ (more exactly, for $|\lambda|^{-1} > r_{K^-V}$ in the space $H_\alpha(\Gamma)$).

Note 8.6. The operator $-K_-$ is a projector in $L_p(\Gamma)$, $p > 1$, in the sense that $(-K^-)^2 = -K^-$ (this follows from the Poincaré-Bertrand formula). Consequently, $K^-(L_p(\Gamma))$ is a closed subspace in $L_p(\Gamma)$, where the operator K^- is bounded. If for any $\mu_\pm > 0$ and any $g \in L_p(\Gamma)$ there exists a solution of problem (8.4.1) and (8.4.2) with $F_- \in L_p(\Gamma)$, then the number λ, $|\lambda| < 1$, is not a characteristic number of the operator K^-V in the space $K^-(L_p(\Gamma))$.

Basic interest attaches to the behaviour of the stresses in the neighbourhood of an angle of the curve Γ which are linked to the torsion function $F(z)$ by the formula

$$\sigma_{xz}(z) - i\sigma_{yz}(z) = \mu_-\kappa[F'(z) - i\bar{z}], \qquad (8.4.11)$$

with $z \in \Gamma^-$, κ is the torque. In order to obtain for all boundary values such a series of successive approximations, consider the condition which will ensure the truth of the formula

$$F'_-(t) = \sum_{n=0}^{\infty} \lambda^{n+1}[(K^-V)^n(K^-g)]'(t), \ t \in \Gamma. \qquad (8.4.12)$$

The differentiability of F_- along Γ and equation (8.4.12) follow, as is known, from the differentiability along Γ of the terms of the series in (8.4.12). Therefore, it will be proved that F_- enters in narrower sense than $H_\alpha(\Gamma)$, $\alpha \in (0,1)$ of the Banach space of smooth functions, firstly, satisfying conditions 1 - 3, where convergence in the norm of the sequence of the functions $\{f_n\}$ follows the uniform convergence of $\{f_n\}$ and $\{f_n'\}$ on Γ. For the proof ([21]) will now be essential the absence on the sectionally smooth contour of angular points, not being points of sharpness (point of rotation), and likewise sharpness with a low order of contact, as, for example, "logarithmic peaks", yielded in polar coordinates (ρ, θ) by the inequality $-c_1|\log \rho|^{-1} < \theta < c_2|\log \rho|^{-1}$, where c_1, c_2 are positive constants.

Formulate now the following particular case of the result [21], relating to the contour with sharpness degrees of order

$$\begin{cases} y_1 = c_1 x^{1+\varepsilon_1}, \\ y_2 = c_2 x^{1+\varepsilon_2}, \end{cases} \qquad (8.4.13)$$

where $x \geq 0$; $\varepsilon_1, \varepsilon_2 > 0$; $c_1, c_2 \geq 0$ are fixed for every sharpness number.

Theorem 8.8. Let Γ be a sectionally smooth contour without angular points. If every sharpness of the contour Γ in a local coordinate system with origin at the point of sharpness and a real axis, corresponding to the non-oriented tangent at this point, has a degree of order (8.4.13), then for sufficiently small $|\lambda|$ the derivative of the torsion function is continuous up to the border Γ, and its contraction on Γ is represented by the, uniformly convergent with respect to $t \in \Gamma$, series (8.4.12).

In particular, if the contour Γ satisfies the condition of theorem 8.8, then for all sufficiently small values of $|\lambda|$ the stresses are continuous up to the boundary and (8.4.11) holds in $G^- \cup \Gamma$.

Note 8.7. An analogous assertion holds also for the boundary values outside Γ.

Note 8.8. In plane problems of the theory of elasticity for regions with elastic, peaked inclusions, an absence of special stress features at points of sharpness has been established by asymptotic methods in [50].

As an example of the employment of these formulae, consider ([22]) a rod with hypotrochoidal section G^+, reinforced by a cylinder the diameter of which is much larger than the diameter of the region G^+. The boundary G^+ is the hypotrochoid Γ, i.e., the region of the unit circle $T = \{|\tau| = 1\}$ for the transformation $\varphi(\tau) = \tau + m\tau^{-m}$, $n = 1, 2, 3, \ldots$, $0 \leq m \leq n^{-1}$ conformal outside the unit circle. Denote by ψ the inverse mapping of the region onto the outside of the unit circle. For $m = n^{-1}$, the contour Γ has $n+1$ points of sharpness and satisfies the condition of theorem 8.8.

The zero approximation of the boundary value of F_- of the torsion function

$$\lambda(Sg)(t) = \lambda \lim_{G^- \ni z \to t} \frac{1}{2\pi i} \int_\Gamma \frac{i|w|^2}{w - z}\, dw =$$

$$= i\lambda \lim_{G^- \ni z \to t} \frac{1}{2\pi i} \int_T \frac{|\varphi(\tau)|^2 \varphi(\tau)}{\varphi(\tau) - z}\, d\tau =$$

$$= i\lambda \lim_{G^- \ni z \to t} \frac{1}{2\pi i} \int_T \frac{1 + m^2 + m\left(\tau^{n+1} + \tau^{-n-1}\right)}{\varphi(\tau) - z}\, \varphi'(\tau)\, d\tau =$$

$$= i\lambda\{1 + m^2 + m[S(\psi^{n+1}) + S(\psi^{-n-1})]\},\quad t \in \Gamma,$$

reduces to the evaluation of the integrals $S(\psi^\nu)$, $\nu = \pm 1, \pm 2, ...$, which are given on the basis of the formula

$$(S\psi^\nu)(t) = a_\nu(t) - \psi^\nu(t), \tag{8.4.14}$$

where $a_\nu(t)$ is the limit as $z \to t$ less the value of the function $f(w) = w^\nu \varphi'(w)/(z - \varphi(w))$ at infinity.

Thus, the zero approximation F_- equals

$$\lambda(Sg)(t) = i\lambda\{1 + m^2 + m[a_{n+1}(t) - \psi^{n+1}(t) - \psi^{-n-1}(t)]\}.$$

Applying to this expression the operators V and S and using again (8.4.14), one obtains the following term of the series

$$\lambda^2(SVSg)(t) = -i\lambda^2\{1 + m^2 + m[S\bar{a}_{n+1}(t) + \psi^{-n-1}(t) - a_{n+1}(t) + \psi^{n+1}(t)]\}.$$

For numerical purposes, it is convenient to transfer the function, entering into this expression, to the unit circle, setting $t = \varphi(\tau)$, $\tau \in T$. Then $\varphi^\nu(t) = \tau^\nu$, and the computation of the successive approximations reduces to the determination of the terms of the iterative sequence $a_{n+1}(\varphi(\tau))$, $[(SV)a_{n+1}](\varphi(\tau))$, $[(SV)^2(a_{n+1})](\varphi(\tau)),...$.

The successive approximations for the boundary values of the stresses likewise can be represented at functions $\tau \in T$, noting that $\psi'(t) = (1 -$

$mn\tau^{-n-1}$). Only the first two terms of the series will be given here:

$$\tau_{xz}^-(t) - i\tau_{yz}^-(t) =$$
$$= \mu_-\kappa[\lambda(Sg)'(\varphi(\tau)) + \lambda^2(SVSg)'(\varphi(\tau)) - i\varphi(\tau)] =$$
$$= i\mu_-\kappa\left\{\lambda m\left[a'_{n+1}(\varphi(\tau)) + (n+1)\frac{1-\tau^{2n+2}}{\tau^{n+2}-mn\tau}\right] - \right.$$
$$- \lambda^2 m\left[(S\bar{a}_{n+1})'(\varphi(\tau)) + \right.$$
$$\left.\left. + (n+1)\frac{\tau^{2n+1}-1}{\tau^{n+2}-mn\tau} - a'_{n+1}(\varphi(\tau))\right] - \varphi(\tau)\right\} + O(\lambda^3). \tag{8.4.15}$$

Finally, the formulas will be presented for the case $n = m^{-1} = 2$, when Γ is the hypercycloid with three sharp points. Then

$$a'_3(\varphi(\tau)) = 3\tau^2 + 3\tau^{-1} + 3/4\tau^{-4},$$

$$(S\bar{a}_3)'(\varphi(\tau)) = 1/32(39\tau^{-4} + 15/2\tau^{-7} + 3/4\tau^{-10})$$

and (8.4.15) assumes the form

$$\sigma_{xz}^-(t) - i\sigma_{yz}^-(t) =$$

$$= i\mu_-\kappa[3/8\lambda\tau^{-4} - 1/64\lambda^2(15\tau^{-4} + 15/2\tau^{-7} + 3/4\tau^{-10}) - \tau^{-1} - 1/2\tau^2] + O(\lambda^3).$$

The results of the computations are shown in Fig. 20 as curves of $\sigma_{xz}^-(\varphi(\tau))$, $\sigma_{yz}^-(\varphi(\tau))$, corresponding to $\lambda = 0,1; 0,4$ (for the sake of simplicity, $\mu_-\kappa = 1$).

§5. Space problems of the theory of cracks

The need to analyse the state of strain and stress of bodies with cracks has led to the appearance of a number of papers in which different models of three-dimensional problems are treated. In particular, the solutions of the basic problems of the theory of elasticity are constructed for spaces with disk-shaped cracks and with cracks in the form of half-planes [31, 72]. As regards the case of cracks, lying on arbitrary smooth surfaces, numerical results have been obtained by use of the classical equation of potential theory as well as of pseudo-differential (for example, hyper-singular) boundary integral equations [33,16]. Such equations have been analysed for the first time for the basic problems of the theory of elasticity for spaces with cracks of arbitrary shape by E. Stephan and M. Costabel [87, 101]. They considered an equation of the first kind when displacements (Dirichlet type problem) and a hyper-singular equation when stresses (Neumann type problem) were specified on the boundary of the crack. They also studied the asymptotic properties of the solution of the above mentioned equations near crack boundaries and constructed asymptotic, quasi-optimal boundary elements. A brief exposition of these problems is presented in this section.

Let Σ be a smooth, closed surface in R^3 space and Δ the surface of a crack with smooth edge γ, where $\Delta \subset \Sigma$. The functional class (Sobolev space) necessary for the formulation of the results will now be determined. For $s \geq 0$, let

$$H^s(R^3) = \{u \in L_2(R^3) : \|u\|_s^2 = \int\limits_{R^3} |Fu|^2 (1 + |x|^2)^{s/2} \, dx < \infty\},$$

where $(Fu)(x)$ is the Fourier transform of the function $u(x)$. Moreover, let $H^3(\Sigma)$ be the contraction of functions of the class $H^{s+1/2}(R^3)$ onto the surface Σ. For $s < 0$, this class is defined with the aid of the expression

$$< u, v >= \int\limits_{\Sigma} u(x)\, v(x)\, dx \, .$$

More exactly, for $s < 0$, the class $H^s(\Sigma)$ is this manifold of linear, continuous functionals, defined on $H^{-s}(\Sigma)$ with norm

$$\|u\|_s = \sup\{< u, v >: \|v\|_{-s} \leq 1\} \, .$$

This space is utilized for the definition of the classes of distributions on Δ; let

$$H^{\circ s}(\Delta) = \{u \in H^s(\Sigma) : \operatorname{supp} u \subset \Delta\},$$

$$H^s(\Delta) = \{u\big|_\Delta : u \in H^s(\Sigma)\} \, .$$

The basic problems of the theory of elasticity are considered in terms of fields of displacements $u = (u_1, u_2, u_3)$ of a homogeneous, isotropic, elastic medium $\Omega_\Delta = R^3/\overline{\Delta}$.

Problem II of the theory of elasticity reduces to finding the displacement field $u(x)$, satisfying the Lame equations

$$\mu \Delta u + (\lambda + \mu)\operatorname{grad} \operatorname{div} u = 0, \; x \in \Omega_\Delta, \tag{8.5.1}$$

and the boundary conditions

$$\sigma^{(n)} u\big|_{\Delta_+} = \Psi_+ , \; \sigma^{(n)} u\big|_{\Delta_-} = \Psi_- , \tag{8.5.2}$$

where

$$\Psi_+, \Psi_- \in H^{1/2}(\Delta), \ \Psi = \Psi_+ - \Psi_- \in H^{\circ -1/2}(\Delta) \qquad (8.5.3)$$

and

$$\int_\Delta \Psi \, ds = 0, \qquad (8.5.4)$$

with $n = n(x)$ the normal vector, external to the surface Σ. Besides, one has at infinity the conditions

$$u(x) = o(1), \ \frac{\partial}{\partial x_j} u(x) = o(|x|^{-1}), \ j = 1, 2, 3, \ |x| \to \infty. \qquad (8.5.5)$$

The first boundary value problem of the theory of elasticity is formulated in an analogous manner. For given $u_+, u_- \in H^{1/2}(\Delta)$, $u_+ - u_- \in H^{\circ 1/2}(\Delta)$, find u, satisfying (8.5.1), the boundary conditions

$$u\big|_{\Delta_+} = u_+, \ u\big|_{\Delta_-} = u_- \qquad (8.5.6)$$

and conditions (8.5.5). In (8.5.2) and (8.5.6), the symbols Δ_+ and Δ_- denote the two "shores" of the surface of the cracks in a three-dimensional body.

A variation of the formulation of the problems of the theory of elasticity follows. Denote by H^1_Δ the energy space of the vector functions, i.e., the closure of all C^∞ functions f in $R^3/\overline{\Delta}$ which have on Δ the limits f_+ and f_- satisfying (8.5.5) and for which the finite norm is

$$\|f\|^2 = \int_{\Omega_\Delta} E(f, f) \, dx + \int_\Delta (|f_+|^2 + |f_-|^2) \, dS,$$

where $E(f, f)$ is the density of elastic energy. In addition, let $\widetilde{H}^1_\Delta$ be a subspace of H^1_Δ of functions f such that $f_+ = f_- = 0$ on Δ.

The varied formulation of the second boundary value problem of the theory of elasticity is written in the following form: Find $u \in H^1_\Delta$ such that for all $v \in H^1_\Delta$ and Ψ_+, Ψ_- as also above, the condition

$$\int_{\Omega_\Delta} E(u, v) \, dx = \int_\Delta (\Psi_+ - \Psi_-) v \, dS. \qquad (8.5.7)$$

is fulfilled. Now let $W \in H^1_\Delta$ be functions for which $W\big|_\Delta = \Phi = u_+ - u_-$, then problem I is equivalent to the problem: For given $W \in H^1_\Delta$, find $u \in H^1_\Delta$ such that $u - W \in \widetilde{H}^1_\Delta$ and for all $v \in \widetilde{H}^1_\Delta$

$$\int\limits_{\Omega_\Delta} E(u,v)\, dx = 0 . \tag{8.5.8}$$

The proof of the existence and uniqueness of the solutions of this problems is give in [101].

Consider now the derivation of the integral equations. Using Betti's formula, one finds

$$u(x) = \int\limits_\Delta \{\Gamma(y,x)t(y) - (\sigma^{(n)}\Gamma(y,x))^* \Phi(y)\}\, dS_y , \tag{8.5.9}$$

where

$$\Phi = [u] = u\big|_{\Delta_+} - u\big|_{\Delta_-} , \quad t = [\sigma^{(n)}u] = \sigma^{(n)}u\big|_{\Delta_+} - \sigma^{(n)}u\big|_{\Delta_-} , \tag{8.5.10}$$

$$\Gamma^k_j(y,x) = -\frac{\lambda + 3\mu}{8\pi\mu(\lambda + 2\mu)} \left\{ \frac{1}{|x-y|}\, \delta_{jk} + \frac{\lambda+\mu}{\lambda+3\mu}\, \frac{(x_j - y_j)(x_k - y_k)}{|x-y|^3} \right\}.$$

Applying known results on the behaviour of the double layer potential on the boundary Δ (cf. chapter 5, §4), one finds an analogue to Somigliani's formula

$$1/2(u_+ + u_-)(x) = \int\limits_\Delta \{\Gamma(y,x)t(y) - (\sigma^{(n)}\Gamma(y,x))\Phi(y)\}\, dS_y , \quad x \in \Delta .$$

$$\tag{8.5.11}$$

Moreover, acting with the operator $\sigma^{(n)}$ on the formula (8.5.9) and going to the limit on the boundary, one obtains

$$1/2(\sigma^{(n_x)}u_+ + \sigma^{(n_y)}u_-)(x) =$$

$$= \int\limits_\Delta \{\sigma^{(n_x)}\Gamma(x,y)t(y) - (\sigma^{(n_x)}(\sigma^{(n_y)}\Gamma(y,x)))\Phi(y)\}\, dS_y , \quad s \in \Delta . \tag{8.5.12}$$

The formulation of the second boundary value problems of the theory of elasticity in (8.5.12) leads to the integral equation

$$D\Phi(x) \equiv -\int\limits_\Delta \sigma^{(n_x)}(\sigma^{(n_y)}\Gamma(y,x))^* \Phi(y)\, dS_y = f(x), \quad x \in \Delta , \tag{8.5.13}$$

that of the first problem in (8.5.11) to the integral equation

$$V t(x) \equiv \int_{\Delta} \Gamma(y, x) t(y) \, dS_y = g(x), \ x \in \Delta. \qquad (8.5.14)$$

where

$$f(x) = 1/2(\Psi_+ + \Psi_-)(x) - \Lambda(\Psi_+ - \Psi_-)(x),$$

$$g(x) = 1/2(u_+ + u_-)(x) - \Lambda^*(u_+ - u_-)(x),$$

and Λ^* is the operator, adjoint to the integral operator

$$\Lambda v(x) := \int_{\Delta} \sigma^{(n_x)} \Gamma(y, x) \, v(y) \, dS_y, \ x \in \Delta.$$

From the method of derivation of the integral equation (8.5.13) and (8.5.14) follows the existence of the solutions $\Phi \in H^{\circ 1/2}(\Delta)$ and $t \in H^{\circ -1/2}(\Delta)$, respectively. It will now be shown that the field of displacements $u(x)$, defined by (8.5.19), is the energy solution of the Dirichlet and Neumann problems, respectively. In fact, since $\Phi \in H^{\circ 1/2}(\Delta)$ and $t \in H^{\circ -1/2}(\Delta)$, then (by definition) $\Phi \in H^{\circ 1/2}(\Sigma)$ and $t \in H^{\circ -1/2}(\Sigma)$, where $\operatorname{supp}\Phi \subset \Delta$ and $\operatorname{supp}t \subset \Delta$. Therefore (8.5.9) may be rewritten as

$$u(x) = \int_{\Sigma} \left\{ \Gamma(y, x) t(y) - \left(\sigma^{(n_y)} \Gamma(y, x) \right)^* \Phi(y) \right\} dS_y, \ x \notin \Delta. \qquad (8.5.15)$$

The surface Σ is smooth, and the functions $u(x)$, defined by (8.5.15), are the energy solution of the problem

$$\mu \Delta u + (\lambda + \mu) \operatorname{grad} \operatorname{div} u = 0, \ x \notin \Delta \qquad (8.5.16)$$

with the boundary conditions

$$[u]\big|_{\Delta} = \Phi \in H^{\circ 1/2}(\Delta), \ [\sigma^{(n_y)} u]\big|_{\Delta} = t \in H^{\circ -1/2}(\Delta) \qquad (8.5.17)$$

and the conditions at infinity (8.5.5). This problem is a particular case of the contact problem for sectionally homogeneous media with the same elastic constants. It has been studied in detail in particular, the existence and uniqueness of the energy solution of this problem has been proven.

The integral operators D and V of equations (8.5.13) and (8.5.14) differ from the classical operators of potential theory, however, they (like also the classical operators, appear in a wide class of pseudo-differential operators). The proof of the single-valued solubility of equations (8.5.13) and (8.5.14) is based on the strict ellipticity of the operators D and V and requires an introduction to the concept of the symbol of pseudo-differential operator.

Let for smooth functions $u(x)$ with compact support in R^2 an operator A be defined by

$$Au(x) = \int\limits_{R^2} \int\limits_{R^2} \exp(i(x-y)\xi)\sigma_A(x,\xi)u(y)\,dy\,d\xi, \qquad (8.5.18)$$

where the integral is understood to be repeated, and the function $\sigma_A(x,\xi)$ is expandable in the asymptotic series

$$\sigma_A(x,\xi) \sim \sum_{j=0}^{\infty} |\xi|^{x_j} \sigma_j(x,\xi), \qquad (8.5.19)$$

with the sequence of real terms r_j decreasing monotonically $r_j \to -\infty$ for $j \to \infty$, the smooth functions $\sigma_j(x,\xi)$ defined for any $x, \xi \neq 0$ and homogeneous of degree 0 with respect to ξ. The number r_0 is called the order of the operator A.

If some class of functions of the form (8.5.19) forms a ring, then $\sigma_A(x,\xi)$ is called the symbol of the operator A for this ring of operators (cf. part I).

A linear operator B, given on smooth functions with compact support in Σ, is called a pseudo-differential operator on Σ, if for any coordinate diffeomorphism $\chi : \Sigma_1 \to \Sigma_2$, $\Sigma_1 \subset \tilde{\Sigma}$, $\Sigma_2 \subset R^2$ the operator, defined by

$$Au(x) = (Bu^*)(y), \; u^*(p) = u(\chi(p)), \; y = \chi^{-1}(x), \qquad (8.5.20)$$

admits a representation of the form (8.5.18).

For the pseudo-differential operator B on a closed, smooth surface Σ is defined its symbol $\sigma_A(x,\xi)$, given by (8.5.20). However, in general, this symbol depends on the choice of the coordinate diffeomorphism χ. It turns out that a quantity, not depending on the choice of χ, is the principal symbol

$$\sigma_B^0(x,\xi) = \sigma_A^0(x,\xi) = |\xi|^{r_0} \sigma_0(x,\xi).$$

An operator B is said to be elliptic, if

$$\sigma_B^0(x,\xi) \neq 0, \; x \in \Sigma, \; \xi \in R^2/\{0\}.$$

Extend the operator B continuously into $H^s(\Sigma)$, so that $B : H^s(\Sigma) \to H^{s-r_0}(\Sigma)$. It is known that the index of an elliptic operator depends only on the principal symbol (cf. [84]).

The determination of the principal symbols of the operators D and V at an arbitrary point $x \in \Sigma$ formally reduces to the replacement of Σ on the tangent plane Σ at the point $x \in \Sigma$ and application of the Fourier transformation to the thus derived operators. One has (cf. [102]):

$$\sigma_D^0(x,\xi) = \frac{\mu^2}{|\xi|} \begin{pmatrix} |\xi|^2 + \varepsilon\xi_1^2 & \varepsilon\xi_1\xi_2 & 0 \\ \varepsilon\xi_1\xi_2 & |\xi|^2 + \varepsilon\xi_2^2 & 0 \\ 0 & 0(1+\varepsilon)|\xi|^2 \end{pmatrix}, \qquad (8.5.21)$$

$$\sigma_V^0(x,\xi) = \frac{(\lambda + 3\mu)|\xi|^{-3}}{2(\lambda + 2\mu)\mu} \begin{pmatrix} |\xi|^2 + k\xi_2^2 & -k\xi_1\xi_2 & 0 \\ -k\xi_1\xi_2 & |\xi|^2 + k\xi_1^2 & 0 \\ 0 & 0 & |\xi|^2 \end{pmatrix}. \qquad (8.5.22)$$

where $0 < \varepsilon = \lambda(\lambda + 2\mu)^{-1} < 1$ and $0 < k = (\lambda + \mu)(\lambda + 3\mu)^{-1} < 1$. By the limitations imposed on ε and k, the principal symbols are positively defined for $|\xi| = 1$, $\xi \in R^2$, which entails the strict ellipticity of D and V. Formulae (8.5.21) and (8.5.22) imply that the operator D is of order $+1$, and the operator V of order -1; in particular, the transformation $D : H^{\circ 1/2}(\Delta) \to H^{-1/2}(\Delta)$ and $V : H^{\circ -1/2}(\Delta) \to H^{1/2}(\Delta)$ are continuous. From these formulae follows the correctness of Fredolm's alternative, i.e., the vanishing of the indices of the operators D and V.

Theorem 8.9. Let $g(x) \in H^{1/2}(\Delta)$ and $f(x) \in H^{-1/2}(\Delta)$ be given functions, then the integral equations (8.5.14) and (8.5.13) have the unique solutions $t(x) \in H^{\circ -1/2}(\Delta)$ and $\Phi(x) \in H^{\circ 1/2}(\Delta)$, respectively.

Proof. Since the indices of the operators D and V vanish, then it is sufficient to verify that the homogeneous equations corresponding to (8.5.13) and (8.5.14) have only the zero solution. Let this not be so and there exist $t(x) \in H^{\circ -1/2}(\Delta)$ and $\Phi(x) \in H^{\circ -1/2}(\Delta)$, where $t(x) \neq 0$ and $\Phi \neq 0$. Then, for $x \in R^2/\Delta$, define the functions

$$u(x) = \int_\Delta \Gamma(y, x) t(y) \, d_y S \,,$$

$$v(x) = -\int_\Delta \Phi(y) \sigma^{(n_y)} \Gamma(y, x)' \, d_y S \,,$$

which are variational solutions of problems I and II of the theory of elasticity, respectively. The potential of the simple layer is continuous for transition through the boundary Σ, and by the strength of the homogeneous equation, corresponding to (8.5.14), one has $u|_\Delta = 0$. It follows now from the uniqueness of the variational solution that the simple layer potential vanishes identically, i.e., $t(y) = 0$ for $y \in \Delta$. This verification proves the single-valued solubility of equation (8.5.14). Analogous reasoning proves the single-valued solubility of equation (8.5.13). Hence the theorem has been proved.

Note 8.9. The solution $t(x) \in H^{\circ-1/2}(\Delta)$ of the integral equation (8.5.14) of the first boundary value problem of the theory of elasticity depends only on a linear combination of the displacements u_+ and u_- on the boundary Δ, more accurately, on the functions $g(x) = 1/2(u_+ + u_-) + \Lambda^*(u_+ - u_-)$.

Note 8.10. The solution $\Phi(x) \in H^{\circ 1/2}(\Delta)$ of the integral equation (8.5.13) of the second boundary value problem of the theory of elasticity depends only on the linear combination of the stresses Ψ_+ and Ψ_- on the boundary Δ, more accurately on the functions $f(x) = 1/2(\Psi_+ + \Psi_-) + \Lambda(\Psi_+ - \Psi_-)$.

The results of the asymptotic behaviour of the solutions of equations (8.5.13) and (8.5.14) in the neighbourhood of the boundary γ of a crack surface Δ will be formulated next. Let s be the arc length parameter of the curve γ, $\rho(x)$ the distance from the point $x \in \Delta$ to γ (on the surface Δ) and $\eta(x)$ the "cutting" functions $\eta \in C_0^\infty(R^1)$, $0 \le \eta \le 1$, $\eta(x) \equiv 1$ for x belongs to a neighborhood of zero.

Theorem 8.10. 1) Let Ψ_- and $\Psi_+ \in H^{3/2+\sigma}(\Delta)$, $|\sigma| < 1/2$ where $\Psi_+ - \Psi_- \in H^{3/2+\sigma}(\Delta)$. Then the solution of the integral equation (8.5.13) has the form

$$\Phi = \alpha(s)\rho^{1/2}\eta(\rho) + \nu_0 \, ,$$

where $\alpha \in H^{3/2+\sigma}(\gamma)$ and $\nu_0 \in H^{3/2+\sigma}(\Delta)$.

2) Let u_- and $u_+ \in H^{3/2+\sigma}(\Delta)$, where $u_+ - u_- \in H^{\circ 3/2+\sigma}(\Delta)$. Then the solution of the integral equation (8.5.14) has the form

$$t = \beta(s)\rho^{-1/2}\eta(\rho) + \Psi_0 \, ,$$

where $\beta \in H^{1/2+\sigma}(\gamma)$ and $\Psi_0 \in H^{\circ 1/2+\sigma'}(\Delta)$ for any $\sigma' < 0$.

3) Correct apriori estimates with positive constants c and $s = 3/2 + \varepsilon$, $\varepsilon > 0$ are

$$\|\alpha\|_{H^s(\gamma)} + \|\nu_0\|_{H^s(\Delta)} \le c\big\{\|\Psi_+ + \Psi_-\|_{H^s(\Delta)} + \|\Psi_+ - \Psi_-\|_{H^s(\Delta)}\big\} \, ,$$

$$\|\beta\|_{H^{s-1}(\gamma)} + \|\Psi_0\|_{H^{s-1-\varepsilon}(\Delta)} \le c\big\{\|u_+ + u_-\|_{H^s(\Delta)} + \|u_+ - u_-\|_{H^s(\Delta)}\big\} \, .$$

The proof of this theorem is based on an investigation of the model problem on a crack in the form of a half-plane (where the solution applies) and the positive definiteness of the principal symbols of the operators D and V (cf. [87,101]).

Two types of boundary elements have been constructed for the numerical realization of the boundary integral equations (8.5.13) and (8.5.14). Standard Galerkin approximations, used normally for smooth surfaces Δ, have maximal accuracy of order $O(h^{1-\varepsilon})$, $\varepsilon > 0$. The modified Galerkin approximation, based on special approximations to the solution of equations (8.5.13) and (8.5.14) near the boundaries of the surface Δ (theorem 8.10) have maximal accuracy (for the same norm) of order $O(h^{2-\varepsilon})$.

Let $x = X(v)$, $v \in U$ be a regular parametrization of the surface Δ, and U a region in R^2, the boundary of which maps onto $\gamma = \partial\Delta$. Select, following [87], a sequence of regular, triangulated regions U with maximal cell dimensions $h > 0$ and let $S^{\alpha+1,m}$ $0 \le m \le d$ be finite elements (sectionally polynomial functions, d the degree of the polynomials, m their smoothness). Formulas $W_h(x) = \mu(X^{-1}(x))$, $\mu \in S^{d+1,m}$ determine boundary elements on the surface Δ, where, in particular, $S^{2,1}(\Delta)$ is a piecewise linear functions, continuous on Δ, and $S^{0,1}(\Delta)$ a piecewise constant functions.

If $H_h^1 \subset H^{\circ 1/2}(\Delta)$ and $H_h^2 \subset H^{\circ -1/2}(\Delta)$ are finite dimensional spaces, comprising finite numbers of elements, then the standard Galerkin method , in application to equations (8.5.13) and (8.5.14) is formulated in the following manner.

Find $\Phi_h \in H_h^1 \subset H^{\circ 1/2}(\Delta)$ such that for all $v \in H_h^1$ is fulfilled

$$< D\Phi_h, v >=< f, v >, \qquad (8.5.23)$$

find $t_h \in H_h^2 \subset H^{\circ -1/2}(\Delta)$ such that for all $v \in H_h^2$ is fulfilled

$$< Vt_h, v >=< g, v > \qquad (8.5.24)$$

The simplest space of finite elements is

$$H_h^1 = \left\{ W_h \in S^{2,1}(\Delta) | w_h = 0 \text{ on } \gamma \right\}, \; H_h^2 = S^{1,0}(\Delta).$$

In their paper [87], E. Stephan and M. Costabel proved the following assertion:

Theorem 8.11. There exists $h_0 > 0$ such that for all $h \le h_0$ there exists a unique solution $\Phi_h \in H_h^1$ for (8.5.23) and a unique solution $t_h \in H_h^2$ for (8.5.24). If Φ and t are exact solution of equations (8.5.13) and (8.5.14), respectively, then

$$\|\Phi - \Phi_h\|_H^{\circ 1/2+\eta}(\Delta) \le Ch^{\sigma-\eta}\|f\|_{H^{-1/2+\sigma}(\Delta)},$$

$$\|t - t_h\|_H^{o-1/2+\eta}(\Delta) \leq Ch^{\sigma-\eta}\|g\|_{H^{1/2+\sigma}(\Delta)},$$

where $|\eta| < 1/2$, $|\sigma| < 1/2$.

The asymptotic convergence may be improved by introduction of the special space of boundary elements

$$H_h^1 = \left\{\Phi_h = \alpha_h(s)\rho^{1/2}\eta(\rho) + V_{0h} \,|\, \alpha_h \in S^{2,1}(\gamma),\; V_{0h} \in S^{2,1}(\Delta)\right\},$$

$$H_h^2 = \left\{t_h = \beta_h(s)\rho^{-1/2}\eta(\rho) + \Psi_{0h} \,|\, \beta_h \in S^{1,0}(\gamma),\; \Psi_{0h} \in S^{1,0}(\Delta),\; \Psi_{0h}|_\gamma = 0\right\}.$$

Theorem 8.12. Problems (8.5.23) and (8.5.24) with spaces H_h^1 and H_h^2 together with H_h^1 and H_h^2, respectively, are single-valuedly soluble for all $h \leq h_0$ and one has the estimates

$$\|\Phi - \Phi_h\|_H^{o-1/2+\eta}(\Delta) \leq Ch^{2+\sigma-\eta-\varepsilon}\|f\|_{H^{-3/2+\sigma}(\Delta)},$$

$$\|t - t_h\|_H^{o-3/2+\eta}(\Delta) \leq Ch^{2\sigma-\eta-\varepsilon}\|f\|_{H^{3/2+\sigma}(\Delta)},$$

where $-1/2 \leq 1$, $|\sigma| < 1/2$, $\varepsilon > 0$ and the constant $c > 0$ does not on $h > 0$.

APPENDIX I
COSSERAT SPECTRUM

Introduction

1^0. Consider the spectrum of the bundle of the operators of the static theory of elasticity for homogeneous media. Originally, this spectrum was investigated by the known French mathematicians and mechanicians Eugene and Francis Cosserat who published on this topic nine papers between 1898 and 1901 (cf. [12-20]). Their work was continued during the Sixties and Seventies of this century by S.G. Mikhlin and, partly, by V.G. Maz'ja whose paper [5] contains a sufficiently complete bibliography of the subject.

The term operator bundle is given to operators A_ω which depend of a numerical, real or complex parameter ω. As it is well known, the differential equations of the static theory of elasticity for homogeneous, isotropic media may be presented in the form

$$\Delta^* u := \Delta u + \omega \operatorname{grad} \operatorname{div} u = -\mu^{-1} F(x)\,, \ x \in \Omega\,, \qquad (A1.0.1)$$

where Ω is a region of Euclidean space, u is the displacement vector, F is the body force vector, μ is the shear modulus of the elastic medium, $\omega = (1 - 2\nu)^{-1}$ and ν is Poisson's constant for the same medium. Equations (A1.0.1) will be studied either for the boundary conditions of problem I

$$u\big|_{\partial\Omega} = \varphi(x)\,, \qquad (A1.0.2)$$

or for those of problem II $\sigma^{(n)}\big|_{\partial\omega} = \psi$ or otherwise

$$\left[(\omega-1)\alpha_j \operatorname{div} u + \sum_{k=1}^{3} \left(\frac{\partial u_j}{\partial x_k} + \frac{\partial u_k}{\partial x_j}\right)\alpha_k\right]_{\partial\Omega} = \mu^{-1}\psi_j(x)\,; \ j = 1,2,3\,, \ (A1.0.3)$$

where u_j, x_j, ψ_j are the components of the vectors u, x and ψ, respectively, α_j are the direction cosines of the external normal to the boundary $\partial\Omega$ of the region Ω, and $\psi(x)$ is the stress vector, acting at the point $x \in \partial\Omega$.

The left hand sides of equations (A1.0.1) and (A1.0.2) (or, correspondingly, of (A1.0.1) and (A1.0.3)) which determine the bundle of the operators with parameter ω, will be denoted by Δ_ω^*. For real media, Poisson's constant has values in the ranges $-1 < \nu < 1/2$ or $1/3 < \omega < \infty$; however, in what follows, all possible real as well as complex values, including ω, will be considered $\omega = \infty$.

Let A_ω be an arbitrary bundle of operators which are linear for all values of ω. The manifold of the values of the parameter ω for which the inverse operator A_ω^{-1} either does not exist or exists, but is unbounded, or is only defined on a non-compact set will be referrred to as the spectrum of this bundle. The spectrum of the bundle Δ_ω^* of the operators of the theory of elasticity will be called Cosserat spectrum. It follows immediately from the existence and uniqueness theorems of the solution of the theory of elasticity that the Cosserat spectrum is located in the complex plane ω outside the interval $1/3 < \omega < \infty$ of the real axis.

2^0. The problem of the Cosserat spectrum may be reduced to that of the spectrum of a certain integral equation, as will be explained by the example of problem I. Let $G(x,y)$ be the Green function of the Dirichlet problem for the Laplace operator in the region Ω. In order to somewhat simplify the manipulations, assume that in (A1.0.2) the function $\varphi(x) \equiv 0$. Recall that

$$G(x,y) = \frac{1}{4\pi}\left(\frac{1}{r} - g(x,y)\right),$$

where $r = |x - y|$, and $g(x,y)$ is the function which is harmonic in terms of one of the variables x and y in the region Ω, if the second point is fixed in the same region, and, in addition, that this function becomes $1/r$, if one of the points x, y lies on $\partial\Omega$. Equations (A1.0.1) and (A1.0.2) yield the integral equation

$$u(x) = \omega \int_\Omega G(x,y)\,\mathrm{grad}\,\mathrm{div}\,u(y)\,dy + \Phi(x), \qquad (A1.0.4)$$

where

$$\Phi(x) = \mu^{-1} \int_\Omega G(x,y)\,F(y)\,dy\,, \quad x \in \Omega\,.$$

Integrate (A1.0.4) by parts. Denoting by $x = i_k$ the unit vectors of the coordinate axes x_k, one finds

$$\int_\Omega G(x,y)\,\mathrm{grad}\,\mathrm{div}\,u(y)\,dy = -(4\pi)^{-1}\sum_{j,k=1}^{3} i_k \int_\Omega \frac{\partial}{\partial y_k}r^{-1}\frac{\partial u_j}{\partial y_j}\,dy +$$

$$+ (4\pi)^{-1}\sum_{j,k=1}^{3} i_k \int_\Omega \frac{\partial g}{\partial y_k}\frac{\partial u_j}{\partial y_j}\,dy, \qquad (A1.0.5)$$

The second integral on the right hand side may again be integrated by parts, and, since $u|_{\partial\Omega} = 0$, one finds

$$\int\limits_{\Omega} \frac{\partial g}{\partial y_k} \frac{\partial u_j}{\partial y_j} \, dy = \int\limits_{\Omega} u_j(y) \frac{\partial^2 g}{\partial y_j \partial y_k} \, dy \; .$$

In order to take by parts the first integral, one has to exercise some caution (cf. chapter 3, §4). Surround the point x by a sphere of radius ε with centre at x; choose ε to small that this sphere together with its boundary lies entirely in Ω. Consider the integral

$$- \int\limits_{\Omega/(r<\varepsilon)} \frac{\partial}{\partial y_k} r^{-1} \frac{\partial u_j}{\partial y_j} \, dy \; ; \qquad\qquad (A1.0.6)$$

which, as $\varepsilon \to 0$, tends to the first integral on the right hand side of (A1.0.5) uniformly in any, closed, internal sub-region of Ω. The integral (A1.0.6) may be integrated by parts:

$$- \int\limits_{\Omega/(r<\varepsilon)} \frac{\partial}{\partial y_k} r^{-1} \frac{\partial u_j}{\partial y_j} dy = \int\limits_{\Omega/(r<\varepsilon)} \frac{\partial^2}{\partial y_j \partial y_k} r^{-1} u_j(y) \, dy -$$

$$- \int\limits_{S} \frac{\partial}{\partial y_k} r^{-1} u_j \cos(n, y_j) \, dS \; ; \qquad (A1.0.7)$$

where S is the sphere $r = \varepsilon$. If $\varepsilon \to 0$, then the first integral on the right hand side has as limit the three-dimensional, singular integral;

$$\int\limits_{\Omega} \frac{\partial^2}{\partial y_j \partial y_k} r^{-1} u_j(y) \, dy \; ;$$

the limit of the second integral will be found next. The normal n, external to the sphere S, is directed in the opposite direction to the vector r, hence

$$\cos(n, y_j) = -\cos(r, y_j) = -(y_j - x_j)/r \; .$$

It is readily seen that the limit of the second integral of (A1.0.7) vanishes for $j \neq k$, and equals $-4\pi u(x)/3$ for $j = k$. Substitution of these results into (A1.0.5) reduces equation (A1.0.4) to the form

$$(1 - \omega/3)u(x) - \omega(4\pi)^{-1} \sum_{j,k=1}^{3} \int\limits_{\Omega} \left[\frac{\partial^2}{\partial y_j \partial y_k} r^{-1} - \frac{\partial^2 g}{\partial y_j \partial y_k} \right] u_j(y) \, dy = \Phi(x)$$

$$(A1.0.8)$$

a system of three integral equations in the three unknowns $u_j(x)$, $j = 1, 2, 3$. Clearly the Cosserat spectrum coincides with the spectrum of this system. Every element of the matrix of the kernel of system (A1.0.8) is a sum of two terms the first of which is the singular kernel $\partial^2 r^{-1}/\partial y_j \partial y_k$. As regards the second term $\partial^2 g/\partial y_j \partial y_k$, it will be continuous, if one of the points x or y lies inside Ω; however, if both points lie on $\partial\Omega$, then, if they coincide, this term becomes infinite so that it is not summable in Ω. The theory of such equations has not been developed fully; there exist only partial results for analogous one-dimensional integral equations the first of which were obtained by Tricomi [9]. It is possible that the following study of Cosserat spectrum presents an opportunity for the construction of a theory of such a kind of multi-dimensional integral equations at least in certain cases.

§1. The investigations of Eugene and Francis Cosserat

The earlier referred to papers [12-20] are brief notes almost without proofs; their contents, in chronological order , will now be presented.

1^0. In [12], the authors consider the first boundary value problem for the sphere in the absence of body forces:

$$\Delta u + \omega \operatorname{grad} \operatorname{div} u = 0\,, \quad |x| < a\,, \qquad (A1.1.1)$$

$$u\big|_{|x|=a} = \varphi(x)\,. \qquad (A1.1.2)$$

The solution of this problem was known at that time. The authors reduced it to the following form. Let $u_0(x)$ be a vector, harmonic in the sphere $|x| < 0$ and satisfying condition (A1.1.2). Expand the harmonic function $\operatorname{div} u_0$ in the series

$$\operatorname{div} u_0 = \sum_{n=0}^{\infty} F_n(x)\,, \qquad (A1.1.3)$$

where $F_n(x)$ are homogeneous, harmonic polynomials of degree n. For $n > 0$, set

$$u_n(x) = \frac{1}{2(2n+1)}(|x|^2 - a^2)\operatorname{grad} F_n(x)\,, \quad \omega_n = -\frac{2n+1}{n} \qquad (A1.1.4)$$

The solution of problem (A1.1.1) and (A1.1.2) may now be reduced to the form

$$u(x) = u_0(x) + \omega \sum_{n=1}^{\infty} \frac{\omega_n u_n(x)}{\omega - \omega_n}\,. \qquad (A1.1.5)$$

Obviously, $u_n(x)\big|_{|x|=a} = 0$; it is also easily verified that $u_n(x)$ satisfies the homogeneous equation of the theory of elasticity for $\omega = \omega_n$, hence it is seen that the number $\omega_n = -(2n+1)/n$ is the characteristic number of the bundle of operators of the theory of elasticity for problem I in the case of the sphere; the corresponding characteristic vectors $u_n(x)$ are determined by (A1.1.4). The multiplicity of the characteristic number ω_n equals $2n+1$, the number of linearly independent, harmonic, homogeneous polynomials of degree n three-dimensional space.

Note that formulas (A1.1.4) and (A1.1.5) are easily derived without assuming beforehand that the solution of the problem of the theory of elasticity for the sphere is known. Consider the general case of an m-dimensional, $m \geq 3$ space. It is directly verified that the number $\omega_n = -(2n+m-2)/n$ and the vector functions

$$u_n(x) = c_n(|x|^2 - a^2)\operatorname{grad} F_n(x), \quad c_n = const, \tag{A1.1.6}$$

are the characteristic number and characteristic functions of Cosserat of the first boundary value problem for the sphere. Seek the solution of problem (A1.1.1) in the form

$$u(x) = u_0(x) + \sum_{n=1}^{\infty} a_n u_n(x). \tag{A1.1.7}$$

where the vector $u_0(x)$ has been defined above. This solution satisfies the boundary condition (A1.1.2), and there remains to choose the coefficients a_n in such a manner that equation (A1.1.1) is satisfied. Substitution from (A1.1.7) into (A1.1.1) yields

$$\omega \operatorname{div} u_0 + \sum_{n=1}^{\infty} a_n(\omega - \omega_n)\operatorname{div} u_n = const.$$

Expand the harmonic function $\operatorname{div} u_0$ in a series of homogeneous, harmonic polynomials. Choose the polynomials $F_n(x)$ in the formula (A1.1.6) in such a way that this expansion has the form (A1.1.4), with the divisor $2n+1$ replaced by $2n+m-2$. Then $c_n^{-1} = 2(2n+m-2)$ and $a_n = \omega\omega_n/(\omega - \omega_n)$, which leads to (A1.1.5). The convergence of the series, generalizing the series (A1.1.5), will be studied in the corresponding metric in the next section.

2^0. Paper [13] takes one more important step in the investigation of the spectrum of the bundle Δ_ω^*: the class of solutions of equation (A1.1.1) is given for arbitrary regions for the value $\omega = -1$. This solution has the form

$$u = x \times \operatorname{grad} \varphi + \operatorname{grad} \psi. \tag{A1.1.8}$$

where the cross denotes the vector product and φ and ψ are harmonic functions. The vectors (A1.1.8) comprise an infinity of such vectors which, being linearly independent, satisfy the boundary condition $u|_{\partial\Omega} = 0$, hence it follows that in every case of problem I the significance of the parameter $\omega = -1$ is that it is a characteristic number of infinite multiplicity.

Note that the class (A1.1.8) does not exhaust all the solutions of the homogeneous equations of the theory of elasticity for $\omega = -1$.

3^0. Paper [14] plans the construction of the spectrum of problem I for the ellipsoid. A method of construction of the characteristic vectors is presented which have the form of polynomials of any given degree and the corresponding characteristic number is determined. In fact, let

$$\Phi(x) = \frac{x_1^2}{a_1^2} + \frac{x_2^2}{a_2^2} + \frac{x_3^2}{a_3^2} - 1$$

and $\Phi(x) = 0$ be the equation of a given ellipsoid. The characteristic vectors are sought in the form of a product of $\Phi(x)$ and a polynomial of degree $n-1$. Substitution into the differential equation leads to a homogeneous, algebraic system, the determinant of which for $n > 1$ has one root $\omega = -1$; all remaining roots being real and different (including also for $n = 1$). If one remains on the side of the root $\omega = -1$, then its is asserted that the remaining roots yield $2n + 1$ polynomials of degree $n + 1$, which are also the characteristic vectors of problem I for the ellipsoid. For $n = 1$ and $n = 2$, the characteristic numbers and vectors were computed. Thus, if $n = 1$, then

$$u_1(x) = (\Phi(x), 0, 0), \quad \omega_1 = -a_1^2\left(\frac{1}{a_1^2} + \frac{1}{a_2^2} + \frac{1}{a_3^2}\right).$$

4^0. Paper [15] contains two important results which deal with problem I for arbitrary, finite regions. It is proved there that, if $u_k(x)$ and $u_n(x)$ are the characteristic vectors of problems I, corresponding to different characteristic Cosserat numbers, then

$$\int_\Omega \operatorname{div} u_k \operatorname{div} u_n \, dx = 0. \tag{A1.1.9}$$

Furthermore, for certain boundaries, which will not be discussed here, the authors prove a theorem according to which the solution of problem (A1.0.1) and (A1.0.2) is a single-valued analytic function of ω; its characteristic points are real and lie on the segment $[-\infty, -1]$. For the proof of this theorem is essential the establishment of the fact that for $|\omega| < 1$ the solution of problem (A1.0.1) and (A1.0.2) can be expanded in series of powers of ω. At this point, the authors made a mistake, later corrected by S.G. Mikhlin [5].

Note the consequences of the Cosserat theorems:

1) The Cosserat spectrum of problem I is located on the segment $[-\infty, -1]$;

2) if $|\omega| < 1$, the problem (A1.0.1) and (A1.0.2) may be solved by iteration according to

$$\Delta u^{(0)} = F(x), \; u^{(0)}\big|_{\partial\Omega} = g(x),$$
$$n \geq 1, \; \Delta u^{(n)} = -\omega \operatorname{grad} \operatorname{div} u^{(n-1)}, \; u^{(n)}\big|_{\partial\Omega} = 0, \tag{A1.1.10}$$

where $g(x)$ is a function, given on $\partial\Omega$.

If ν is real, then the condition $|\omega| < 1$ signifies that either $\nu > 1$ or $\nu < 0$. Only the second possibility may correspond to the law of conservation of energy, but elastic materials with negative Poisson constant are unknown.

Some strengthening of the second consequence is given in paper [16], where it is proved that the solution of problem I of the theory of elasticity may be expanded in a certain series which converges, if $K := (3\sqrt{3}\,|\omega| + 2)|\omega + 2|^{-1} < 3$. If $\omega < 0$, then $\nu > 1/2$, and this does not correspond to any elastic material. However, if $\omega > 0$, then the condition $K < 3$ yields $\omega < 2(\sqrt{3} + 1)/3$ or $\nu < (7 - 3\sqrt{3})/8 \approx 0.225$.

5^0. In paper [17], problem II of the $|x| < a$ theory of elasticity is studied for the sphere and the following formulas are presented for the characteristic numbers and vectors:

$$\tilde{u}_n(x) = x F_n(x) + \frac{a^2 - 3|x|^2}{2(2n + 1)} \operatorname{grad} F_n(x),$$
$$\tilde{\omega}_n = \frac{2n + 1}{2n^2 + 4n + 3}; \; n = 0, 1, 2, \ldots; \tag{A1.1.11}$$

where, as above, $F_n(x)$ denotes a homogeneous, harmonic polynomial of degree n. The solution of problem II for the sphere

$$\Delta u + \omega \operatorname{grad} \operatorname{div} u = 0, \; |x| < a;$$
$$(\omega - 1)\alpha_j \operatorname{div} u + \sum_{k=1}^{3} \left(\frac{\partial u_j}{\partial x_k} + \frac{\partial u_k}{\partial x_j} \right) \alpha_k = h_j(x); \; j = 1, 2, 3; \; |x| = a, \tag{A1.1.12}$$

is obtained in the form

$$u = \tilde{u}_0(x) + \omega \sum_{n=0}^{\infty} \frac{\tilde{\omega}_n \tilde{u}_n(x)}{\omega - \tilde{\omega}_n}, \qquad (A1.1.13)$$

where $u_0(x)$ is a harmonic vector-function which satisfies on the sphere $|x| = a$ some boundary condition which depends on the vector $h(x)$.

6^0. Paper [18] is devoted to problems of the theory of elasticity for a spherical shell. The authors obtain for the characteristic numbers of problem I the quadratic equation

$$\left(\omega_n + \frac{2n+1}{n}\right)\left(\omega_n + \frac{2n+1}{n+1}\right) = \frac{(2n-1)(2n+3)}{4} \cdot$$
$$\cdot \frac{(a'^2 - a^2)\omega_n^2}{(a'^{2n+3} - a^{2n+3})(a'^{-2n+1} - a^{-2n+1})}, \qquad (A1.1.14)$$

where a and a' are the radiuses of the boundaries of the shell. For $n \to \infty$, this equation becomes $(\omega_\infty + 2)^2 = 0$, hence it follows that $\omega_n \to n \to \infty - 2$ as $n \to \infty$. Moreover, the denominator in (A1.1.14) is negative, hence the roots lie in the segment $-2(2n+1)/n < \omega_n < -(2n+1)/(n+1)$ and it is readily seen that all finite multiplicity characteristic numbers lie in the interval $-3 < \omega_n < -3/2$.

The characteristic numbers of problem II satisfy a somewhat more complicated equation from which it may be seen that with a growth of the number n these numbers tend to zero, and that all of them are included in the segment $[-2, 1/3]$.

Certain results have also been obtained for the more general problem when there is given on the boundary of the region a linear combination of stresses and displacements. This problem is considered in paper [19] for the ellipsoid for which are computed the first few characteristic numbers and vectors.

7^0. In their last note [20], the Cosserat's prove that the number $\omega = 1/3$ is characteristic for problem II for any finite region. These results will be summarized next.

For certain regions (the sphere, spherical shell, ellipsoid), the existence of a finite-fold multiplicity spectrum of the bundle of operators of problems I and II of the three-dimensional theory of elasticity is proved; several results are also obtained for the mixed problem. It is explained for the stated regions that the finite-fold characteristic numbers of problem I gather at the point $\omega = -2$, those of problem II at the point $\omega = 0$; the authors state finite intervals which contain all finite-fold characteristic numbers.

Representations of the solution of problems I and II are given for the sphere in terms of characteristic vectors, corresponding to finite-fold characteristic numbers.

A proof is given of the result that for problem I the number $\omega = -1$ is the characteristic number of infinite-foldness.

The characteristic vectors of problem I, corresponding to different finite-fold characteristic numbers, satisfy the orthogonality relations (A1.1.9).

The solution of problem I for arbitrary finite regions is an analytic function of ω, the singular points of which lie in the interval $[-\infty, -1]$.

If $|\omega| < 1$, then problem I may be solved by iteration by means of the formulas (A1.1.10).

§2. Problem I for finite regions

1^0. Singular solutions of the vector equation of the three-dimensional theory of elasticity (cf. chapter 6)

$$\Delta u + \omega \operatorname{grad} \operatorname{div} u = \delta(x - y)e^j \qquad (A1.2.1)$$

where δ is the Dirac function, $e^1 = \begin{pmatrix} 1 \\ 0 \\ 0 \end{pmatrix}$ and $e^2 = \begin{pmatrix} 0 \\ 1 \\ 0 \end{pmatrix}$ are the unit vectors,

are constructed with the aid of the so-called Kelvin-Somigliani tensor $\Gamma(x, y)$, the components $\Gamma_i^j(x, y)$ of which are determined by the formulas

$$\Gamma_i^j(x, y) = \frac{1}{8\pi(\omega + 1)} \left[\frac{(-\omega + 2)\delta_{ij}}{r} - \frac{\omega(y_j - x_j)(y_i - x_i)}{r^3} \right]. \qquad (A1.2.2)$$

The vector $\Gamma^j = \begin{pmatrix} \Gamma_1^j \\ \Gamma_2^j \\ \Gamma_3^j \end{pmatrix}$ solves equation (A1.2.1).

Formula (A1.2.2) shows that for all values of ω, except for $\omega = -1$, and for $x \neq y$ this singular solution is analytic in $x = (x_1, x_2, x_3)$. The singular solution (A1.2.2) is not unique - every other singular solution differs from the Somigliani tensor by a solution of the homogeneous equation

$$\Delta u + \omega \operatorname{grad} \operatorname{div} u = 0. \qquad (A1.2.3)$$

Each such solution is analytic in the neighbourhood of any point, as follows, for example, from (A1.0.4). Thus, for $x \neq y$ and for $\omega \neq -1$, any solution of equation (A1.2.1) is analytic, hence follows (cf. [7], p. 288) that for all finite values of ω, except $\omega = -1$, the operator Δ_ω^* is elliptic. It is likewise readily shown, according to I.G. Petrovskii [cf. [4], p. 225), that for all those ω the operators of the theory of elasticity are elliptic.

It will now be explained for what values of ω the "condition of completeness" of Agmon-Douglas-Nierenberg [11] are violated. Consider the non-homogeneous equation of the theory of elasticity for non-homogeneous boundary conditions (equation (A1.0.1) and (A1.0.2)); assume that the Ω is finite and its boundary $\partial\Omega$ sufficiently smooth.

Recall the procedure for verification of completeness. Let Ω be a region of m-dimensional Euclidean space with a sufficiently smooth boundary $\partial\Omega$. Let there be given in Ω the differential operator Lu, where u is a vector with N components and $L = L(x, D) = \|L_{jk}(x; D)\|$ is a square matrix of order N, which depends polynomially on the differentiation operator $D = \left(\frac{\partial}{\partial x_1}, \frac{\partial}{\partial x_2}, ..., \frac{\partial}{\partial x_m}\right)$. Let the boundary conditions have the form $G(x; D) = \varphi(x)$, $x \in \partial\Omega$, where, in the general case, G is a matrix of order $N \times N_1$ which likewise depends polynomially on D. Formed the principal parts [18] L_0 and G_0 of the operators L and G, construct the corresponding characteristic matrices $L_0(x; \xi)$ and $G_0(x; \xi)$, $\xi \in R_m$ and denote by $\hat{L}_0(x; \xi)$ the matrix, adjoint to $L_0(x; \xi)$ (i.e., the matrix of the algebraic complements of the elements of the matrix $L_0(x; \xi)$).

Select an arbitrary point $x \in \partial\Omega$ and construct there a local coordinate system with the m-th axis directed along the normal to $\partial\Omega$; denote the coordinates in this system by $\xi = (\xi_1, \xi_2, ..., \xi_{m-1}, \tau)$ and write $\xi = (\xi', \tau)$, where $\xi' = (\xi_1, \xi_2, ..., \xi_{m-1})$. Construct the equation $\mathrm{Det} L_0(x; \xi', \tau) = 0$ with the unknown τ. If the operator L is elliptic, this equation does not have real roots; assume that this equation has the same number of roots as with positive as also with negative imaginary parts.

Introduce the notation

$$M^+(x; \xi', \tau) = \prod_{j=1}^{k}(\tau - \tau_j(x; \xi')), \qquad (A1.2.4)$$

where $\tau_j(x; \xi')$, $j = 1, 2, ..., k$ are the roots with positive, imaginary parts. The condition of completeness consists of the fact that for $\xi' \neq 0$ the columns of the matrix $G_0(x; \xi'\tau)\hat{L}_0(x; \xi', \tau)$ are linearly independent with respect to the modulus $M^+(x; \xi', \tau)$.

In the sequel, only the case $N = N_1 = 3$ will be considered, when the matrices L and G do not depend on x and coincide with their principal parts; verification of completeness at any point then reduces to the fact that the determinant, which consists of the remainders of the division of the elements of $G(\xi',\tau) \times L(\xi',\tau)$ by $M^+(\xi',\tau)$, must be non-zero.

In the case of problem (A1.0.1) and (A1.0.2), which is of interest here, one has $L_0 = \Delta + \omega\,\mathrm{grad\,div}$ and G is the unit matrix. The characteristic matrix for the differential operator of the theory of elasticity is now

$$\begin{pmatrix} (1+\omega)\xi_1^2 + \xi_2^2 + \tau^2 & \omega\xi_1\xi_2 & \omega\xi_1\xi_3 \\ \omega\xi_1\xi_2 & \xi_1^2 + (1+\omega)\xi_2^2 + \tau^2 & \omega\xi_2\xi_3 \\ \omega\xi_1\xi_3 & \omega\xi_1\xi_2 & \xi_1^2 + \xi_2^2 + (1+\omega)\tau^2 \end{pmatrix}.$$

$$(A1.2.5)$$

Setting its determinant equal to zero, one finds $(1+\omega)(\xi_1^2 + \xi_2^2 + \tau^2)^3 = 0$; for $\xi' \neq 0$, this equation has in the upper half-plane one three-fold root $\tau = i(\xi_1^2 + \xi_2^2)^{1/2} = i|\xi|^2$, hence $M^+(\xi',\tau) = (\tau - i|\xi|^2)$.

The following observation will facilitate the further manipulations. First of all, the matrices L_0, $\hat{L}_0$, G_0 are homogeneous with respect to ξ' and τ, hence it is sufficient to verify the condition of completeness for the unit vectors ξ'. Moreover, since the differential equation as also the boundary condition in problem (A1.0.1) and (A1.0.2) are invariant with respect to rotation of the coordinate axes, then the local coordinate system, referred to above, may be chosen such that $\xi_1 = 1$, $\xi_2 = 0$, when $M^+(\xi',\tau) = (\tau - i)^3$ and

$$L_0 = \begin{pmatrix} 1 + (\omega)\tau^2 & 0 & -\omega\tau \\ 0 & (1+\omega)(1+\tau^2)^2 & 0 \\ -\omega\tau & 0 & 1+\omega+\tau^2 \end{pmatrix} (1+\tau^2)^2 . \quad (A1.2.6)$$

In the present case, G is the unit matrix, it is sufficient to construct the determinant from the remainders of the division of the elements of the preceding matrix by $(\tau - i)^3$. A study of the central element of this matrix yields the sole value $\omega = -1$ which, however, has already been excluded earlier. There remains to consider the analogous determinant for the second order matrix

$$\begin{pmatrix} 1 + (1+\omega)\tau^2 & -\omega\tau \\ -\omega\tau & 1 + (1+\omega)\tau^2 \end{pmatrix} .$$

The value of the corresponding determinant is simply found to be $4(\omega + 2)^2$. Thus, except for the earlier excluded values $\omega = -1, \infty$ for which the ellipticity of the operator of the theory of elasticity is lost, there exists still the single value $\omega = -2$ at which the condition of completeness for problem I is lost. Note that for the same value $\omega = -2$, there vanishes the symbolic determinant of the system of singular integral equations, to which problems I and II of the theory of elasticity may be reduced [4].

3^0. Consider problem I for the homogeneous boundary condition

$$\Delta u + \omega \,\mathrm{grad}\,\mathrm{div}\,u = F(x), \ x \in \Omega,$$
$$u = 0, \ x \in \partial\Omega. \tag{A1.2.7}$$

Denote the operator of this problem by P_ω, so that the problem may be rewritten in the form $P_\omega u = F(x)$. It follows from the results of [1] and [11] that for $\omega \neq -1, -2, \infty$ the operator P_ω is Noetherian, i.e., that it is normally soluble and has a finite index, like the operator from $E(\Omega) = \mathbf{W}_2^2(\Omega) \cap \mathbf{W}^{\circ 1}_2(\Omega)$ in $\mathbf{L}_2(\Omega)$.

This assertion may be made essentially more precise. Let H, H_1, H_2 be Hilbert spaces, where each of the following is bounded and dense contained in the preceding one: $H_2 \subset H_1 \subset H$. Furthermore, let P act on H_2 in H; let still $D(P^*) \subset H_1$ and $\overline{D(P^*)} = H_1$. Denote by H_{-1} the space, conjugate to with H respect to the scalar product in H. Construct the extension $\tilde{P}$ of the operator P, acting from H_1 into H_{-1} and defined as follows: Let $u \in H_1$ be such that $|(u, P^*v)| \leq C_u \|v\|_{H_1}$, $\forall v \in D(P^*)$. Then (u, P^*v) determines a functional on v which admits extension with pect to continuity over the entire H_1. Denote this functional by $\tilde{P}u$:

$$(\tilde{P}u, v) = (u, P^*v). \tag{A1.2.8}$$

It is seen from this definition that $\tilde{P}$ is an extension of P, acting from H_1 into H_{-1}. In an analogous manner, an extension $\tilde{P}^*$ of the operator P^* may be constructed; if $P^{**} = P$, then $\tilde{P}^* = \tilde{P}^*$.

Lemma A1.1. If an operator P is, then, under the conditions described above, the operator $\tilde{P}$ likewise in Noetherian. The zeroes of the operators P and $\tilde{P}$, and likewise those of P^* and $\tilde{P}^*$, coincide.

Let φ_j $(j = 1, 2, ..., n)$ be the base of the subspace $N(P)$ of zeroes of the operator P. Obviously, the φ_j are likewise the zeroes of the operator $\tilde{P}$. Just as, if ψ_k $(k = 1, 2, ..., n^*)$ is the base of the zeroes of the operator P^*, then the ψ_k are likewise the zeroes of the operator $\tilde{P}^*$. It will be shown that the operators $\tilde{P}$ and $\tilde{P}^*$ do not have zeroes which are linearly independent with φ_j and ψ_k, respectively. The subspace of zeroes, introduced above, may be assumed to be orthonormal. For example, consider the operator $\tilde{P}$. Let $\tilde{P}u = 0$ then $(u, P^*v) = 0$. Considering u as a functional in H, it is seen that it is annulled on the subspace $\mathrm{Im}P^*$ by the values of the operator P^*. There remains to find the value of this functional on the orthogonal extension space $H \ominus \mathrm{Im}P^* := \Phi$. Let

$$(u, \varphi_j) = c_j, \ \sum_{j=1}^{n} c_j \varphi_j = \varphi_0,$$

then $(u - \varphi_0, \varphi) = 0$, $\forall \varphi \in \Phi$ and, clearly, $u - \varphi_0$ is annulled on the entire space H, hence $u = \varphi_0$, as it has to be shown.

Next, it will be proved that the equation

$$\tilde{P}u = g\,, \; g \in H_{-1}\,, \qquad\qquad (A1.2.9)$$

is soluble, if $g \perp \psi$. The solution of this equation is the functional, satisfying the relation

$$(u, P^*v) = (g, v)\,, \; \forall v \in \mathrm{Im}P^*\,,$$

which determines the functional u on the subspace $\mathrm{Im}P^*$. In order to prove this result, it is sufficient to verify that, if v_1 and v_2 are two solutions of the equation $P^*v = g$, then $(g, v_1 - v_2) = 0$. However, this is obvious, because $(v_1 - v_2) \perp \psi$ and $g \perp \psi$. To fix, for example, $(u, \varphi) = 0$, $\varphi \in \Phi$ and the solution of equation (A1.2.9) has been constructed. Thereby the assertion above has been proved.

A consequence of this result is that the indices of the operators P and $\tilde{P}$ coincide. Setting $P = P_\omega$, $H = \mathbf{L}_2(\Omega)$, $H_1 = \mathbf{W}^{\circ 1}_2(\Omega)$, $H_2 = E(\Omega)$, one arrives at the following statement: If $\omega \neq -1, -2, \infty$, then the operator of problem I of the theory of elasticity is Noetherian by being an operator from H_1 into H_{-1}.

In the sequel, the indices 1 and -1 will be referred to as scalar products and norms in the spaces H_1 and H_{-1}, respectively.

4^0. Denote by A the vector operator $-\Delta$ for the boundary condition $u|_{\partial\Omega} = 0$. The problem $P_\omega u = 0$ is equivalent to the problem

$$A^{-1}\mathrm{grad}\,\mathrm{div}u - \eta u = 0\,, \; \eta = 1/\omega\,.$$

Introduce in the space H_1 the scalar product and norm according to the formulas

$$[u, v] = \int_\Omega \sum_{k=1}^{3} \frac{\partial u}{\partial x_k} \frac{\partial v}{\partial x_k}\, dx\,, \; |u|^2 = [u, u]\,. \qquad\qquad (A1.2.10)$$

Note that these definitions coincide with the energy product and energy norm of the operator A. It will be shown that for such a choice of scalar product the operator $B = -A^{-1}\mathrm{grad}\,\mathrm{div}$ is in the space H_1 symmetric, non-negative and bounded. To start with, assume that $u, v \in E(u)$. Then $[Bu, v] = -[A^{-1}\mathrm{grad}\,\mathrm{div}u, v] = -(\mathrm{grad}\,\mathrm{div}u, v) = (\mathrm{div}u, \mathrm{div}v)$. In particular, $[Bu, v] = \|\mathrm{div}u\|^2 \geq 0$, where $(,)$ and $\|\cdot\|$ denote the scalar product and norm in $L_2(\Omega)$ or in $\mathbf{L}_2(\Omega)$, depending on the fact that one has either to deal with a scalc or with a vector product.

Multiply the known identity $\Delta u = \operatorname{grad}\operatorname{div}u - \operatorname{rot}^2 u$ scalarly (in the sense of vector algebra) by u and integrate over Ω. Next, integrating by parts and using the boundary condition $u|_{\partial\Omega} = 0$, one arrives at

$$|u|^2 = D(u) = \|\operatorname{div}u\|^2 + \|\operatorname{rot}u\|^2 ; \qquad (A1.2.11)$$

where $D(u)$ stands for the Dirichlet integral for the vector u; let $D(u,v)$ denote the corresponding bi-linear form. The last result then yields

$$\|\operatorname{div}u\| \le |u|^2 , \ \forall u \in H_1 . \qquad (A1.2.12)$$

The equality sign is attained: it is sufficient to set $u = \operatorname{grad}u$, where u is a function which is sufficiently smooth in Ω and vanishes near $\partial\Omega$. Hence one finds that $|B| = 1$. The operator B, defined on a dense set in H_1, the manifold $E(\Omega)$ admits extension with respect to continuity over the entire H_1; the thus extended operator B will be self-adjoint, with a spectrum contained in the segment $[0,1]$.

Let $\omega \neq -1, -2, \infty$. The operator $\Delta + \omega\operatorname{grad}\operatorname{div}$ is elliptic, and problem I satisfies the condition of completeness. It follows then from lemma A1.1 that the characteristic vectors of the bundle P_ω coincide with the ordinary characteristic vectors of the operator B; the last, in particular, are analytic. However, if $\omega = -1, -2, \infty$, then the characteristic vectors of the bundle $P\omega$ are called the characteristic vectors of the operator B; in that case, one can only assert regarding the characteristic Cosserat vectors that they belong to the space H_1. It follows from the above that the Cosserat spectrum of the bundle P_ω lies on the infinite segment $[-\infty, -1]$.

Let $\omega \neq -1, -2, \infty$. Consider the operator $P_\omega - \zeta I$, where ζ is a numerical parameter and I the unit operator in H_1. Consider the question regarding the ordinary spectrum of the operator P_ω, i.e., regarding those values of ζ for which for the given ω the operator $(P_\omega - \zeta I)^{-1}$ either does not exist, or is unbounded, or is defined on a non-dense manifold. Special interest attaches to the value $\zeta = 0$. In view of what has been said above, it may belong to the ordinary spectrum of the operator P_ω only if $\omega \in [-\infty, -1]$. These values of ω will be considered below. For any $F \in H_{-1}$, the equations

$$P_\omega u = F , \qquad (A1.2.13)$$

$$I + \omega B = A^{-1}F \qquad (A1.2.14)$$

are equivalent, hence it follows that $\zeta = 0$ belong simultaneously or does not at all belong to the spectra of both operators P_ω and $I + \omega B$. Problem

(A1.2.13) is Noetherian, hence the operator P_ω has a finite number of zeroes, and this equation is soluble, if its free term satisfies a finite number of orthogonality conditions of the form $l_j F = 0$, $j = 1, 2, ..., k$, where the l_j are bounded functionals in H_{-1}. Let $\|l_j\|_{-1} = c_j$. Then $|l_j F| \leq c_j \|F\|_{-1}$. Moreover, the operator A is bounded as operator from H_1 into H_{-1}. Let $\|A\|_{H_1 \to H_{-1}} = c$, when $\|F\|_{-1} = \|AA^{-1}F\|_{-1} \leq c\|A^{-1}F\|_1$, and hence $|l_j F| \leq cc_j \|A^{-1}F\|_1$. Thus, l_j is a functional on $A^{-1}F$, bounded in H_1. Denoting this new functional by m_j, one has the sufficient condition for the solubility of equation (A1.2.13) $m_j(A^{-1}F) = 0$, $j = 1, 2, ..., k$. However, these conditions are also sufficient for the solubility of equation (A1.2.14), hence the operator $I + \omega B$ is normally soluble, and the conjugate operator $(I + \omega B)^*$ has a finite number of zeroes. Moreover, the operator $I + \omega B$ likewise has a finite number of zeroes - as many as has the operator P_ω. These results show that $I + \omega B$ is a Noctherian operator; in particular, the region of its values is closed. However, this operator is yet self-adjoint, hence $\eta = -1/\omega$ is either a regular point or the characteristic number of the operator B has finite multiplicity. This means that the points of the Cosserat spectrum of problem I of the theory of elasticity, other than $-1, -2, \infty$, are characteristic numbers of finite multiplicity.

The essential spectrum of the operator B (and with it, that of the spectrum-operator P_ω) cannot comprise more than the three points $\omega = -1, -2, \infty$. In such a case, the spectral expansion of the operator B does not contain an integral term, and the system of characteristic vectors of this operator is complete and orthogonal in H_1. However, the characteristic vectors of the operator B and the bundle P_ω coincide, hence it follows that the system of the characteristic Cosserat vectors of problem I of the theory elasticity is orthogonal (it may be assumed to be ortho-normal) and complete in $H_1 = \mathbf{W}^{\circ 1}_2(\Omega)$. Especially, it is complete in $H = \mathbf{L}_2(\Omega)$.

The character of the orthogonality of the characteristic Cosserat vectors of problem I will now be explained in greater detail. As it has already been stated

$$[u_k, u_n] = 0, \ k \neq n. \tag{A1.2.15}$$

Besides, u_k and u_n are characteristic vectors of the self-adjoint operator B, hence $[Bu_k, u_n] = 0$. However,

$$[Bu_k, u_n] = (ABu_k, u_n) = -(\operatorname{grad} \operatorname{div} u_k, u_n).$$

Integrating by parts and taking into consideration the boundary condition $u_k\big|_{\partial\Omega} = 0$, one finds that

$$(\operatorname{div} u_k, \operatorname{div} u_n) = 0, \ k \neq n. \tag{A1.2.16}$$

That equality which was obtained already by the Cosserat's (cf. (A1.1.9)). Next, it follows from (A1.2.11) that

$$[u_k, u_n] = (\operatorname{div} u_k, \operatorname{div} u_n) + (\operatorname{rot} u_k, \operatorname{rot} u_n),$$

hence

$$(\operatorname{rot} u_k, \operatorname{rot} u_n) = 0, \; k \neq n. \tag{A1.2.17}$$

5^0. It will be shown that the points $\omega = \infty$ and $\omega = -1$ are characteristic Cosserat numbers of problem I of infinite multiplicity. Recall that for the point $\omega = -1$ this assertion follows likewise from the Cosserat's results.

If $\omega = \infty$, then the homogeneous equation of the theory of elasticity has the form $\operatorname{div} u = 0$. Let u be a characteristic vector, then $(\operatorname{grad} \operatorname{div} u, u) = 0$. Integrating by parts and using the boundary condition, one finds $\|\operatorname{div} u\|^2$. Thus, any characteristic Cosserat vector of problem I, corresponding to the characteristic number $\omega = \infty$, belongs to the space H_1 and satisfies the relation

$$\operatorname{div} u = 0. \tag{A1.2.18}$$

The converse is also obvious: The vector $u \in H_1$ which satisfies equation (A1.2.18), is the characteristic Cosserat vector of problem I, corresponding to the characteristic number $\omega = \infty$. The space of these vectors has infinite dimension; let it be denoted by $H_1^{(\infty)}$.

If $\omega = -1$, then the homogeneous equation of the theory of elasticity assumes the form $\operatorname{rot}^2 u = 0$. If u of the corresponding characteristic vector, then $0 = (\operatorname{rot}^2 u, u) = \|\operatorname{rot} u\|^2$, and, consequently,

$$\operatorname{rot} u = 0. \tag{A1.2.19}$$

The converse if obvious. Thus, the characteristic Cosserat vectors of problem I, corresponding to the characteristic number $\omega = -1$, form in H_1 the infinite-dimensional sub-space $H_1^{(-1)}$, defined by (A1.2.19).

6^0. Denote by $\{u_n^{(-1)}\}$ and $\{u_n^{(\infty)}\}$ any bases of the sub-spaces $H_1^{(-1)}$ and $H_1^{(\infty)}$, orthonormalized in the metric of H_1. Moreover, let $\{u_n\}$ be the system of the remaining characteristic Cosserat vectors of the bundle P_ω, orthonormalized in the same metric, and let ω_n be the corresponding characteristic Cosserat numbers. Finally, let $u(x)$ be the solution of the problem $P_\omega u = F$, where, for example, $F \in H$, and ω does not belong to the Cosserat spectrum of the bundle P_ω. One finds by ordinary means that

$$u(x) = \sum_{n=1}^{\infty} \left\{ \frac{\omega_n f_n}{\omega_n - \omega} u_n(x) + \frac{f_n^{(-1)}}{1 + \omega} u_n^{(-1)}(x) + f_n^{(\infty)} u_n^{(\infty)}(x) \right\}; \tag{A1.2.20}$$

where $f_n = u^{(-2)}(F, u_n)$, $f_n^{(-1)} = u^{(-1)}(F, u_n^{(-1)})$, $f_n^{(\infty)} = u^{(-1)}(F, u_n^{(\infty)})$.

In particular, consider the homogeneous equation of the theory of elasticity

$$\Delta w + \omega \operatorname{grad} \operatorname{div} u = 0 \qquad (A1.2.21)$$

for the non-homogeneous boundary condition

$$u\big|_{\partial\Omega} = g(x). \qquad (A1.2.22)$$

Assume that ω does not belong to the Cosserat spectrum of problem I. Construct the harmonic vector $u_0(x)$ which satisfies this condition, setting $u - u_0 \neq w$. Then

$$\Delta w + \omega \operatorname{grad} \operatorname{div} w = F(x) := -\omega \operatorname{grad} \operatorname{div} u_0. \qquad (A1.2.23)$$

Assume that $u_0 \in H_1$, when, obviously, also $w \in H_1$. Formula (A1.2.20) applies to the vector w. It will be shown that $f_n^{(-1)} = f_n^{(\infty)} = 0$. In fact,

$$f_n^{(\infty)} = -\omega(\operatorname{grad} \operatorname{div} u_0, u_n^{(\infty)}) = \omega(\operatorname{div} u_0, \operatorname{div} u_n^{(\infty)}) = 0,$$
$$f_n^{(-1)} = -\omega(\operatorname{grad} \operatorname{div} u_0, u_n^{(-1)}) = \omega(\operatorname{rot}^2 u_0, u_n^{(-1)}) = \omega(\operatorname{rot} u_0, \operatorname{rot} u_n^{(-1)}) = 0.$$

Simplify this expression into

$$f_n = -\omega(\operatorname{grad} \operatorname{div} u_0, u_n) = \omega(\operatorname{div} u_0, \operatorname{div} u_n).$$

One has now

$$u(x) = u_0(x) + \omega \sum_{n=1}^{\infty} \frac{\omega_n}{\omega_n - \omega}(\operatorname{div} u_0, \operatorname{div} u_k)u_n(x). \qquad (A1.2.24)$$

This is a formula which generalizes (A1.1.5), obtained by the Cosserat for the sphere, to the case of an arbitrary finite region with sufficiently smooth boundary.

7^0. **Theorem A1.1.** Let $\{u_n(x)\}$ be the union of all characteristic Cosserat vectors of problem I, corresponding to characteristic numbers other than $\omega = -1, \infty$. Then the system of functions $\{\operatorname{div} u_n\}$, completed by constants, is complete in the space $G_2(\Omega)$ of functions which are harmonic and quadratically summable in Ω.

First of all, it will be shown that $\operatorname{div} u_n \in G_2(\Omega)$. Since $u_n \in H_1$, then $\operatorname{div} u_n \in L_2(\Omega)$. Moreover, the function $u_n(x)$ is a solution of the homogeneous, elliptic equation $\Delta u_n + \omega_n \operatorname{grad} \operatorname{div} u_n = 0$, and therefore it is analytic in Ω. Applying to the last equation the operation div, one finds $(1 + \omega_n)\Delta\mathrm{div}u_n = 0$; since $\omega_n \neq -1$, one has $\Delta\mathrm{div}u_n = 0$.

Now let $\varphi_0 \in G_2(\Omega)$ and

$$(\varphi_0, \mathrm{div}u_n) = 0\,,\ n = 1, 2, \ldots. \tag{A1.2.25}$$

It will be shown that one can construct a harmonic vector $v_0(x)$ such that $v_0 \in \mathbf{W}_2^1(\Omega)$ and $\mathrm{div}v_0 = \varphi_0$. The vector v_0 will be sought in the form of the simple layer potential

$$v_0(x) = \int_{\partial\Omega} \frac{\mu(\xi)n}{r}\, d_\xi\partial\Omega\,;\ r = |x - \xi|\,,\ u = \partial\Omega\,,$$

where $\mu(\xi)$ is a scalar function and n is the unit external normal to $\partial\Omega$ at the point $\xi \in \partial\Omega$. Then

$$\mathrm{div}v_0 = \int_{\partial\Omega} \mu(\xi)\frac{\partial}{\partial n}\, r^{-1}\, d_\xi\partial\Omega$$

and the density $\mu(\xi)$ is determined from the integral equation of the internal Dirichlet problem for Laplace's equation

$$\mu(x) - (2\pi)^{-1}\int_{\partial\Omega} \mu(\xi)\frac{\partial}{\partial n}r^{-1}\, d_\xi\partial\Omega = (2\pi)^{-1}\varphi(x)\,,\ x \in \partial\Omega\,.$$

Using the methods of [2], it may be shown that, if $\varphi \in L_2(\Omega)$, then $v_0 \in \mathbf{W}_2^1(\Omega)$.

Proceed now to the boundary value problem

$$\Delta u + \omega\mathrm{grad}\,\mathrm{div}u = 0\,,\ u\big|_{\partial\Omega} = v_0\,, \tag{A1.2.26}$$

it will be assumed that ω does not belong to the Cosserat spectrum of the bundle Δ_ω^*. For example, it may be assumed that $\omega > 0$. It follows from (A1.2.24) and (A1.2.25) that the solution of the last problem is $u(x) = v_0(x)$. Its substitution into (A1.2.26) yields $\mathrm{grad}\,\mathrm{div}v_0 = \mathrm{grad}\varphi_0$ and $\varphi_0 = const$, as it was required to show.

The results of this section permit the formulation of

Theorem A1.2. The bundle of operators of problem I of the theory of elasticity has a countable system of characteristic vectors which satisfy each of the orthogonality conditions (A1.2.15)—(A1.2.17). This system is complete in each of the spaces $H_1 = \mathbf{W}^{\circ 1}_2(\Omega)$ and $H = \mathbf{L}_2(\Omega)$. The characteristic numbers of the above bundle are located on the segment $-\infty \leq \omega \leq -1$ and may accumulate only at the points $\omega = -1, -2, \infty$. The numbers $\omega = -1$ and $\omega = \infty$ are characteristic values of infinite multiplicity.

8^0. In conclusion of this section, a few words will be devoted to the solubility of problem (A1.2.7) by iteration. It has been noted in §1 that iteration in accordance with (A1.1.11) converges, according to results of the Cosserat's, but that this convergence had only been proved for values of Poisson's constant which do not correspond to any known elastic medium. However, it is not difficult to reconstruct the iteration process so that it converges for all values ν, corresponding to the law of conservation of energy. Using the identity $\Delta = \operatorname{grad} \operatorname{div} - \operatorname{rot}^2$, give the boundary value problem the form

$$\Delta w + \omega^* \operatorname{rot}^2 w = -\mu^{-1} F, \quad w = (1+\omega)u, \quad w\big|_{\partial\Omega} = \varphi(x), \qquad (A1.2.27)$$

where $\omega^* = \omega/(1+\omega) = 1/2(1-\nu)$. Let $|\omega^*| < 1$. It follows from the identity (A1.2.11) that the homogeneous problem

$$\Delta w + \omega^* \operatorname{rot}^2 w = 0, \quad w\big|_{\partial\Omega} = 0$$

has only the zero solution. To the circle $|\omega^*| < 1$ corresponds the half-plane $\operatorname{Re}\omega > -1/2$ which is free of the Cosserat spectrum. In such a case, the homogeneous problem is soluble in the stated circle, and then its solution is unique. Likewise it is obvious that in the same circle the solution represents an analytic function of ω^*, and therefore it can be expanded in a power series in ω^*. This means that for $|\omega^*| < 1$ problem (A1.2.7) can be solved by the following iteration process

$$\Delta w^{(0)} = \mu^{-1} F, \quad w^{(0)}\big|_{\partial\Omega} = \varphi,$$
$$n \geq 0, \quad \Delta w^{(n+1)} = -\omega^* \operatorname{rot}^2 w^{(n)}, \quad w^{(n+1)}\big|_{\partial\Omega} = 0. \qquad (A1.2.28)$$

The condition $|\omega^*| < 1$ is fulfilled, if $\nu > 3/2$ or $\nu < 1/2$.

§3. Isolated points of the essential Cosserat spectrum

1^0. **Theorem A1.3** The points $\omega = \infty$ and $\omega = -1$ are isolated points of the Cosserat spectrum.

The proof depends on the integral equation to which Lichtenshtein [21] reduced problem I of the theory of elasticity. This equation has the form

$$\theta(x) - \frac{\omega^+}{4\pi} \int_{\partial\Omega} \rho \frac{\partial^2 G}{\partial\rho\partial n} \theta(y)\, d_y\partial\Omega = \frac{2}{2+\omega}\theta_0(x), \quad x \in \partial\Omega; \qquad (A1.3.1)$$

it has been derived under the assumption that there are no body forces. In this equation, $\theta = \mathrm{div}\,u$, $\theta_0 = \mathrm{div}\,u_0$, where u is the unknown displacement vector, u_0 is a vector which is harmonic in Ω and coincides with u on $\partial\Omega$, $\omega^+ = \omega/(\omega + 2)$, $G(x,y)$ is the Green function of the Dirichlet problem for Laplace's equation, n is the external normal to $\partial\Omega$, and, finally, ρ is the radius vector from the point x to the point y as well as the length of this radius vector.

Two assumptions are essential for the derivation of equation (A1.3.1): 1) there exists a mutually single-valued relationship between the stresses and displacements; 2) since it is assumed that body forces are absent, one has $\Delta\mathrm{div}\,u = 0$. The first assumption means that all values of ω are admissible, except, possible, $\omega = 1/3, \infty$. In fact, the determinant of the equations of Hooke's law vanishes for $\mu = 0$ and for $3\lambda + 2\mu = 0$, where λ and μ are the Lame constants; remember that $\nu = \lambda/2(\lambda + \mu)$. The second assumption requires that $\lambda + 2\mu \neq 0$. This condition excludes the values $\nu = 1$ or $\omega = -1$. Finally, one must exclude the value $\omega = -2$ for which in equation (A1.3.1) $\omega^+ = \infty$.

Thus, the values $\omega = -2, -1, 1/3, \infty$ are excluded from the ensuing considerations. For all remaining values of ω, equation (A1.3.1) and problem I of the theory of elasticity are equivalent: If a vector u solves the boundary value problem, then $\theta = \mathrm{div}\,u$ solves equation (A1.3.1); conversely, if θ solves equation (A1.3.1), then (cf., for example, [8]) the vector

$$u(x) = u_0(x) + \frac{\omega}{8\pi} \int\limits_{\partial\Omega} (y - x)\frac{\partial G(x,y)}{\partial n}\,\theta(y)\,d_y\,\partial\Omega$$

solves problem I.

The kernel of equation (A1.3.1) has for small ρ the estimate $O(\rho^{-1})$, so that the Fredholm theory is applicable to this equation.

Denote by ω_n the finite-fold, characteristic Cosserat number of problem I and by ω_n^+ the characteristic number of equation (A1.3.1). The equivalence, explained above, yields the relation

$$\omega_n^+ = -\frac{\omega_n}{\omega_n + 2}. \tag{A1.3.2}$$

If it could be shown that equation (A1.3.1) has only a finite number of characteristic numbers, then the bundle P_ω would have a finite number of finite-fold characteristic numbers; then theorem A1.3 would be obvious. However, if equation (A1.3.1) has a countable manifold of characteristic numbers, then

$\omega_n^+ \to \infty$, and, by relation (A1.3.2), $\omega_n \to -2$. Theorem A1.3 is proved for this case.

It follows from section A1.3 that equation (A1.3.1) is soluble by Liechtenstein iteration.

2^0. Note the following two consequences of theorem A1.3

1) It has been explained in §2, 7^0 that there exists an infinite manifold of characteristic Cosserat numbers of problem I, other than $\omega = -1$ and $\omega = \infty$. Since they may accumulate only at points of the essential spectrum, then their single point of accumulation is $\omega = -2$, and one of the following two possibilities arises:

a) $\omega = -2$ is a characteristic number of infinite multiplicity;

b) $\omega = -2$ is the limit point of finite -fold characteristic numbers.
The authors are not aware of the existence of a finite region for which the first possibility arises, but it occurs, for example, for the half-space [3].

2) Consider the contraction of B_0 into the operator B on the subspace $H_1 \ominus H_1^{(\infty)}$. By theorem A1.3, the spectrum of the operator B_0 does not contain the point $\eta = \infty$, and therefore it is bounded positively from below. Let k_0^2 be the lower bound of this spectrum, when $(B_0 u, u) \geq k_0^2 \|u\|^2$. A simple transformation yields

$$\|\mathrm{div} u\| \geq k_0 |u|, \quad u \in H_1 \ominus H_1^{(\infty)}. \tag{A1.3.3}$$

The inequality

$$\|\mathrm{rot} u\| \geq k_1 |u|, \quad k_1 > 0, \quad u \in H_1 \ominus H_1^{(-1)}. \tag{A1.3.4}$$

may be proved in the same manner.

3^0. Knowing the characteristic Cosserat numbers and characteristic Cosserat vectors of problem I of the theory of elasticity, one can solve the following problems:

a) Find the vector $u \in H_1 \ominus H_1^{(\infty)}$ which satisfies the equation $\mathrm{div} u = \varphi(x)$, $x \in \Omega$, where $\varphi(x)$ is a given function, and the boundary condition $u|_{\partial\Omega} = 0$.

b) Find the vector $u \in H_1 \ominus H_1^{(-1)}$ such that $\mathrm{rot} u = \psi(x)$, $u|_{\partial\Omega} = 0$.

4^0. This section will be closed with the following observation. The homogeneous equation of the theory of elasticity may be given the form

$$(\Delta - 2\mathrm{grad}\,\mathrm{div})u + \kappa\,\mathrm{grad}\,\mathrm{div} u = 0, \quad \kappa = \omega + 2, \tag{A1.3.5}$$

Consider the solution of this equation which belongs to the infinitely dimensional space $H_1^{(0)} = H_1 \ominus H_1^{(-1)} \ominus H_1^{(\infty)}$; in this sub-space, the spectrum of equation (A1.3.5) consists only of the characteristic numbers $\kappa_n = \omega_n + 2 \to 0$ as $n \to \infty$. It will be shown that the elements of the sub-space $H_1^{(0)}$ are generalized solutions of the equation

$$\mathrm{rot}^3 u = 0 \,. \qquad (A1.3.6)$$

In fact, any characteristic vector $u_n \in H_1^{(0)}$ satisfies the equation $\mathrm{rot}^2 u_n = (1+\omega_n)\mathrm{grad}\,\mathrm{div} u_n$, and, consequently, also equation (A1.3.6). The last is true also for linear combinations of vectors u_n, hence it is true in the generalized sense for any vector from the sub-space $H_1^{(0)}$ which, obviously, is stiff on the vectors u_n, $n = 1, 2, \dots$.

§4. The second boundary value problem

1^0. Consider the general non-homogeneous second boundary value problem of the theory of elasticity

$$\Delta u + \omega \,\mathrm{grad}\,\mathrm{div} u = \mu^{-1} F(x), \; x \in \Omega; \qquad (A1.4.1)$$

$$(\omega - 1)\alpha_j \mathrm{div} u + \sum_{k=1}^{3} \left(\frac{\partial u_j}{\partial x_k} + \frac{\partial u_k}{\partial x_j} \right) \alpha_k = \Psi_j(x); \; j = 1, 2, 3, \; x \in \partial\Omega. \quad (A1.4.2)$$

To start with, assume that the region Ω is finite, and that its boundary $\partial\Omega$ is sufficiently smooth. In the boundary condition (A1.4.2), $\alpha_k = \cos(n, x_k)$ and n is the normal to $\partial\Omega$. It will now be shown for what values of ω the completeness condition for problem (A1.4.1) and (A1.4.2) is violated; the values $\omega = -1$ and $\omega = \infty$ for which equation (A1.4.1) fails to be elliptic will be excluded beforehand.

For $x \in \partial\Omega$, equation (A1.4.1) and the boundary conditions (A1.4.2) are invariant with respect to translation and rotation of the coordinate axes, hence it may be assumed that they are described in a local coordinate system with origin at the point x. Then $\alpha_1 = \alpha_2 = 0$, $\alpha_3 = 1$, and the boundary conditions assumes the form

$$\begin{cases} \dfrac{\partial u_1}{\partial x_3} + \dfrac{\partial u_3}{\partial x_1} = \psi_1(x), \; \dfrac{\partial u_2}{\partial x_3} + \dfrac{\partial u_3}{\partial x_2} = \psi_2(x), \\[2mm] (\omega - 1)\mathrm{div} u + 2\dfrac{\partial u_3}{\partial x_3} = 0 \,. \end{cases} \qquad (A1.4.3)$$

As in §2, it may be assumed that $\xi_1 = 1$, $\xi_2 = 0$. The matrix $G = G_0$ has the form

$$\begin{pmatrix} \tau & 0 & 1 \\ 0 & \tau & 1 \\ \omega - 1 & 0 & \omega + 1 \end{pmatrix} \,.$$

Multiply it from the right by the matrix (A1.2.6) and multiply the remainders of the division by $M^+(\xi',\tau) = (\tau - i)^3$. The determinant of these remainders has the form $m = \tau - i$

$$m^3 \begin{vmatrix} 2[2\omega - i(5\omega + 2)m] & 0 & 2[2\omega + (3\omega - 2)m] \\ 0 & -2[(\omega + 1)(1 - 3i\,m)] & 0 \\ 2[2i\omega + (3\omega + 2)m] & 0 & -2[2\omega - i(\omega - 2)m] \end{vmatrix}$$

and equals

$$64\omega(\omega + 1)(\omega + 2)(\tau - i)^3[1 - 3i(\tau - i)] . \qquad (A1.4.4)$$

Thus, if it does not assume the earlier excluded values $\omega = -1, \infty$, then the completeness condition for problem II is violated at the points $\omega = 0$ and $\omega = -2$, hence it follows that problem II of the theory of elasticity is Noetherian for all values of ω, except possibly for $\omega = -2, -1, 0, \infty$.

Assume now that the vector of the body forces as well as the vector of the surface stresses satisfy the conditions of static equilibrium; this means that these vectors are orthogonal (respectively, in the metric $L_2(\Omega)$ and $L_2(\partial\Omega)$) to arbitrary, rigid displacement.

Denote by $\widetilde{W}_2^1(\Omega)$, $\widetilde{L}_2(\Omega)$ and $\widetilde{L}_2(\partial\Omega)$ the sub-spaces of the corresponding spaces, orthogonal (the first two in the metric of $L_2(\Omega)$, the third in the metric $L_2(\partial\Omega)$) to arbitrary, rigid displacement. Introduce the function pair $U = (u, v)$ of the Hilbert space $\Pi(\Omega)$, with the norm

$$\|U\|_\Pi^2 = \|u\|_{W_2^{-1}} + \|v\|_{L_2} . \qquad (A1.4.5)$$

Denote by N_1 and N_2 the operators which act from $W_2^1(\Omega)$ into $\Pi(\Omega)$ and are defined by the formulas

$$N_1(u) = (-\Delta u - \operatorname{grad} \operatorname{div} u; \delta u) , \qquad (A1.4.6)$$

where the vector $\delta u = (\delta_1 u, \delta_2 u, \delta_3 u)$ has the form

$$\delta_j u = \sum_{k=1}^{3} \left(\frac{\partial u_j}{\partial x_k} + \frac{\partial u_k}{\partial x_j} \right) \alpha_k , \; j = 1, 2, 3 \qquad (A1.4.7)$$

and, with the normal unit vector to $\partial\Omega$,

$$N_2(u) = (-\operatorname{grad} \operatorname{div} u, n \operatorname{div} u) . \qquad (A1.4.8)$$

Problem (A1.4.1) and (A1.4.2) may now be written in the form

$$(N_1 + \eta N_2)u = (-\mu^{-1}F, \psi)\,; \quad \eta = \omega - 1\,. \qquad (A1.4.9)$$

Note that N_1 is the operator of problem II of the theory of elasticity for $\omega = 1$, or $\nu = 0$.

Denote by $E(u)$ the density of potential strain energy, corresponding to the values of the elastic constants $\nu = 0$ and $\mu = 1$:

$$2E(u) = 1/2 \sum_{j,k=1}^{3} \left(\frac{\partial u_j}{\partial x_k} + \frac{\partial u_k}{\partial x_j}\right)^2\,.$$

The norm in $\widetilde{\mathbf{W}}_2^1(\Omega)$ will be given by the relation

$$\|u\|_{\overline{\mathbf{W}}_2^{(1)}} = 2 \int_\Omega E(u)\,dx\,. \qquad (A1.4.10)$$

The operator N_1^{-1} exists and is defined on the entire space $\Pi(\Omega)$. Applying it to both sides of equation (A1.4.9), one finds

$$(I + \eta N_1^{-1} N_2)u = N_1^{-1}(-\mu^{-1}F, \psi) := f\,. \qquad (A1.4.11)$$

where I is the identity operator and f some known function; in the problem of characteristic values, one has $f = 0$.

It will be shown that the operator $N_1^{-1} N_2$ is symmetric, non-negative and bounded in $\widetilde{\mathbf{W}}_2^1$. Denote by Δ_1^* the differential operator of the theory of elasticity for $\omega = 1$ and by $p_1(u)$ the corresponding vector of the surface stresses

$$\Delta_1^* = \Delta u + \operatorname{grad} \operatorname{div} u\,,$$

$$(p_1(u))_j = \sum_{k=1}^{3} \left(\frac{\partial u_j}{\partial x_k} + \frac{\partial u_k}{\partial x_j}\right)\alpha_k\,, \quad j = 1, 2, 3\,, \qquad (A1.4.12)$$

so that $N_1(u) = (-\Delta_1^* u, p_1(u))$. By Betti's formula

$$-(\Delta_1^* u, u) = 2 \int_\Omega E(u)\,dx - \int_{\partial\Omega} u p_1(u)\,d\partial\Omega\,.$$

Hence

$$(u, v)_{\widetilde{W}_2^1} = -(\Delta_1^* u, v) + \int_{\partial\Omega} v p_1(u)\, d\partial\Omega. \qquad (A1.4.13)$$

Moreover,

$$(N_1^{-1} N_2 u, v)_{\widetilde{W}_2^1} = -(\Delta_1^* N_1^{-1} N_2 u, v) + \int_{\partial\Omega} v p_1(N_1^{-1} N_2 u)\, d\partial\Omega =$$
$$\qquad (A1.4.14)$$
$$= -(\operatorname{grad}\operatorname{div} u, v)_{L_2} + \int_{\partial\Omega} v(\operatorname{div} u) n\, d\partial\Omega = (\operatorname{div} u, \operatorname{div} v)_{L_2}.$$

It follows from the last identity that the operator $N_1^{-1} N_2$ is symmetric. Setting in this identity $u = v$, one finds that the operator $N_1^{-1} N_2$ is non-negative:

$$(N_1^{-1} N_2 u, u) = \|\operatorname{div} u\|_{L_2(\Omega)}^2 \geq 0.$$

Finally, the boundedness of this operator follows from the inequality

$$\|\operatorname{div} u\|_{L_2(\Omega)}^2 = \int_\Omega (\operatorname{div} u)^2\, dx \leq 3 \int_\Omega \sum_{k=1}^3 \left(\frac{\partial u_k}{\partial x_k}\right)^2 dx \leq$$
$$\leq 3/2 \int_\Omega 2E_1(u)\, dx = 3/2\|u\|_{\widetilde{W}_2^1(\Omega)}.$$

There remains now to repeat the reasoning of §2 which leads to

Theorem A1.4. For problem II of the theory of elasticity, there exists a system of characteristic Cosserat vectors which is orthogonal in the metric (A1.4.10). This system is complete in $\widetilde{W}_2^1(\Omega)$ and in $\widetilde{L}_2(\Omega)$. The corresponding characteristic numbers are located on the segment $-\infty \leq \omega \leq 1/3$ and can accumulate only at the points $\omega = 0, -1, -2, \infty$.

2^0. **Theorem A1.5.** For problem II of the theory of elasticity, the numbers $\omega = -\infty$ and $\omega = -1$ are isolated characteristic Cosserat numbers of infinite multiplicity.

If $\omega = \infty$, then the homogeneous differential equation of the theory of elasticity and the homogeneous boundary condition of problem II assume, respectively, the form $\operatorname{grad}\operatorname{div} u = 0$ and $\operatorname{div} u = 0$. Hence it is seen that the manifold of the characteristic Cosserat vectors of problem II, corresponding to the characteristic number $\omega = \infty$, coincides with the manifold of the vectors of class $\widetilde{\mathbf{W}}_2^1(\Omega)$ with zero divergence; obviously, this manifold is infinite-dimensional. If $\omega = -1$, one will not succeed to write down such a simple manifold of characteristic Cosserat vectors; however, it is not difficult to prove that this manifold is infinite-dimensional. The equation of the theory of elasticity in the given case assumes the form $\mathrm{rot}^2 u = 0$. Let $\Phi(x)$ be an arbitrary function of the class $C^3(\Omega)$ which vanishes in the neighbourhood of the boundary $\partial\Omega$. The vector $u = \mathrm{grad}\Phi$ satisfies the homogeneous equations (A1.4.1) and (A1.4.2). if one sets there $\omega = -1$.

It will now be shown that the points $\omega = -1$ and $\omega = \infty$ are isolated points of the Cosserat spectrum of problem II. For this purpose, use will be made of the singular integral equation of problem II (problem (A1.4.1) and (A1.4.2)) which has the form

$$u(x) + 2 \int_{\partial\Omega} P(x,y)u(y)\,d_y\partial\Omega = 2g(x)\,. \qquad (A1.4.15)$$

The notation will now be explained. Let $V = -\Gamma(x,y)$, where Γ is the Somigliani tensor and Γ^j, $j = 1,2,3$ its columns. If $v(x)$ is any vector of elastic displacements in Ω, then $p(v)$ will denote its corresponding vector of elastic stresses, acting on a surface area element $\partial\Omega$. The kernel $P(x,y)$ of equation (A1.4.15) is the matrix with the columns $p(-\Gamma^j)$, $j = 1,2,3$; it is readily seen that the elements of this matrix are

$$P_{jk}(x,y) = \frac{1}{8\pi(1-\nu)}\left[\frac{1-2\nu}{r^3}(\xi_j\delta_{kl} - \xi_k\delta_{jl} - \xi_l\delta_{jk}) - 3r^{-5}\xi_j\xi_k\xi_l\right]\alpha_l\,,$$
$$(A1.4.16)$$

where $\xi = y - x$, $r = |\xi|$, and the α_l are the direction cosines of the external normal to $\partial\Omega$ at the point y; summation with respect to the subscript l occurs between the limits 1 and 3. Finally, the function $g(x)$ is determined by

$$g(x) = \int_{\Omega} V(x,y)\mu^{-1}F(y)\,dy + \int_{\partial\Omega} V(x,y)\psi(y)\,d_y\partial\Omega\,. \qquad (A1.4.17)$$

A derivation of equation (A1.4.15) is given in [4] as well as in chapter 6 of this book.

Problem (A1.4.1) and (A1.4.2) and equation (A1.4.15) are equivalent: if a boundary value problem has the solution $u(x)$, then the restriction on $\partial\Omega$ of this solution satisfies equation (A1.4.15); conversely, if this equation has the solution $u(x)$, $x \in \partial\Omega$, then substitution of this solution into

$$u(x) = -\int_{\partial\Omega} P(x,y)u(y)\,d_y\partial\Omega + g(x)\,, \qquad (A1.4.18)$$

as follows readily from Betti's formulas, yields the solution of the boundary value problem.

In a local coordinate system, the symbolic matrix of equation (A1.4.15) has the form

$$\Phi = \begin{pmatrix} 1 & 0 & -i\delta\cos\Theta \\ 0 & 1 & -i\delta\sin\Theta \\ -i\delta\cos\Theta & -i\delta\sin\Theta & 1 \end{pmatrix}, \ \delta = \frac{1}{\omega+1}; \qquad (A1.4.19)$$

its determinant equals $1 - \delta^2 = \omega(\omega+2)/(\omega+1)^2$, hence the symbol of equation (A1.4.15) degenerates only for $\omega = 0$ and $\omega = -2$. For other values of ω, the stated equation admits regularization, hence it follows readily that for $\omega \neq 0$ the solution of equation (A1.4.15) satisfies a system of Fredholm equations the matrix kernel of which is an analytic function of ω with the isolated, singular points $\omega = 0, -2$. The last assertion is easily verified with the aid of the expression for the inverse symbolic matrix

$$\Phi^{-1} = \frac{(\omega+1)^2}{2\omega(\omega+2)} \begin{pmatrix} 1 - \delta^2\sin^2\Theta & -\delta^2\sin\Theta\cos\Theta & i\delta\cos\Theta \\ -\delta^2\sin\Theta\cos\Theta & 1 - \delta^2\cos^2\Theta & i\delta\sin\Theta \\ -i\delta\cos\Theta & i\delta\sin\Theta & 1 \end{pmatrix}.$$
$$(A1.4.20)$$

Note still that for the matrix kernel mentioned here the Fredholm system the points $\omega = -1, \infty$ are regular. It follows from these observations (cf., for example, [6], p. 34) that the points $\omega = -1, \infty$ may only be isolated singularities of the spectrum of the above mentioned system, and in such a case these points are isolated points of the Cosserat spectrum of problem II of the theory of elasticity.

Note that one can also study by such methods the spectrum of problem I.

3^0. It will be shown that the characteristic Cosserat vectors of problem II satisfy the condition of orthogonality (A1.1.9). Denote these vectors by $\tilde{u}_n$, and their corresponding Cosserat numbers by $\tilde{\omega}_n$. It has been shown in 1^0 above that the vectors $\tilde{u}_n$ are orthogonal in the metric, generated by the quadratic form (A1.4.10), and that their system is complete in $\mathbf{W}_2^1(\Omega)$.

Let $\nu_0 \neq 0$ be any fixed value of Poisson's constant from the segment $1 < \nu_0 < 1/2$; for such values of ν_0, twice the density of potential strain energy (it is assumed that $\mu = 1$) equals

$$2E_{\nu_0}(u) = \frac{2\nu_0}{1 - 2\nu_0}(\operatorname{div}u)^2 + 1/2 \sum_{j,k=1}^{3} \left(\frac{\partial u_j}{\partial x_k} + \frac{\partial u_k}{\partial x_j}\right)^2 \qquad (A1.4.21)$$

which represents a positive quadratic form of the strains. It will be shown that the characteristic vectors $\tilde{u}_n$ are orthogonal in the metric, generated by this quadratic form:

$$\|u\|_{\widetilde{W}_2^1}^2 = \int\limits_\Omega \left[\frac{2\nu_0}{1 - 2\nu_0}(\operatorname{div}u)^2 + 1/2 \sum_{j,k=1}^3 \left(\frac{\partial u_j}{\partial x_k} + \frac{\partial u_k}{\partial x_j} \right)^2 \right] dx . \quad (A1.4.22)$$

Set $\omega_0 = (1 - 2\nu_0)^{-1}$. Give the differential equation and boundary conditions of problem II the form

$$\Delta u + \omega_0 \operatorname{grad} \operatorname{div}u + (\omega - \omega_0)\operatorname{grad} \operatorname{div}u = -\mu^{-1}F(x), \quad x \in \Omega ;$$

$$(\omega_0 - 1)\alpha_j \operatorname{div}u + \sum_{k=1}^3 \left(\frac{\partial u_j}{\partial x_k} + \frac{\partial u_k}{\partial x_j} \right)\alpha_k + (\omega - \omega_0)\alpha_j \operatorname{div}u = \psi_j , \quad x \in \partial\Omega .$$

$$(A1.4.23)$$

Denote by N_0 the operator, acting from $\widetilde{\mathbf{W}}_2^1(\Omega)$ into $\Pi(\Omega)$ according to the formulas

$$N_0 u = (-\Delta u - \omega_0 \operatorname{grad} \operatorname{div}u, \varepsilon u) ,$$

$$\varepsilon_j u = (\omega_0 - 1)\alpha_j \operatorname{div}u + \sum_{k=1}^3 \left(\frac{\partial u_j}{\partial x_k} + \frac{\partial u_k}{\partial x_j} \right)\alpha_k , \quad j = 1, 2, 3 . \qquad (A1.4.24)$$

where N_0 is, exactly apart from the factor μ, the operator of problem II of the theory of elasticity for the value ν_0 of Poisson's constant. Problem (A1.4.23) may be given the form

$$(N_0 + \kappa N_2)u = (-F/\mu, \psi), \quad \kappa = \omega - \omega_0 . \qquad (A1.4.25)$$

Repeating with corresponding changes the reasoning of 1^0, one finds that in the metric (A1.4.22) the operator $N_0^{-1}N_2$ is non-negative, bounded and self-adjoint; for this, the characteristic Cosserat vectors of problem II are the characteristic vectors of the operator $N_1^{-1}N_2$. Then they are orthogonal in the metric (A1.4.22); simultaneously, as has been shown in 1^0, they are orthogonal in the metric (A1.4.10). Hence it is seen that the characteristic Cosserat vectors are orthogonal in the sense of formula (A1.1.9).

4^0. The representation of the solution of problem II of the theory of elasticity in terms of the characteristic vectors and characteristic numbers of Cosserat will now be derived. Let ω not belong to the Cosserat spectrum of problem II. Then the solution of problem (A1.4.1) and (A1.4.2) exists, is unique and can be expanded in a series with respect to the characteristic Cosserat vectors of problem II:

$$u(x) = \sum_{n=1}^{\infty} a_n \tilde{u}_n(x)\,.$$

By (A1.1.9),

$$a_n = \frac{(\mathrm{div}\,u, \mathrm{div}\,\tilde{u}_n)}{\|\mathrm{div}\,\tilde{u}_n\|^2}\,. \tag{A1.4.26}$$

Let the characteristic vectors be orthonormal in the metric (A1.4.10). Then

$$a_n = \left(u, \tilde{u}_n\right)_{\widetilde{W}_2^1} = 2 \int_{\Omega} E_1(u, u_n)\,dx\,. \tag{A1.4.27}$$

Multiply equation (A1.4.1) scalarly in the metric of $\mathbf{L}_2(\Omega)$ by $\tilde{u}_n$. Transforming the right hand side of the result using Betti's formula, one finds

$$a_n[1 + (\omega - 1)\|\mathrm{div}\,\tilde{u}_n\|^2] = (f, \tilde{u}_n)_{\Pi}\,; \tag{A4.28}$$

where $f = (-\mu^{-1}F, \psi)$ is the union of the given data for problem II.

Multiply the identity $\Delta\tilde{u}_n + \omega_n\mathrm{grad}\,\mathrm{div}\,\tilde{u}_n = 0$ scalarly in the metric $\mathbf{L}_2(\Omega)$ by $\tilde{u}_n$ and transform the result by Betti's formula. Using the boundary condition and normalization of $\tilde{u}_n$, one finds that $\|\mathrm{div}\,\tilde{u}_n\|^2 = (1 - \tilde{\omega}_n)^{-1}$. By (A1.4.28), one now obtains

$$u(x) = \sum_{n=1}^{\infty} \frac{1 - \tilde{\omega}_n}{\omega - \omega_n} \left(f, \tilde{u}_n\right)_{\Pi} \tilde{u}_n(x)\,. \tag{A1.4.29}$$

The terms corresponding to the characteristic numbers $\omega = -1$ and $\omega = \infty$ may be separated; if one denotes their corresponding characteristic vectors by $\tilde{u}_n^{(-1)}$ and $\tilde{u}_n^{(\infty)}$, then

$$u(x) = \sum_{n=1}^{\infty} \left\{ \frac{2}{\omega + 1} \left(f, \tilde{u}_n^{(-1)}\right)_{\Pi} \tilde{u}_n^{(-1)}(x) + \left(f, \tilde{u}_n^{(\infty)}\right)_{\Pi} \tilde{u}_n^{(\infty)}(x) + \frac{1 - \tilde{\omega}_n}{\omega - \tilde{\omega}_n} \tilde{u}_n(x) \right\}\,.$$
$$\tag{A1.4.30}$$

where $\tilde{\omega}_n$ denotes the characteristic Cosserat numbers other than -1 and ∞, and $\tilde{u}_n(x)$ are their corresponding characteristic vectors.

5^0. The third boundary value problem differs from the second by the fact that there appear on the left hand side of the boundary condition (A1.4.2) terms hu_j, where h is a given, say, continuous function of the point $x \in \partial\Omega$. This addition changes neither the matrix G_0 (§2, 2^0) nor the singular part of equation (A1.4.15), hence it follows that the basic results of this section can be transferred without change to problem III of the theory of elasticity. In particular, theorems A1.4 and A1.5 remain in force.

§5. The case of infinite regions

1^0. Let Ω be an infinite region of three-dimensional space, bounded by a finite, closed, sufficiently smooth surface $\partial\Omega$. Consider problem I of the theory of elasticity

$$\Delta u + \omega \operatorname{grad} \operatorname{div} u = -\mu^{-1} F(x), \quad x \in \Omega, \qquad (A1.5.1)$$

$$u\big|_{\partial\Omega} = 0; \qquad (A1.5.2)$$

for the sake of simplicity, consideration will be restricted to a homogeneous boundary condition. The generalized solution of this problem will be sought in the energy space H_0 of the Dirichlet problem for the vector Laplace equation. Recall (cf. [3]) that this space consists of vectors with finite Dirichlet integral which vanish on $\partial\Omega$, and that the Dirichlet integral equals the quadratic norm in this space.

The left hand side of equation (A1.5.1) together with the boundary condition (A1.5.2) determines a bundle of operators which, as before, will be denoted by P_ω. As above, the Cosserat spectrum of the problem under consideration will be referred to as the spectrum of the bundle P_ω.

2^0. The generalize solution of problem (A1.5.1) and (A1.5.2), by definition, satisfies the integral identity

$$D(u,\zeta) + \omega(\operatorname{div} u, \operatorname{div}\zeta) = (\mu^{-1}F, \zeta), \quad \forall \zeta \in H_0. \qquad (A1.5.3)$$

In the sequel, the energy product and norm in H_0 will be denoted by the usual symbols $[,]$ and $|\cdot|$ so that $[u,v] = D(u,v)$ and $|u|^2 = D(u)$. Let the brackets () denote an integral of the form

$$(p,q) = \int\limits_\Omega p(x)\, q(x)\, dx\,,$$

where p and q do not necessarily belong to $\mathbf{L}_2(\Omega)$.

In order that the generalized solution of problem (A1.5.1) and (A1.5.2) will exist, it is necessary and sufficient to fulfill the following conditions: The integral (F, ζ) exists whether or whether not the vector $\zeta \in H_0$ and it represents itself a functional, bounded in H_0. It is not difficult to see that this, in its turn, is necessary and sufficient for the existence of the symmetric tensor $T = [\sigma_{jk}]_{j,k=1}^{j,k=3}$, $\sigma_{jk} = \sigma_{kj}$ which satisfies the equilibrium equations

$$\frac{\partial \sigma_{kj}}{\partial x_k} + F_j = 0, \; j = 1, 2, 3\,.$$

and the inequality

$$\int\limits_{\Omega} \sum_{j,k=1}^{3} \sigma_{jk}^2 \, dt < \infty .$$

In fact, if the generalized solution of problem (A1.5.1) and (A1.5.2) exists, then one may take for T the stress tensor which corresponds to the displacement vector u according to Hooke's law. Conversely, if the tensor T with the stated properties exists, then, integrating by parts, one obtains

$$(\mu^{-1}F,\zeta) = - \int\limits_{\Omega} \zeta \operatorname{div} T \, dx = - \int\limits_{\Omega} \sum_{j,k=1}^{3} \frac{\partial \sigma_{kj}}{\partial x_k} \, \zeta_j \, dx = \int\limits_{\Omega} \sum_{j,k=1}^{3} \sigma_{kj} \frac{\partial \zeta_j}{\partial x_k} \, dx .$$

Moreover, by Bunyakovskii's inequality,

$$|(\mu^{-1}F,\zeta)| \le \mu^{-1} \Big[\int\limits_{\Omega} \sum_{j,k=1}^{3} \sigma_{kj} \, dx \Big] \|\zeta\|_{L_2}^2 \le C \|\zeta\|^2 ,$$

which signifies that the functional $(\mu^{-1}F,\zeta)$ is bounded in H_0 and, consequently, the generalized solution exists.

By Riesz' theorem, there exists one and only one vector $f \in H_0$ for which for any $\zeta \in H_0$ the identity $(\mu^{-1}F,\zeta) = [f,\zeta]$ holds, or, what is the same thing,

$$[u,\zeta] + \omega(\operatorname{div} u, \operatorname{div}\zeta) = [f,\zeta] . \tag{A1.5.4}$$

The space H_0 may be obtained as the closure of the manifold of the finite vectors in the metric of the Dirichlet integral [8]; here the word "finite" indicates that an infinitely differentiable function, defined in Ω, equals zero in the neighbourhood of $\partial\Omega$ and near the point at infinity. Let $u \in H_0$ and $\{u^{(n)}\}$ be a sequence of finite vector functions, approximating u in the metric of the Dirichlet integral. Multiply the identity $\Delta u^{(n)} = \operatorname{grad} \operatorname{div} u^{(n)} - \operatorname{rot}^2 u^{(n)}$ scalarly, in the sense of the metric of $L_2(\Omega)$, on $u^{(n)}$, integrating the result parts to arrive at $|u^{(n)}|^2 = \|\operatorname{div} u^{(n)}\|^2 + \|\operatorname{rot} u^{(n)}\|^2$, and, finally, proceed to the limit $n \to \infty$, to obtain

$$\forall u \in H_0 , \quad |u|^2 = \|\operatorname{div} u\|^2 + \|\operatorname{rot} u\|^2 . \tag{A1.5.5}$$

3^0. Consider now identity (A1.5.4). By Cauchy's inequality and identity (A1.5.5), one has

$$(\operatorname{div} u, \operatorname{div}\zeta) \le \|\operatorname{div} u\| \cdot \|\operatorname{div}\zeta\| \le |u| \cdot |\zeta| . \tag{A1.5.6}$$

It will be assumed that u is a fixed and ζ a variable element of the space H_0. The last inequality shows that the functional $(\operatorname{div}u, \operatorname{div}\zeta)$, linear with respect to ζ, is bounded in H_0 and that its norm does not exceed $|u|$. By the strength of the theorem of Riesz, referred to above, there exists one and only one element of the space H_0; this element will be denoted by Ru, such that

$$(\operatorname{div}u, \operatorname{div}\zeta) = [Ru, \zeta] . \qquad (A1.5.7)$$

Substituting this result into identity (A1.5.4), one sees that the generalized solution of problem (A1.5.1) and (A1.5.2) satisfies the equation

$$\omega^{-1}u + Ru = \omega^{-1}f ; \qquad (A1.5.8)$$

the converse is obvious.

Certain simple properties of the operator R will now be noted. Setting in (A1.5.7) $\zeta = u$, one finds $[Ru, u] = \|\operatorname{div}u\|^2$; this result shows that the operator R in H_0 is self-adjoint and non-negative. Setting next in (A1.5.7) $\eta = Ru$, one obtains

$$|Ru|^2 = (\operatorname{div}Ru, \operatorname{div}u) \leq \|\operatorname{div}Ru\| \cdot \|\operatorname{div}u\| \leq |Ru| \cdot |u| \qquad (A1.5.9)$$

and, consequently, $|R| \leq 1$.

The Cosserat spectrum of problem I coincides with the manifold of those values ω for which $-\omega^{-1}$ belongs to the spectrum of the operator R, hence it follows that the Cosserat spectrum of problem I lies on the segment $-\infty \leq \omega \leq -1$. It is readily verified, as in the case of finite regions, that the points $\omega = \infty$ and $\omega = -1$ are characteristic Cosserat numbers of infinite multiplicity: Characteristic vectors are any vectors of H_0 for which $\operatorname{div}u = 0$ and $\operatorname{rot}u = 0$, respectively.

4^0. **Theorem A.6.** In the case of infinite regions with sufficiently smooth boundaries, the Cosserat spectrum of problem I of the theory of elasticity consists of the points $-1, -2, \infty$ and, besides, of characteristic numbers of finite multiplicity which may only accumulate at the point $\omega = -2$. The system of all characteristic Cosserat vectors is complete in H_0.

The proof requires only that part of the theorems which relates to characteristic numbers. The assertion of the completeness of this system of characteristic vectors follows from the first assertion of the theorem and from self-adjointness of the operator R. The theorem will be proved below in the following two sections.

5^0. Consider the differential properties of the Characteristic Cosserat vectors. Exclude from the study the values $\omega = -1$ and $\omega = \infty$. For the remaining values of ω, the system of equations of the theory of elasticity is elliptic and, in the absence of body forces, $\Delta \operatorname{div} u = 0$. Exclude, in addition, the value $\omega = -2$; for the remaining values of ω in any finite region the condition of completeness is fulfilled for problem I.

Let ω be a characteristic Cosserat number, other than -2, -1 and ∞, and let u be the corresponding characteristic vector. The system of equations of the theory of elasticity is elliptic and therefore the vector u is analytic inside Ω. Consider a finite region Ω_R, bounded by the surface $\partial\Omega$ and the sphere S_R, the radius R of which is sufficiently large and the centre of which coincides with the coordinate origin. It follows from the results of [1] that for a sufficiently smooth surface $\partial\Omega$ the vector u is as smooth as desired in the region $\bar\Omega_R$. For any elastic displacement vector which belongs to the space H_0 and is twice continuously differentiable in the closed region $\bar\Omega$, the following formula, following from Betti's formula applies:

$$u(x) = -\int_\Omega V(x,y)\Delta^* u(y)\,dy - \int_{\partial\Omega} [P(x,y)u(y) - V(x,y)P(u(y))]\,d_y\,\partial\Omega\,.$$

$$(A1.5.10)$$

$6^0.$ Consider problem $(A1.5.1)$ and $(A1.5.2)$. Formula $(A1.5.10)$ yields

$$u(x) = \Phi(x) - \int_{\partial\Omega} V(x,y)p(u(y))\,d_y\,\partial\Omega\,, \qquad (A1.5.11)$$

where

$$\Phi(x) = \mu^{-1}\int_\Omega V(x,y)F(y)\,dy \qquad (A1.5.12)$$

is a known vector function. Performing on both sides of $(A1.5.11)$ the operation p and letting $x \to \partial\Omega$, one obtains the vector, singular integral equation

$$p(u(x)) + 2\int_{\partial\Omega} P^*(y,x)p(u(y))\,d_y\,\partial\Omega = 2p(\Phi(x))\,, \qquad (A1.5.13)$$

where $P^*(y,x)$ is the matrix, conjugate to the matrix $P(x,y)$, introduced in §4. In the derivation of this equation, the following limit formula (cf. [4]) and likewise results of chapter 6 of this book have been employed: If

$$v(x) = \int_{\partial\Omega} V(x,y)\rho(y)\,d_y\,\partial\Omega\,,$$

and the point $x \in \Omega$, then

$$\lim_{x \to \partial\Omega} p(v(x)) = (1/2)\rho(x) + \int_{\partial\Omega} P^*(y, x)\rho(y)\, d_y\, \partial\Omega\,.$$

It is not difficult to see that the singular equation (A1.5.13) is equivalent to the boundary value problem (A1.5.1) and (A1.5.2).

This vector equation is equivalent to a system of three scalar equations; the symbolic matrix of this system is very close to the symbolic matrix of the system studied in §4. The same analysis, applied in the given case, demonstrates the following. For all values of ω, except for the values $\omega = -1, -2, \infty$, the system (A1.5.13) reduces to an equivalent Fredholm system. This has a unique solution for all values of ω, except for the earlier excluded values and besides still a finite or countable manifold of values which only may accumulate at the points $\omega = -2$ and $\omega = 0$; the values of ω from the last manifold are finite-fold characteristic Cosserat numbers. It has been seen above (3^0) that the point $\omega = 0$ does not belong to the Cosserat spectrum of problem I, hence follows that the finite-fold characteristic Cosserat numbers of problem I for infinite regions with finite, sufficiently smooth boundaries can only accumulate at the point $\omega = -2$.

The final results demonstrate the same as also in the case of finite regions. The essential Cosserat spectrum of problem I of the theory of elasticity for infinite regions contains the three numbers: -1, -2, ∞. The first and last of these numbers are isolated characteristic numbers of infinite multiplicity, the second is the limit of finite-fold characteristic numbers. The system of the characteristic vectors, corresponding to the characteristic numbers of finite as well as infinite multiplicity is complete in H_0.

7^0. For problem II, the same results as for the case of finite regions may be established for infinite regions. They will not be formulated here in detail.

§6. Plane problems

1^0. Consider to start with the second plane problem which in many respecvts is the simpler one; in particular, the smoothness of the contour does not play a special role — it is sufficient to assume that the contour is sectionally smooth. Consider a finite simpliconnected region Ω of the plane (x_1, x_2). The stresses and displacements in this region satisfy the equilibrium equations and Hooke's law, from which follows the well known relation

$$(\lambda + 2\mu)\Delta\theta + \mathrm{div}F = 0\,, \quad \theta = \mathrm{div}u\,, \qquad (A1.6.1)$$

where F is the vector of the body forces and u is the displacement vector. It will be useful in the sequel to note that the Lame constants λ and μ are linked to Poisson's constant ν and Young's modulus E by the relations

$$\lambda = \frac{E\nu}{(1+\nu)(1-2\nu)}, \ \mu = \frac{E}{2(1+\nu)}. \qquad (A1.6.2)$$

It likewise follows from Hooke's law that

$$2(\lambda + \mu)\theta = \sigma_{11} + \sigma_{22}. \qquad (A1.6.3)$$

Thus, for any values of the Lame constants, as long as $\lambda + 2\mu \neq 0$, it is true that

$$\Delta(\sigma_{11} + \sigma_{22}) + \frac{2(\lambda + \mu)}{\lambda + 2\mu}\operatorname{div}F = 0. \qquad (A1.6.4)$$

The equilibrium equations, equation (A1.6.4) and the boundary conditions of problem II define uniquely the stress tensor; for the existence of the solution, it is necessary and sufficient that the given forces are in static equilibrium. The displacement are uniquely determined, if one demands, for example, that they are orthogonal to an arbitrary, rigid displacement.

It follows from these statements that the Cosserat spectrum of the plane second problem arises either when equation (A1.6.4) is not true, i.e., for $\lambda + 2\mu = 0$, which yields $\omega = -1$, or when to zero stresses correspond non-zero displacements. The latter situation can arise in two cases: 1) the determinant of the Hooke system vanishes, i.e.,

$$4\mu^2(\lambda + \mu) = \frac{4E^3\omega^4}{(3\omega - 1)^2},$$

which yields the value $\omega = 0$; 2) one of the Lame constants becomes infinite, so that Hooke's law has nosense. Relation (A1.6.2) shows that for a finite value of Young's modulus E the Lame constants may become infinite either for $\nu = -1$, i.e., for $\omega = 1/3$, or for $\nu = 1/2$, i.e., for $\omega = \infty$.

The value $\nu = -1$ may occur also for the finite values of the Lame constants $\lambda = -2/3$ and $\mu = 1$; then the determinant of the equations of Hooke's law is non-zero. Hence it follows that $\omega = 1/3$ does not belong to the Cosserat spectrum of the second plane problem, and this spectrum consists only of the three points $\omega = 0, -1, \infty$. It will be shown that these points are characteristic values of infinite multiplicity; with this aim in mind, the correpsonding characteristic vectors will be found.

For the characteristic number $\omega = \infty$, the characteristic vectors, as in the three-dimensional case, are determined by the single condition

$$\mathrm{div}\, u = \frac{\partial u_1}{\partial x_1} + \frac{\partial u_2}{\partial x_2} = 0 \,. \qquad (A1.6.5)$$

Consider the case $\omega = -1$ in more detail. The differential equations of the theory of elasticity in terms of displacements assume in this case the form

$$\frac{\partial}{\partial x_2}\left(\frac{\partial u_1}{\partial x_2} + \frac{\partial u_2}{\partial x_1}\right) = 0, \; \frac{\partial}{\partial x_2}\left(\frac{\partial u_1}{\partial x_2} + \frac{\partial u_2}{\partial x_1}\right) = 0 \,. \qquad (A1.6.6)$$

hence $\frac{\partial u_1}{\partial x_2} + \frac{\partial u_2}{\partial x_1} = c$, where c is an arbitrary constant. The general solution of the last equation is

$$u_1 = \frac{\partial \Phi}{\partial x_1} + cx_2 \,, \; u_2 = \frac{\partial \Phi}{\partial x_2} \,, \qquad (A1.6.7)$$

where Φ is an arbitrary, twice differentiable function. The boundary conditions become

$$\alpha_2 \frac{\partial u_2}{\partial x_1} + \alpha_1 \frac{\partial u_2}{\partial x_2} = 0, \; \alpha_2 \frac{\partial u_1}{\partial x_1} + \alpha_1 \frac{\partial u_1}{\partial x_2} = 0 \,,$$

where α_1 and α_2 are the direction cosines of the normal to the boundary. Substituting here the expressions for u_1 and u_2 from (A1.6.7), one finds

$$\left.\frac{\partial \Phi}{\partial x_1}\right|_{\partial\Omega} = -cx_2 + c_1 \,, \; \left.\frac{\partial \Phi}{\partial x_2}\right|_{\partial\Omega} = c_2 \,, \; c_2 = const \,. \qquad (A1.6.8)$$

This condition is not contradicted, if

$$\int\limits_{\partial\Omega} \left[(-cx_2 + c_1)\, dx + c_2\, dx_2\right] = 0 \,,$$

hence $c = 0$; the characteristic Cosserat vectors, corresponding to the characteristic number $\omega = -1$, are the gradients of functions the first derivatives of which are constant on the boundary of the region. The converse assertion is obvious.

In the case $\omega = 0$, the differential equations and boundary conditions of problem II assume the form

$$\Delta u_1 = 0 \,, \; \Delta u_2 = 0 \,, \qquad (A1.6.9)$$

$$\begin{cases} \left(\dfrac{\partial u_1}{\partial x_1} - \dfrac{\partial u_2}{\partial x_2}\right)\alpha_1 + \left(\dfrac{\partial u_1}{\partial x_2} + \dfrac{\partial u_2}{\partial x_1}\right)\alpha_2 = 0\,, \\[4mm] \left(\dfrac{\partial u_1}{\partial x_1} - \dfrac{\partial u_2}{\partial x_2}\right)\alpha_2 + \left(\dfrac{\partial u_1}{\partial x_2} + \dfrac{\partial u_2}{\partial x_1}\right)\alpha_1 = 0\,. \end{cases} \qquad (A1.6.10)$$

It follows from the last equation that

$$\frac{\partial u_1}{\partial x_1} - \frac{\partial u_2}{\partial x_2} = 0\,, \quad \frac{\partial u_1}{\partial x_2} + \frac{\partial u_2}{\partial x_1} = 0\,.$$

The functions u_1 and u_2 are harmonic, hence the last relations are also true inside the region. Thus, the vector (u_1, u_2) is then and only then a characteristic Cosserat vector of the second plane problem , corresponding to the characteristic number $\omega = 0$, when $u_1 + iu_2$ is a function of the variable $z = x_1 + ix_2$ which is , holomorphic in the given region.

2^0. The manifold of characteristic Cosserat vectors is complete in the corresponding metric, and the vectors, corresponding to different characteristic numbers, are orthogonal, hence the solution of the second plane problem may be represented in the form

$$u = u^{(\infty)} + u^{(-1)} + u^{(0)}\,, \qquad (A1.6.11)$$

where the terms on the right hand side are projections of the displacement vector onto the subspace of the characteristic Cosserat vectors, corresponding to the characteristic numbers $\omega = \infty, -1, 0$.

If there are no body forces, then the second term on the right hand side in (A1.6.11) disappears. In fact, by the formulas of G.V. Kolosov–N.I. Muskhelishvili (cf., for example, chapter 5 of this book)

$$2\mu(u_1 + iu_2) = 2\omega^{-1}\varphi(z) + [\varphi(z) - z\overline{\varphi'(z)} - \overline{\psi(z)}]\,, \qquad (A1.6.12)$$

where $\varphi(z)$ and $\psi(z)$ are the Goursat functions. The vector

$$u^{(0)} = (\mu\omega)^{-1}(\mathrm{Re}\varphi(z)\,, \ \mathrm{Im}\varphi(z))\,. \qquad (A1.6.13)$$

is the characteristic Cosserat vector, corresponding to the number $\omega = 0$. Moreover, setting

$$2\mu\big(u_1^{(\infty)} + iu_2^{(\infty)}\big) = \varphi(z) + \overline{z\varphi'(z)} - \overline{\psi(z)}\,, \qquad (A1.6.14)$$

one finds that $\mathrm{div}\,u^{(\infty)} = 0$ and, consequently, $u^{(\infty)}$ is the characteristic Cosserat vector, corresponding to the characteristic number $\omega = \infty$.

3^0. It has been explained in §2 that the essential Cosserat spectrum of problem I of the three-dimensional theory of elasticity consists of the points $\omega = -1, 2, \infty$. For other values of ω, the system of the equations of the theory of elasticity remains elliptic, and the boundary conditions satisfy the condition of completeness. These results can be transferred without difficulty to the plane problem.

§7. A qualitative study of the solutions for Poisson coefficients with values near 1/2

As it is known, the method of photo-elasticity studies by means of optically sensitive models. For such materials, Poisson's constant is usually close to 1/2, and there arises the question, which is especially important in the case of three-dimensional problems, by how much may the stresses in the models and in nature differ from each other. This question has been studied in the paper [10] by E.I. Edelshtein. An attempt will be made here to investigate the same question, employing decomposition of the problems of the theory of elasticity in accordance with the Cosserat spectrum.

The question is posed regarding the behaviour of the stresses for $\nu < 1/2$ and $\nu \to 1/2$, i.e., for $\omega \to +\infty$. The answer to this question may be obtained from formulas (A1.2.20), (A1.2.24) and (A1.4.30), which yield the solutions of the basic problems of the theory of elasticity, and from the fact that $\omega = \infty$ is an isolated point of the Cosserat spectrum; the last circumstance permits to justify the limiting processes which will be performed below. The displacement vector will be given the form $u(x) = \hat{u}(x) + u_\infty(x)$, where $u_\infty(x) = \lim\limits_{\omega \to \infty} u(x)$

In the case of a rigidly clamped boundary, one has by (A1.2.20)

$$\mu \hat{u}(x) = \sum_{n=1}^{\infty} \left\{ \frac{(F, u_n^{(-1)})}{1+\omega} u_n^{(-1)}(x) + \frac{(F, u_n)\omega_n}{\omega - \omega_n} u_n(x) \right\},$$

$$\mu u^{(\infty)}(x) = \sum_{n=1}^{\infty} (F, u_n^{(\infty)}) u_n^{(\infty)}(x);$$

$$(A1.7.1)$$

where $F(x)$ is the vector of the body forces.

If $\nu \to 1/2$, then the displacement $\hat{u}$ and the corresponding strains tend to zero. Consider the behaviour of the stresses. Assume, for example, that for $\nu \to 1/2$ Young's modulus $E \to E_0$, $0 < E_0 < \infty$.

Formula (A1.6.2) shows that under this condition the stress tensor $[\hat{\sigma}_{jk}]$, corresponding to the displacement $\hat{u}$, tends to the spherical tensor the components of which are determined by the formulas

$$\hat{\sigma}_{jj} = \sum_{n=1}^{\infty} \left[(F, u_n^{(-1)})\mathrm{div}\,u_n^{(-1)}(x) + \omega_n(F, u_n)\mathrm{div}\,u_n(x) \right],$$

$$\hat{\sigma}_{jk} \neq 0, \ j = k,$$

$$(A1.7.2)$$

hence it is clear that during the employment of the photo-elastic method the stresses in the model and in nature may differ to an essential degree.

Thus, if $u^{(\infty)}(x) = 0$, then formula (A1.7.2) shows that in the model the stresses actually are reduced to hydro-static stress, while in nature, these stresses may be quite different.

APPENDIX II

THE CALDERON–SEELY PROJECTORS AND THE REDUCTION OF BOUNDARY VALUE PROBLEMS OF ELLIPTIC OPERATORS TO BOUNDARY EQUATIONS*

The study and solution of the boundary value problems of mathematical physics with the aid of the potential method started with the work of Poisson and Green.

In applications to boundary value problem for elliptic systems, describing the state of stress of an elastic body, the questions of their reduction were studied by V.D.Kupradze et al. [4,5]. For this purpose, results of the theory of singular integral operators which were established by Zygmund, Calderon and Mikhlin [33,31,12,14] were employed.

For general elliptic equations with variable coefficients, methods of reduction of boundary values problems to pseudo-differential equations were developed by Lopatinskii, Calderon, Seely and Hermander [11,31, 32,29]. In the work of the last three authors, potentials are described in a certain operator form which does not uitilize the explicit form of the fundamental solution. In numerical work, an application of the potential method (or the boundary integral equations method) was restricted basically to boundary value problems for equations with constant coefficients. This fact, apparently, is due to the difficulties experienced with approximations of the pseudo–differential operators which enter into the boundary conditions.

In 1969, V.S. Ryaben'kii [23] constructed potentials for systems of difference equations with constant coefficients and corresponding methods for the reduction of such boundary value problems on difference boundary. Subsequently, in the work of Ryaben'kii and his students, the difference potentials method has been effectively developed; to-day, it has been applied with success to the numerical solution of a number of new problems [22] (as far as the potential method is concerned).

This appendix is concerned with the construction of the Calderon-Seely potential as well as the boundary value problem reduction scheme based on this construction. The present construction of Calderon-Seely potentials and projectors permits to consider jointly classical potential methods as well as the theoretical part of the difference potential method.

---

* Appendix II is written by M.I. Lazarev, V.N. Chikin

As a preliminary demonstration of this material, it will be convenient to deal with Laplace's equation.

Let Ω be a bounded domain in R^3 with the smooth boundary $\partial\Omega$; $\Delta \equiv \partial^2_{x1} + \partial^2_{x2} + \partial^2_{x3}$ – the Laplace operator; G – the operator of convolution with the fundamental solution $g(x,y) = (4\pi)^{-1}(|x-y|)^{-1}$ of the Laplace operator: $Gu = \int\limits_{R^3} g(x,y)u(y)\,dy$.

Denote by $H^s_c(R^3)$ the space of functions which belong to the Sobolev space $H^s(R^3)$ ($s > 5/2$, $s \neq 1/2+$integer) and have compact support.

As it is known, the operator G is semi-inverse to the operator Δ, i.e., $\Delta G \Delta u = \Delta u$ for any function $u \in H^s_c(R^3)$.

Define the operator P:

$$Pu(x) \equiv \Theta_\Omega \Big(u(x) - \frac{1}{4\pi} \int\limits_{\Omega} \frac{1}{|x-y|} \Delta u(y)\,dy \Big),$$

where Θ_Ω is the characteristic function of the manifold Ω. Green's formula for the domain Ω can then be given the form

$$Pu(x) = \frac{1}{4\pi} \int\limits_{\partial\Omega} \partial_\nu \Big(\frac{1}{|x-y|} \Big) u(y)\,d_y s - \frac{1}{4\pi} \int\limits_{\partial\Omega} \frac{1}{|x-y|} \partial_\nu u(y)\,d_y s\,, \quad x \in \Omega,$$

$$(A2.0.1)$$

where $\partial_\nu u$, $\partial_\nu \big(\frac{1}{|x-y|} \big)$ are the directional derivatives along the outward normal $\nu(y)$ of the functions $u(y)$ and $(|x-y|)^{-1}$, respectively.

Consider the operator P.

It is readily seen that

1) P is a projector: $P^2 = P$.

2) The image of the projector P coincides with the space of the zeroes of the Laplace operator: $\operatorname{Im} P = \ker \Delta$.

3) Let $\gamma : H^3(R^3) \to DC_1(\partial\Omega) : u \to (u, \partial_\nu u)|_{\partial\Omega}$ be the operator of taking the first order Cauchy data. The function $u_1 = Pu$ depends only on the first order Cauchy data of the function u, i.e., $\ker\gamma \subset \ker P$.

In particular, it follows from 1) and 2) that the equations $\Theta_\Omega \Delta u = 0$ and $u = Pu$ are equivalent. By 3), the relation $p\gamma = \gamma P$ correctly defines the projector p, acting in the space of the first order Cauchy data. In addition, the projector p is a *Calderon-Seely projector*: its image coincides with the first order Cauchy data of functions, satisfying Laplace's equation in the domain Ω, smooth up to the boundary, i.e., $\operatorname{Im} p = \gamma(\ker \Delta)$.

Thus, the equations $\Delta u(x) = 0$, $x \in \Omega$, $u \in H^s(\Omega)$ and $p\psi(x) = \psi(x)$, $x \in \partial\Omega$, $\psi \in DC_1(\partial\Omega)$ are equivalent in the sense that there exists a mutually unique correspondence between their solutions: $\psi = \gamma u$, $u = \pi \psi$ where the operator γ is an operator of taking the first order Cauchy data and π (*the Calderon-Seely potential*) is correctly defined by the relation $P = \pi\gamma$ (by 3)).

It should be emphasized that $p\psi = \psi$ is an equation relative to a function given on the boundary of the domain $\partial\Omega$.

The standard boundary condition states the resriction on the boundary of a differential operator of not higher than the first order. Therefore the initial boundary value problem

$$\begin{cases} \Delta u(x) = 0, \ x \in \Omega \\ lu(x) = f(x), \ x \in \partial\Omega, \end{cases}$$

is equivalent to the system

$$\begin{cases} (I - p)\psi(x) = 0, \ x \in \partial\Omega \\ \tilde{l}\psi(x) = f, \ x \in \partial\Omega \end{cases} \tag{A2.0.2}$$

where $\tilde{l}$ is defined by $l = \tilde{l}\gamma$.

The system (A2.0.2) is in a certain sense canonical: In the case of an arbitrary elliptic operator L of order m the initial boundary value problem is equivalent to the system (A2.0.2), where p is the projector on $\gamma(\ker L)$, γ - the operator of taking the Cauchy data of order $m - 1$.

The construction of the Calderon-Seely projector p in the classical boundary integral equations method is realized by application to both sides of the indentity (A2.0.1) of the limit process $x \to x_0 \in \partial\Omega$:

$$\begin{cases} Pu(x_0) = \dfrac{1}{2}\, u(x_0) + \dfrac{1}{4\pi} \displaystyle\int_{\partial\Omega} \partial_\nu\left(\dfrac{1}{|x_0 - y|}\right) u(y)\, d_y s - \\ \qquad\qquad\qquad\quad - \dfrac{1}{4\pi} \displaystyle\int_{\partial\Omega} \dfrac{1}{|x_0 - y|}\, \partial_\nu u(y)\, d_y s \\[2ex] \partial_\nu Pu(x_0) = \dfrac{1}{2}\, \partial_\nu u(x_0) + \dfrac{1}{4\pi}\, \partial_\nu \displaystyle\int_{\partial\Omega} \partial_\nu\left(\dfrac{1}{|x_0 - y|}\right) u(y)\, d_y s - \\ \qquad\qquad\qquad\quad - \dfrac{1}{4\pi}\, \partial_\nu \displaystyle\int_{\partial\Omega} \dfrac{1}{|x_0 - y|}\, \partial_\nu u(y)\, d_y s\,, \end{cases}$$

$$p(u(x_0), \partial_\nu u(x_0)) = \gamma Pu(x_0) = \begin{cases} (Pu)(x_0) \\ (\partial_\nu P)u(x_0)\,. \end{cases}$$

The representation of the projector p as a limiting value of the surface integrals (and their normal derivatives), entering into Green's formula, is effective when an explicit form of the fundamental solution exists. Otherwise, such a representation ceases to yield a meaningful instruction for the construction of algorithms for the solution of the system (A2.0.2).

An alternative approach – the Ryaben'kii's difference potentials method – employs projectors p in the form $p\psi = \gamma P\gamma_r^{-1}\psi$, where γ_r^{-1} is some right–inverse to γ (the independence on the choice of the right–inverse follows from 3)).

The last formula can be rewritten: $p\psi = \gamma(I - G\Theta_\Omega\Delta)\gamma_r^{-1}\psi$, or, since $G\Delta = I$,

$$p\psi = \gamma G(I - \Theta_\Omega)\Delta\gamma_r^{-1}\psi. \qquad (A2.0.3)$$

The operator G of the convolution with fundamental solutions may be replaced in (A2.0.3) by any operator, semi-inverse to Δ. In particular, such operator may be the operator Γ, inverse to the restriction of the operator Δ on the sub-space of the functions, given on the domain $\Omega' \supset \overline{\Omega}$, selected by the condition $l'u = 0$ on $\partial\Omega'$, such that Γ exists. Note that, by strength of the independence of p on γ_r^{-1}, the region of definition of the projector P may be restricted to functions with support in some arbitrary neighbourhood the boundary.

A projector in the form (A2.0.3) conserves the constructibility also in the case that there does not exist an explicit form of the fundamental solution. The effectiveness of the numerical procedures for this is rendered by the presence of fast algorithms in domains of special shape (for example, the algorithms of Fedorenko's multi-grid method [27]).

Since the effectiveness of algorithms depends essentially on the smoothness of the given data, a significant role in the construction of the projector p is enacted by the extension operator of the function over first order Cauchy data (of order $m - 1$ for an elliptic operator of order m) to functions in the vicinity of the boundary $\partial\Omega$ of maximally possible smoothness. The construction of the extension is based on the following: for any element $(\varphi^0, \varphi^1) \in H^{s-1/2}(\partial\Omega) \oplus H^{s-3/2}(\partial\Omega)$ of the first order Cauchy data space there exist the functions $u \in H^s(\Omega)$ satisfying the conditions

$$u\big|_{\partial\Omega} = \varphi^0, \ \partial_\nu u\big|_{\partial\Omega} = \varphi^1 \qquad (A2.0.4)$$

$$\partial_\nu^j\Delta u\big|_{\partial\Omega} = 0, \ j = 0,...,b-2, \qquad (A2.0.5)$$

where $b = [s - 1/2]$ is the largest integer not exceeding $s - 1/2$. The existence of the functions $u \in H^s(\Omega)$, satisfying (A2.0.4) and (A2.0.5), follows from the non-degeneracy of the symbol of the Laplace operator: in any local system of coordinates (χ_1, χ_2), acting in the vicinity $O \subset \partial\Omega$, the operator Δ may be written in the form: $\Delta = a_0(x)\partial_\nu^2 + a_1(x)\partial_\nu + a_2(x)$, where the $a_j(x)$ are differential operators of order j, containing only derivatives in the directions of χ_1 and χ_2, and $a_0(x) \neq 0$; $x \in O$.

Therefore the "algebraic system"

$$\begin{cases} a_0(x)\partial_\nu^2 u(x) + a_1^0(x)\partial_\nu u(x) + a_2^0(x)u(x) = 0 \\ a_0(x)\partial_\nu^3 u(x) + a_1^1(x)\partial_\nu^2 u(x) + a_2^1(x)\partial_\nu u(x) + a_3^1(x)u(x) = 0 \\ \dotfill \\ a_0(x)\partial_\nu^b u(x) + \ldots + a_b^b(x)u(x) = 0 \end{cases}$$

where $a_j^i(x) = \partial_\nu a_{j-1}^{i-1}(x) + a_j^{i-1}(x)$ has a unique solution with respect to the unknowns $\partial_\nu^2 u(x), ..., \partial_\nu^b u(x)$. There remains to apply the trace theorem according to which for any element $\{\varphi^j\}_{j=1}^b$ of the space of Cauchy data there exists a function $u \in H^s(\Omega)$ such that $\partial_\nu^j u|_{\partial\Omega} = \varphi^j$ (the definition of the trace space will be introduced in §1. 5^0; for more details cf. [10]).

Denote by $H_\Delta^s(\mathrm{R}^n)$ the space of functions of $H_c^s(\mathrm{R}^n)$, which satisfy the conditions (A2.0.5). Note the following properties of the subspace $H_\Delta^s(\mathrm{R}^n)$:

1) For each $\varphi \in DC_1(\partial\Omega)$ there exists a function $u \in H_\Delta^s(\mathrm{R}^n)$ which is unique apart from a function having trace equal to zero on the boundary $\partial\Omega$, and conversely, i.e., $DC_1(\partial\Omega)$ is isomorphic $\tilde\gamma(H_\Delta^s(\mathrm{R}^n))$ where $\tilde\gamma$ is the operator of taking the trace over $\partial\Omega$.

2) Any function which satisfies Laplace's equation in Ω, belongs to the subspace $H_\Delta^s(\mathrm{R}^n)$.

3) The operator $G(I - \Theta_\Omega)\Delta$ is continuous as an operator, acting in $H_\Delta^s(\mathrm{R}^n)$ (since for any function $u \in H_\Delta^s(\mathrm{R}^n)$ of the function $v = \Delta u$ has zero Cauchy data on $\partial\Omega$, and, consequently, the operator $(I - \Theta_\Omega)$ is continuous.)

These properties allow to apply the following method which simplifies the study of the Calderon-Seely projector: coming from the function $\varphi \in DC_1(\partial\Omega)$, the function $u \in H_\Delta^s(R^n)$ is constructed, and the study carries over to the projector $P = G(I - \Theta_\Omega)\Delta$.

In particular, the connection between the projectors on the kernel of an operator, acting inside and outside a domain, is readily established: the first order Cauchy data space is decomposed into a direct sum of traces of functions which satisfy Laplace's equation inside the domain Ω, and of traces of functions which satisfy Laplace's equation outside the domain Ω and which tend to zero at infinity. This result achieved by R.Seely is a generalization of a known fact of the theory of analytic functions regarding expansions of functions, defined on the boundary $\partial\Omega$ of a domain Ω of the complex plane $\mathbb{C}$ into a sum of functions which are analytic inside Ω and outside Ω and tend to zero at infinity.

Within the framework of the present construction, one obtains also a natural generalization of the theorem on involution for a singular Cauchy integral operator. For example, one has for the Laplace operator: $S^2 = I$, where the operator

$$S : H^{s-1/2}(\partial\Omega) \oplus H^{s-3/2}(\partial\Omega) \to H^{s-1/2}(\partial\Omega) \oplus H^{s-3/2}(\partial\Omega)$$

acts according to the rule:

$$S(\varphi^0, \varphi^1) = \begin{cases} -\dfrac{1}{2\pi}\displaystyle\int_{\partial\Omega} \dfrac{1}{|x-y|}\varphi^1(y)\,d_y s + \dfrac{1}{2\pi}\displaystyle\int_{\partial\Omega} \partial_\nu\left(\dfrac{1}{|x-y|}\right)\varphi^0(y)\,d_y s \\[4mm] -\dfrac{1}{2\pi}\displaystyle\int_{\partial\Omega} \partial_{\nu_x} \dfrac{1}{|x-y|}\varphi^1(y)\,d_y s + \\[4mm] +\dfrac{1}{2\pi}\partial_{\nu_x}\displaystyle\int_{\partial\Omega} \partial_\nu\left(\dfrac{1}{|x-y|}\right)\varphi^0(y)\,d_y s\,. \end{cases}$$

Moreover, the operator $\frac{1}{2\pi}\partial_{\nu_x}\int_{\partial\Omega}\partial_\nu\frac{1}{|x-y|}(\cdot)\,d_y s$ is presented in the form of an integro-differential operator with zero non-integral term (cf. §3. 8^0).

The described scheme is generalized in a natural manner to linear elliptic systems; an analogue of potentials and reduction to boundary equations also exists for quasi-elliptic equations [8,9].

The general scheme of reduction on the basis of the Calderon-Seely projector is treated in §1. Its main result is the construction of the Calderon-Seely projector in the form (A2.0.3) for an elliptic differential operator.

A projector of the form (A2.0.3) is utilized in the difference potentials method (DPM). The construction of the difference potential for the Lamé equation is described in §2.

The authors believe it is appropriate to present the basic theorems of the boundary integral equations method (BIEM) in applications to the operator of electro-elasticity at the appendix of this book. The proof of the theorems utilizes an alternative definition of the potential (§3).

Sections 2 and 3 are independent of each other.

The authors wish to express their deep gratitude to the authors of this book for their invitation to write an appendix on this topic as well as for their persistent attention to their work. It will not be an exaggeration to state that the plan and structure of the appendix was worked out in cooperation with them.

Likewise, the authors wish to thank Professor V.S. Ryaben'kii and Doctor B.I. Syleumanov for valuable comments, their assistance in improving the text and their constructive influence which they rendered (and continue to render) on the scientific world view of the authors.

§1. Basic principles of the reduction on a boundary

1^0. **The problem of reduction.** Let $\Omega \subset \mathrm{R}^n$ ($n = 2$ or $n = 3$) be a bounded domain with a smooth boundary $\partial\Omega$, L – a linear, elliptic, differential operator of order m the coefficients of which form a $(k \times k)$ – matrix of smooth functions, l–a linear differential operator of order not larger than $m - 1$.

Consider the boundary value problem

$$(*) \begin{cases} L_\Omega u(x) = 0, \ x \in \Omega \\ lu(x) = f(x), \ x \in \partial\Omega, \end{cases}$$

where L_Ω is the restriction on Ω of the operator L.

It will be assumed in the sequel that the operators L_Ω and l act on functions which belong to a Sobolev space $H^s(\Omega)$ with index $s > m + 1/2$, $s \neq 1/2+$integer.

Regarding the operator L, it will be assumed, in addition, that a domain Ω', enveloping the initial domain Ω (i.e., $\Omega' \supset \overline{\Omega}$) can be found such that the problem

$$\begin{cases} L_\Omega u(x) = v(x), \ x \in \Omega' \\ l'u(x) = 0, \ x \in \partial\Omega', \end{cases} \qquad (A2.1.1)$$

has a unique solution for any function $v \in H^{s-m}(\Omega')$ with compact support. If $\Omega = \mathrm{R}^n$, then the condition l' yields the behaviour of the functions $u \in H^s_{loc}(\mathrm{R}^n)$ as x tends to infinity. The class of the operators which satisfy this assumption is sufficiently wide (cf. [1,16]); in particular, is includes elliptic differential operators with constant coefficients.

Note that by the strength of ellipticity of the operator L, $u \in H^s_{loc}(\Omega')$.

Denote the operator of the solution of problem (A2.1.1) by G.

Reduction of a boundary value problem to the boundary implies replacement of the system $(*)$ by an equivalent system of equations with respect to unknown functions on the boundary. Equivalence is understood in the sense that there exists a continuous, mutually single-valued mapping relating each solution of the reduced system to a solution of the system $(*)$, and conversely.

The next subsection is concerned with the abstract, algebraic reduction scheme; in passing, the properties will be explained which the space must possess in order that the reduced system may be formulated on it.

2^0. **The algebraic reduction scheme.** Let X, Y, F be the Banach spaces and $L : X \to Y$, $l : X \to F$ be continuous, linear operators. In brief: $L \in L(X, Y)$, $l \in L(X, F)$.

It will be assumed that the system of equations of the form

$$Lu = 0, \ lu = f \qquad (A2.1.2)$$

is a model of the boundary value problem, where the the second equation fulfills the role of the boundary condition.

Definition 1. Let Φ be a Banach space. An operator $\pi \in L(\Phi, X)$ will be called the potential of an operator L with density from the space Φ, if $\operatorname{Im} \pi = \ker L$.

Definition 2. Let Ψ be a Banach space. An operator $\gamma \in L(X, \Psi)$ with the image $\operatorname{Im} \gamma$, equal to Ψ, will be called clear trace operator, associated with operator L, if there exist a continuous projector P on $\ker L$ such that

$$\ker \gamma \subset \ker P. \qquad (A2.1.3)$$

In this case, space Ψ will be called clear trace space.

corollary from definitions. For each clear trace operator the potential π determinated by

$$P = \pi\gamma, \qquad (A2.1.4)$$

can be associated. Here P is continuous projector, satisfying (A2.1.3).

Definition 3. Let γ be clear trace operator. The projector $p \in L(\Psi, \Psi)$ such that $\operatorname{Im} p = \gamma(\ker L)$ will be called Calderon–Seely projector.

Let projector P, potential π and clear trace operator γ are related by (A2.1.4). Then operator $p = \gamma\pi$ is Calderon–Seely projector: $\gamma\pi\gamma\pi = \gamma P\pi$, and, since $\pi\varphi \in \operatorname{Im} P = \ker L$ for each φ, $\gamma P\pi = \gamma\pi$. •

Now, everything is available to formulate a problem on clear trace space, which is equivalent to original problem.

Consider a problem

$$\varphi - p\varphi = 0, \ l\pi\varphi = f. \qquad (A2.1.5)$$

Theorem 1. Let Ψ be clear trace space, $\varphi \in \Psi$. Then systems (A2.1.2) and (A2.1.5) are equivalent in sence subsection 1^0. If φ – the solution of system (A2.1.5), then $u = \pi\varphi$ – the solution of system (A2.1.2).

• By definition 2, there exists a projector $P \in L(X, X)$ such that Im $P = \ker L$ and $\ker P \supset \ker \gamma$.

The first of these conditions yields the equivalence of $Lu = 0$ and $Pu = u$, the second condition yields the isomorphism of $\ker L$ and $\gamma(\ker L)$.

Recalling the definitions of the projector p and the potential π, the assertion is proved. •

Examples.

1. The restriction operator on the boundary $\partial\Omega$ of the domain Ω is clear trace operator for the Cauchy-Riemann operator $\partial/\partial\bar{z}$. The corresponding projector is

$$Pu(z) = u(z) - \frac{1}{2\pi i} \int_{\Omega} \frac{\partial u}{\partial\bar{\xi}} \, \frac{d\xi \wedge d\bar{\xi}}{\xi - z} \, ,$$

and the potential π such that $P = \pi\gamma$, is the Cauchy integral

$$\pi u(z) = \frac{1}{2\pi i} \int_{\partial\Omega} \frac{u(\xi)\, d\xi}{\xi - z} \, , \ \xi \in \partial\Omega, \ z \in \Omega .$$

2. The main example: The space DC_{m-1} of Cauchy data of order $m - 1$ is the clear trace space for the elliptic differential operator L of order m. Actually, by Green's formula, written with respect to the fundamental solution of the operator L^*, formally conjugate to L, the projector $Pu(x) = u(x) - \int_{\Omega} g(x, y) Lu(y)\, dy$ is the projector on $\ker L$ and satisfies condition (A2.1.3). In this case, the potential may be expressed in terms of the surface integrals entering into Green's formula: $Pu = \pi(\gamma u)$, where (γu) are the Cauchy data of order $m - 1$ of the function u.

Below, a potential π, given in the space DC_{m-1}, and a Calderon-Seely porjector p, will be defined without surface integrals. Such representations of the potential π and the projector p are employed in the difference potentials method (§2).

In addition, the operator form version of the potential π, in the authors' opinion, turns out to be convenient for the study of the propertries of the potentials of the classical BIEM (§3).

3^0. **Note and announcement.** The concept of clear traces has been introduced by V.S. Ryaben'kii for differential operators [23] and for difference operators [24]. A definition, unifying both cases and equivalent to the one introduced here is given in [7].

Up to the end of this section considerations are restricted to the case when the space of traces is the space DC_{m-1}.

Bellow in this section: the some properties of trace spaces of Sobolev spaces (4^0), the general form of the projector on $\ker L_\Omega$ (5^0), the extension operator of trace of function with respect to its Cauchy data (6^0), the construction of the Calderon-Seely projector (7^0), a connection between the projectors on the kernel inside $(Lu(x) = 0,\ x \in \Omega)$ and on the kernel outside $(Lu(x) = 0,\ x \in \Omega'/\overline{\Omega})$ problem (9^0) and, finally, in 10^0, there will be formulated the reduction problem, concretizing problem (A2.1.5).

4^0. **The spaces of traces and Cauchy data.** First of all, the trace theorem on Sobolev spaces will be recalled (cf., for example, [10]).

Let $\Gamma(\Omega)$ (respectively, $\Gamma(\partial\Omega)$) be the space of infinitely differential real vector functions, given on Ω (and $\partial\Omega$, respectively).

The mapping $\tilde\gamma : \Gamma(\Omega) \to \oplus_{j=0}^{b}\Gamma(\partial\Omega)$, acting by the rule

$$\tilde\gamma(u) = (u, \partial_\nu u, ..., \partial_\nu^b u)\big|_{\partial\Omega},$$

where $b = [s - 1/2]$ is the largest integer which does not exceed $s - 1/2$, is extended to the bounded linear mapping

$$\tilde\gamma : H^s(\Omega) \to \bigoplus_{j=0}^{b} H^{s-j-1/2}(\partial\Omega).$$

Besides, there exists the linear bounded extension operator

$$\tilde\gamma_r^{-1} : \bigoplus_{j=0}^{b} H^{s-j-1/2}(\partial\Omega) \to H^s(\Omega),$$

which is right inverse operator, i.e.,

$$\tilde\gamma\tilde\gamma_r^{-1}((u, \partial_\nu u, ..., \partial_\nu^b)|_{\partial\Omega}) = (u, \partial_\nu u, ..., \partial_\nu^b u)|_{\partial\Omega}.$$

The kernel of the operator $\tilde\gamma$ is the space $H_0^s(\Omega)$, which is the closure in the topology $H^s(\partial\Omega)$ of the subspace $\Gamma_0(\Omega)$, consisting of functions from $\Gamma(\Omega)$ equal to zero on $\partial\Omega$ together with their derivatives along the normal ν to $\partial\Omega$.

An analogous assertion holds in the case when $\partial\Omega$ is the boundary of a domain Ω, lying in the domain Ω', and the operatopr $\tilde\gamma$ is considered to be an operator which acts on the functions (of the corresponding class) given on Ω'.

The mapping $\tilde\gamma : H^s(\Omega) \to \oplus_{j=0}^{b}H^{s-j-1/2}(\partial\Omega)$ is said to be *the operator of taking the trace on $\partial\Omega$*, and the space $DC = \oplus_{j=0}^{b} H^{s-j-1/2}(\partial\Omega)$ – *the trace space*.

The space DC_{m-1} of Cauchy data of order $m-1$:

$$DC_{m-1} \equiv \overset{m-1}{\underset{j=0}{\oplus}} H^{s-j-1/2}(\partial\Omega)$$

will also be defined. Since, by assumption, $s > m+1/2$, one has the mapping

$$\gamma : H^s(\Omega) \to DC_{m-1} : u \to (u, \partial_\nu u, ..., \partial_\nu^{m-1} u)\big|_{\partial\Omega} ,$$

which may be represented as a composition $\gamma = \gamma_m \tilde{\gamma}$, where γ_m is the projection on the first m terms of the direct sum $\oplus_{j=0}^b H^{s-j-1/2}(\partial\Omega)$.

5^0. **Projectors on the kernel of the operator L_Ω.** The construction of the Calderon-Seely projector begins with the construction of the projector P on $\ker L_\Omega$ with a kernel containing the kernel of the operator of taking the trace.

Note, first of all, that if G_Ω is semi-inverse to L_Ω, then the operator $I - G_\Omega L_\Omega$ is the projector on $\ker L_\Omega$, and conversely, any projector P, acting in the space $H^s(\Omega)$ such that $\operatorname{Im} P = \ker L_\Omega$ has the form $I - G_\Omega L_\Omega$, where G_Ω is some semi-inverse to L_Ω.

• In fact, $P^2 = (I - G_\Omega L_\Omega)(I - G_\Omega L_\Omega) = I - G_\Omega L_\Omega$, as long as, by definition of the semi-inverse, $L_\Omega G_\Omega L_\Omega = L_\Omega$. Consequently, P is a projector. If $u \in \ker L_\Omega$, then $u \in \operatorname{Im} P$. If $u \in \operatorname{Im} P$, then $G_\Omega L_\Omega u = 0$, and, since $L_\Omega G_\Omega L_\Omega u = L_\Omega u$, $u \in \ker L_\Omega$. Consequently, $\operatorname{Im} P = \ker L_\Omega$. Conversely, if P is a projector and $\operatorname{Im} P = \ker L_\Omega$, then the restriction L_Ω on $\ker P$ has the trivial kernel, and, consequently, there exists a left–inverse operator which may be taken as G_Ω. •

Thus, the projector on the kernel of an operator is predetermined by the semi-inverse operator to the given one.

Let Ω' be a domain enveloping the domain Ω. Introduce the notation: $\Omega^+ = \Omega$, $\Omega^- = \Omega'/\overline{\Omega}$.

Denote by $H_0^t(\Omega')$ the kernel of the operator $\tilde{\gamma}$ taking the trace on $\partial\Omega$ ($t = s$ or $t = s - m$). Let $\tilde{\gamma}_r^{-1}$ be some right inverse to the operator $\tilde{\gamma}$ such that for any $u \in H^t(\Omega')$ $\operatorname{supp} \tilde{\gamma}_r^{-1} u \subset \Omega''$, where Ω'' is some bounded domain which lies in Ω' (the existence of a continuous right–inverse with the stated property follows from the existence of some continuous right–inverse and the continuity of the product operator on a finite function).

The restriction operator $\rho^+ \in L(H^t(\Omega'), H^t(\Omega^+))$ (respectively, $\rho^- \in L(H^t(\Omega'), H^t(\Omega^-))$) has a continuous right–inverse χ^+ (respectively, χ^-) such that the restriction χ^+ on $H_0^t(\Omega^+)$ (χ^- on $H_0^t(\Omega^-)$) coincides with the operator χ_0^+ (χ_0^-, respectively) of continuation by zero of the functions from $H_0^t(\Omega^+)$ ($H_0^t(\Omega^-)$, respectively) up to the functions of $H_0^t(\Omega')$.

• In fact, since $\ker \rho^+ \subset \ker \tilde{\gamma}$, the relation $\tilde{\gamma} = \tilde{\gamma}^+ \rho^+$ correctly defines the operator $\tilde{\gamma}^+ : H^t(\Omega^+) \to DC$.

The operator $\pi_t = \tilde{\gamma}^{-1}\tilde{\gamma}$ is a continuous projector, acting in the space $H^t(\Omega')$. Consequently, $H^t(\Omega')$ expands the direct sum: $H^t(\Omega') = H_0^t(\Omega') + \operatorname{Im} \pi_t$.

Analogously, $H^t(\Omega^+) = H_0^t(\Omega^+) + \operatorname{Im} \pi_t^+$, where $\pi_t^+ = \rho^+ \tilde{\gamma}^{-1}\tilde{\gamma}^+$.

Then, $\rho^+(H_0^t(\Omega')) = H_0^t(\Omega^+)$ and $\rho^+(\operatorname{Im} \pi_t) = \operatorname{Im} \pi_t^+$.

The operator of continuation by zero χ_0^+ is continuous right–inverse to the operator $\rho^+|_{H_0^t(\Omega')}$ (cf. [10]). The operator $\tilde{\gamma}^{-1}\tilde{\gamma}^+$ is continuous right–inverse (and even inverse) to the operator $\rho^+|_{\operatorname{Im} \pi_t^+}$. Consequently, the operator, acting according to the law

$$\chi^+ u = \chi_0^+(I - \pi_t^+)u + \tilde{\gamma}^{-1}\tilde{\gamma}^+ \pi_t^+ u \,,$$

satisfies the required property. •

Proposition 1. Let G be the operator, defined in 1^0, and χ_{s-m}^+ – right–inverse to the restriction operator $\rho_{s-m}^+ \in \mathrm{L}(H^{s-m}(\Omega'), H^{s-m}(\Omega^+))$ such that

i) the restriction χ_{s-m}^+ on $H_0^{s-m}(\Omega^+)$ coincides with the operator χ_0^+ of continuation by zero of functions of $H_0^+(\Omega^+)$;

ii) for any $u \in H^{s-m}(\Omega^+)$ $\operatorname{supp} \chi_{s-m}^+ u \subset \Omega''$, where Ω'' is some bounded domain, lying in Ω'.

Define the operator G_Ω: $G_\Omega \equiv \rho_s^+ G \chi_{s-m}^+$.

Then the operator $P = I - G_\Omega L_\Omega$: $H^s(\Omega^+) \to H^s(\Omega^+)$ is a continuous projector on $\ker L_\Omega$, whence is true the inclusion

$$\ker P \supset \ker \tilde{\gamma}^+ \,. \tag{$A2.1.6$}$$

• The operator G_Ω is semi-inverse to the operator L_Ω, acting on the functions of $H^s(\Omega^+)$. In fact, by the strength of the locality of the differential operator, the operator L_Ω is correctly defined by the relation $L_\Omega \rho_s^+ = \rho_{s-m}^+ L$. Therefore $L_\Omega G_\Omega L_\Omega = L_\Omega \rho_s^+ G \chi_{s-m}^+ L_\Omega = \rho_{s-m}^+ L G \chi_{s-n}^+ L_\Omega = \rho_{s-m}^+ \chi_{s-m}^+ L_\Omega = L_\Omega$. Consequently, $P = I - G_\Omega L_\Omega$ is the projector on $\ker L_\Omega$.

The inclusion (A2.1.6) is fulfilled, since, if $u \in \ker \tilde{\gamma}$, then $\chi_s^+ u$ is the continuation by zero of the functions u, whenece $G_\Omega L_\Omega = \rho_{s-m}^+ G L \chi_s^+ = I$. •

In order that the projector P could be defined a Calderon-Seely projector, it is necessary to strengthen the inclusion:

$$\ker P \supset \ker \gamma \,, \tag{$A2.1.7$}$$

where γ is the operator of taking the Cauchy data of order $m - 1$.

The final step in the construction of the Calderon-Seely projector involves the following.

Let $\gamma_r^{-1}\colon DC \to H^s(\Omega)$ be the operator of extension with respect to Cauchy data such that for any $u \in \operatorname{Im}\gamma_r^{-1}$ one has

$$\partial_\nu^j L u = 0,\ 0 \le j \le b - m. \tag{A2.1.8}$$

Then the image of restriction of the projector P on $\operatorname{Im}\gamma_r^{-1}$ coincides with the image of the projector P, since, if $u \in \ker L_\Omega \cap H^s(\Omega)$, then $\partial_\nu^j L u = 0$, $0 \le j \le b - m$. Besides, as it will be shown, if $u \in \operatorname{Im}\gamma_r^{-1}$ and $\gamma(u) = 0$, then also $\tilde{\gamma}(u) = 0$. Therefore for functions $u \in \operatorname{Im}\gamma_r^{-1}$ fulfillment of (A2.1.7) follows from inclusion (A2.1.6). Consequently, the restriction of the projector P on $\operatorname{Im}\gamma_r^{-1}$ is a Calderon-Seely projector.

6^0. **Construction of the operator of extension with respect to Cauchy data.** Define the operator L_γ by the condition: $L_\gamma\tilde{\gamma} = \tilde{\gamma}L$, where $\tilde{\gamma}$ is the operator of taking the trace. Since the differential operators are local, the oparator L_γ is defined correctly, i.e., for any $\tilde{\gamma}_r^{-1}$ such that $\tilde{\gamma}\tilde{\gamma}_r^{-1}\varphi = \varphi$, $L_\gamma = \tilde{\gamma}L\tilde{\gamma}_r^{-1}$. From this relation and the existence of a continuous right inverse $\tilde{\gamma}_r^{-1}$ follows that the operator

$$L_\gamma : \bigoplus_{j=0}^{b} H^{s-j-1/2}(\partial\Omega) \to \bigoplus_{j=0}^{b-m} H^{s-m-j-1/2}(\partial\Omega)$$

is continuous as a composition of continuous operators.

Lemma. Let

$$H_L^s(\mathbf{R}^n) = \{u \in H_{loc}^s(\mathbf{R}^n)\,|\,\tilde{\gamma}(u) \in \ker L_\gamma\}.$$

For any $\varphi = (\varphi_0, ..., \varphi_{m-1}) \in DC_{m-1}$ there exists a function $u \in H_L^s(\mathbf{R}^n)$ which is unique up to functions having the trace equal to zero on $\partial\Omega$ such that $\partial_\nu^j u|_{\partial\Omega} = \varphi_j$, $0 \le j \le m-1$.

The operator $\alpha : DC_{m-1} \to \bigoplus_{j=0}^{b} H^{s-j-1/2}(\partial\Omega)$, comparing the function $\varphi \in DC_{m-1}$ to the trace of function $u \in H_L^s(\mathbf{R}^n)$, is continuous, and $\operatorname{Im}\alpha = \ker L_\gamma$.

The proof of this lemma is given in 11^0.

This lemma and the trace theorem in Sobolev space yields directly to the existence of γ_r^{-1} such that $\partial_\nu^j L u = 0$, $0 \le j \le b - m$, for any function $u \in \operatorname{Im}\gamma_r^{-1}$.

7^0. **The Calderon-Seely projector.**

Proposition 2. Let $\alpha : DC_{m-1} \to \bigoplus_{j=0}^{b} H^{s-j-1/2}(\partial\Omega)$ be the extension operator of the trace function with respect to its Cauchy data of order

$m - 1$, defined by the lemma of 6^0, $\tilde{\gamma}_r^{-1} : \oplus_{j=0}^b H^{s-j-1/2}(\partial\Omega) \to H^s(\Omega)$ an extension operator such that functions from $\operatorname{Im}\tilde{\gamma}_r^{-1}$ have a compact support, $\gamma : H^s(\Omega) \to DC_{m-1}$ an operator of taking Cauchy data of order $m - 1$ on $\partial\Omega$, G_Ω – the semi-inverse operator to L_Ω, defined in proposition 1: $G_\Omega \equiv \rho_s^+ G\chi_{s-m}^+$.

Then

i) DC_{m-1} is the clear trace space of the operator L_Ω;

ii) the projector

$$p = \gamma(I - G_\Omega L_\Omega)\tilde{\gamma}_r^{-1}\alpha \qquad (A2.1.9)$$

is a Calderon-Seely projector;

iii) the operator

$$\pi = (I - G_\Omega L_\Omega)\tilde{\gamma}_r^{-1}\alpha \qquad (A2.1.10)$$

is potential.

• The operator $P = (I - G_\Omega L_\Omega)\tilde{\gamma}_r^{-1}\alpha\gamma$ is a projector on $\ker L_\Omega$, moreover, $\ker P \supset \ker\gamma$.

In fact, the operator $\alpha\gamma$ is a projector in the trace space, for all that, $\operatorname{Im}\alpha\gamma = \ker L_\gamma$.

If $u \in \ker L_\Omega$, then $\tilde{\gamma}(u) \in \ker L_\gamma$. Therefore $\alpha\gamma(u) = \tilde{\gamma}(u)$.

Let $\tilde{\gamma}_r^{-1}$ be some right–inverse to the operator $\tilde{\gamma}$. Then $\tilde{\gamma}_r^{-1}\tilde{\gamma}(u) = u + u_0$, where $u_0 \in H_0^s(\Omega)$. However, by proposition 1, $H_0^s(\Omega) \subset \ker(I - G_\Omega L_\Omega)$. Therefore $P(u + u_0) = u$. Consequently $\ker L_\Omega \subset \operatorname{Im} P$.

The inverse inclusion is obvious, since the operator P is the restriction of proector constructed in proposition 1 on the image of the operator $\rho_\Omega\tilde{\gamma}_r^{-1}\alpha\gamma$.

Finally, for the projector P, inclusion (A2.1.7) is fulfilled, since, if $u \in \ker\gamma$, then $\alpha\gamma(u) \equiv 0$.

Therefore the relation $p\gamma = \gamma P$ correctly defines the Calderon-Seely projector p, and the relation $p = \gamma\pi$ correctly defines a potential. •

Theorem 2. Let the operators p and π be defined by the formulae (A2.1.9) and (A2.1.10).

Then the problems

$$L(x) = 0\,, \ x \in \Omega \qquad (A2.1.11)$$

and

$$p\psi(x) = \psi(x)\,, \ x \in \partial\Omega \qquad (A2.1.12)$$

are equivalent.

If $\varphi \in DC_{m-1}$ is the solution of problem (A2.1.12), then $u = \pi\varphi$ is the a solution of problem (A2.1.11). Conversely, if $u \in H^s(\Omega)$ is the solution of problem (A2.1.11), then $\varphi = \gamma u$ is the solution of problem (A2.1.12).

• The assertion of this theorem is a direct consequence of proposition 2 and the definitions of potential and Calderon-Seely projector (2^0). •

8^0. **Representation of the projector P by a volume integral.** Consider the case when the operator G is the convolution with the fundamental solution. Here $\Omega' = \mathbf{R}^n$.

For any function $u \in H^s(\Omega)$ $v = L\rho_s^+ \tilde{\gamma}_r^{-1} \alpha\gamma \in H_0^{s-m}(\Omega)$. Therefore $\tilde{v} = \chi_s^+ v$ is the continuation by zero of the function v. Consequently,

$$\int\limits_{\mathbf{R}^n} g(y,x)\tilde{v}(y)\,dy = \int\limits_{\Omega} g(y,x)v(y)\,dy\,.$$

If $u \in H_L^s(\mathbf{R}^n)$, then $\rho_s^+ \alpha\gamma(u) = u + u_0$, where $u_0 \in H_0^s(\Omega)$. Since $H_0^s(\Omega) \subset \ker(I - G_\Omega L_\Omega)$, for $u \in H_L^s(\mathbf{R}^n)$, one arrives at

$$Pu(x) = u(x) - \int\limits_{\Omega} g(y,x)v(y)\,dy\,,\ x \in \mathbf{R}^n\,.$$

Note that $\operatorname{Im} P \subset H_L^s(\mathbf{R}^n)$.

9^0. **Seely's theorem. The connection between projectors on the kernel of internal and external problems.** As a consequence of theorem 2 of 7^0, a theorem close to Seely's theorem [32] will be obtained.

Denote by L^+ and L^- the operators defined by the relations $L^+\rho_s^+ = \rho_{s-m}^+ L$ and $L^-\rho_s^- = \rho_{s-m}^- L$.

Let $p^+ = \gamma(I - G_\Omega L^+)\tilde{\gamma}_r^{-1}\alpha$, $p^- = \gamma(I - G_{\Omega'/\overline{\Omega}}L^-)\tilde{\gamma}_r^{-1}\alpha$ i.e., the image of the operator p^+ are the Cauchy data of the function from $H^s(\Omega)$, satisfying the equation $L^+u(x) = 0$, $x \in \Omega$, while the image of the projector p^- are the Cauchy data of the functions from $H_{loc}^s(\Omega'/\overline{\Omega})$, satisfying the equation $L^-u(x) = 0$, $x \in \Omega'/\overline{\Omega}$ and conditions l' on Ω' (cf. 1^0).

One has now

Theorem 3. The projectors p^+ and p^- are mutually complementary: $p^+ + p^- = I$.

• For any $\varphi \in DC_{m-1}$ $\psi = \alpha\psi \in \ker L_\gamma$. By the definition of the operator L_γ, the function $L\tilde{\gamma}_r^{-1}\psi$ has a zero trace on $\partial\Omega$ and, consequently, $\chi_{s-m}^+\rho_{s-m}^+(L\tilde{\gamma}_r^{-1}\psi)$ equals to zero for $x \in \Omega'/\overline{\Omega}$.

Therefore one has for functions $v \in \operatorname{Im} \tilde{\gamma}_r^{-1}\alpha\gamma$ the relation $(I - \chi_s^+\rho_s^+)v = \chi_s^- \rho_s^- v$.

Whence $p^- = \gamma(I - G_{\Omega'/\overline{\Omega}}L^-)\tilde{\gamma}_r^{-1}\alpha = I - \gamma G\chi_s^- \rho_s^- L\tilde{\gamma}_r^{-1}\alpha = I - p^+$. •

corollary. Let $\varphi \in DC_{m-1}$. Then there existsthe unique pair of functions (u^+, u^-), where $u^+ \in H^s(\Omega)$, $u^- \in H_{loc}^s(\Omega'/\overline{\Omega})$ and u^- a satisfies condition l', such that $Lu^+ = Lu^- = 0$ and $\gamma u^+ + \gamma u^- = \varphi$.

10^0. **Reduced of a boundary value problem in Cauchy data space.** Consider next the second equation of system $(*)$, yielding the boundary condition.

Let $\tilde{l}^j = \sum\limits_{|k|\le j} b_{jk}\frac{\partial}{\partial x_k}$ be the differential operator of order j, $j \le m-1$, given in some neighbourhood of the boundary $\partial\Omega$.

The boundary operator, acting in DC_{m-1}, is defined by the relation

$$l^j\gamma u = \gamma_0\tilde{l}^j u\,. \tag{A2.1.13}$$

The operator l^j maps the element $\varphi \in DC_{m-1}$ into the element ψ, equal to the trace of zero order of function which is a result of the action of the operator $\tilde{l}^j$ on the function u having a trace coinciding with φ .

If the function u has a zero trace of order $j-1$, then $\gamma_0\tilde{l}^j u = 0$ on $\partial\Omega$, i.e., $\ker\gamma \subset \ker\cdot\gamma_0\tilde{l}^j$, whence follows the independence on the choice of the function u and, thereby, the correctness of the definition of the operator l^j on the relation (A2.1.13).

Let $\chi = (\chi_1,...,\chi_{n-1})$ be a local coordinate system, defined in some neighbourhood of the boundary $\partial\Omega$.

The operator l^j, described in this coordinate system, has the form: $l^j = \sum\limits_{k=0}^{j} \lambda_{jk}(\chi)\partial_\nu u$, where $u \in H_L^s(\mathbf{R}^n)$ such that $\gamma u = \varphi$ and $\lambda_{ij}(\chi)$ are differential operators of order not higher than $(j-i)$ with derivatives only in tangential directions. Consequently, $l^j \in \mathrm{L}(DC_{m-1}, H^{s-j-1/2}(\partial\Omega))$.

It will be convenient to introduce the following terminology:

A system of operators $\{l^j\}_{j=0}^{m-1}$ is called normal, if $\oplus_{j=0}^{m-1}\mathrm{Im}\,l^j$ is isomorphic with DC_{m-1}.

An example of a normal system is the system $\{\partial_\nu^j\}_{j=0}^{m-1}$, since DC_{m-1} is, by definition, $\mathrm{Im}\,\partial_\nu^j$.

It may be shown that the condition which determines the normality of a system of boundary operators is the non-degenracy of the matrix $\lambda_{jj}(\chi)$ for all j, $0 \le j \le m-1$.

Normal systems play the role of the basis in the Cauchy data space: Each element $\varphi \in DC_{m-1}$ is uniquely representable in the form $\{\varphi^j\}_{j=0}^{m-1}$, where $\varphi^j = l^j u$ for functions $u \in H^s(\Omega)$ such that $\gamma u = \varphi$. The transition from one base $\{l^j\}_{j=0}^{m-1}$ to another base $\{l'^j\}_{j=0}^{m-1}$ is given by the rule: $\varphi'^j = \sum\limits_{i=0}^{j} \Lambda_{ji}\varphi^i$, where Λ_{ij} are differential operators of not higher order than $(j-i)$, which have only derivatives in tangential directions.

Considering each operator l^j in a composition with embedding in DC_{m-1}, it may be conceived as a projector in DC_{m-1}. In this case, $\ker l^j = \oplus_{i \neq j}^{m-1} \operatorname{Im} l^i$. In particular, for $m = 2$, each normal system $\{l^0, l^1\}$ defines a pair of mutually complementary projectors.

Let the boundary conditions be given in the form of a system of boundary operators $l = \{l^j i\}_{j=0}^{m/2-1}$. Denote by l_c a system of boundary operators such that $\operatorname{Im} l^j + \operatorname{Im} l_c^j$ is isomorphic to DC_{m-1}.

Since DC_{m-1} is the clear trace space of the operator L_Ω, then, by theorem 1, problem (*) may be rewritten in the equivalent version:

$$\begin{cases} p\psi = \psi \\ l\psi = f, \end{cases}$$

or

$$\begin{cases} lpl\psi + lpl_c\psi = l\psi \\ l_cpl\psi + l_cpl_c\psi = l_c\psi \\ l\psi = f. \end{cases} \qquad (A2.1.14)$$

System (A2.1.14) is the initial object of analysis in the boundary integral equations method as well as in the difference potentials method.

From a most general point of view, the following two approaches to the solution of system (A2.1.14) are possible.

1. The general solution of the second equation has the form: $\psi = f + \varphi$, where $\varphi \in \ker l$. Substitute it into the first equation of system (A2.1.14) and consider the first equation as an equation with respect to φ:

$$(I - p)\varphi = -(I - p)f. \qquad (A2.1.15)$$

If λ is an operator with a domain of definition in DC such that

$$\ker \lambda \cap \operatorname{Im}(I - p)\big|_{\ker l} = \{0\}, \qquad (A2.1.16)$$

then the manifold of solutions (A2.1.15) equations

$$\lambda(I - p)\varphi = -\lambda(I - p)f \qquad (A2.1.17)$$

coincide.

Consequently, if φ is a solution of system (A2.1.17), then $\psi = f + \varphi$ is a solution of system (A2.1.15).

Thus, it will be sufficient for the solution of system (A2.1.15) to solve equation (A2.1.17).

The arbitrariness in the choice of λ may be utilized for the construction of the operator $\lambda(I - p)$ in a more convenient form.

For example, for the Dirichlet problem of the Laplace's equation in a domain $\Omega \subset \mathrm{R}^3$, system (A2.1.15) has the form:

$$\begin{cases} -W^{(1)}(\varphi, x) = (1/2)f(x) - W^{(2)}(f, x) \\ (1/2)\varphi(x) - \partial_\nu W^{(1)}(\varphi, x) = \partial_\nu W^{(2)}(f, x), \end{cases}$$

where

$$W^{(1)}(\varphi, x) = -(4\pi)^{-1} \int_{\partial\Omega} \frac{1}{|x - y|}\, \varphi(y)\, d_y s\,,$$

$$W^{(2)}(\varphi, x) = (4\pi)^{-1} \int_{\partial\Omega} \partial_\nu \frac{1}{|x - y|}\, \varphi(y)\, d_y s\,.$$

The operator ∂_ν may be employed as the operator λ. Condition (A2.1.16) is satisfied, since the kernel of the external Neumann problem (in the class of functions which decrease at infinity and for which the convolution with the fundamental solution is defined and is inverse) is trivial.

Equation (A2.1.17) then assumes the form:

$$(1/2)\varphi(x) - \partial_\nu W^{(1)}(\varphi, x) = \partial_\nu W^{(2)}(f, x)\,.$$

Since the kernel of the outer Dirichlet problem is trivial as well, one can take as operator λ the operator l^0, yielding the Dirichlet condition. Then equation (A2.1.17) becomes:

$$-W^{(1)}(\varphi, x) = (1/2)f(x) - W^{(2)}(f, x)\,.$$

In BIEM boundary equations derived in this manner are referred to as equations of the direct method.

Note that failure to satisfy conditions (A2.1.17) leads to "parasite" solutions.

2. The second method yields the equations of the indirect method.

Interpret the equation

$$lp\chi = f \tag{A2.1.18}$$

as an equation in the unknown χ.

If χ is a solution of this equation, then $\psi = p\chi$ satisfies certainly system (A2.1.15).

Let X_0 be some subspace of the space DC_1. Denote by p_0 the restriction on X_0 of the operator p, i.e., $p_0 = p|_{X_0}$. If there is fulfilled the condition

$$\operatorname{Im} p_0 = \operatorname{Im} p, \qquad\qquad (A2.1.19)$$

then a solution χ of (A2.1.18) which belongs to X_0 may be chosen. Therefore in this case it is sufficient for the construction of the solution of system (A2.1.15) to solve the equation $l p_0 \chi = f$: If χ is a solution of last equation, then $\psi = p_0 \chi = p\chi$ is a solution of system (A2.1.15).

It can be shown that condition (A2.1.19) is satisfied if X_0 is the kernel of the boundary operator l^j, yielding unconditionally soluble the problem in the domain which is outer with respect to the initial domain. For example, since the outer Dirichlet problem for the Laplace operator us unconditionally soluble, then for the internal Neumann problem the kernel of the operator l^0, yielding the Dirichlet condition: $X_0 = \ker l^0 = \operatorname{Im} l^1$, may be select as X_0. In this case, the equation of the indirect method assumes the form:

$$(1/2)\chi(x) + \partial_\nu W^{(1)}(\chi, x) = f\,.$$

If condition (A2.1.19) is not satisfied, then equation (A2.1.18) turns out to be soluble, generally speaking, not for any right hand side, i.e., there occurs a "loss" of solutions. This is just what happens when one attempts to solve the outer Dirichlet problem with the aid of a representation of the solution in the form of a potential of a simple layer (which is equivalent to a restriction on the image of the boundary operator l^0).

In order to avoid a loss of solutions, one has to go to an extension of the domain of definition of the variable ξ (i.e., the space X_0) so that condition (A2.1.19) is satisfied.

A detailed study of the direct and indirect methods is contained in the authors' paper [36].

11^0. **Proof of the lemma of section 6^0.**

• Let $\chi = (\chi_1, ..., \chi_{n-1})$ be the local coordinate system, given in the neighbourhood $O \subset \partial\Omega$, and ν is the normal unit vector.

In the local coordinates $\overline{\chi} = (\chi, \nu)$, the operator $\partial_\nu^j L$ has the form:

$$\Lambda(x)\partial_\nu^{m+j} u + \sum_{i=0}^{m+j-1} \Lambda_j^i(x)\partial_\nu^i u,\ 0 \le j \le b - m\,,$$

where $\Lambda(x)$ is a $(k \times k)$–matrix of smooth functions and the $\Lambda_j^i(x)$ are differential operators of order $l = \min(m, m + j - i)$ which have only derivatives in tangential directions.

The matrix $\Lambda(x)$ is non-degenrate, since, by the strength of the ellipticity of the principal symbol $\sigma_m(x,\xi)$, the operator L does not vanish on any non-zero co-tangential vector ξ, and, in particular, does not vanish on the co-tangential vector $d\nu(x)$. However, $\sigma_m(x, d\nu(x)) = \Lambda(x)$.

Therefore the system

$$\Lambda(x)\varphi^{m+j} + \sum_{i=0}^{m+j-1} \Lambda_j^i(x)\varphi^i = 0 \qquad (A2.1.20)$$

has the unique solution in terms of the unknowns φ^{m+j} $0 \leq j \leq b - m$:

$$\partial_\nu^{m+j} u = -(\Lambda(x))^{-1} \sum_{i=0}^{m+j-1} \Lambda_j^i(\partial_\nu^i u).$$

In this manner, one has determined a mapping α, comparing of the Cauchy data $(\varphi^1, ..., \varphi^{m-1})$ to the element $(\varphi^1, ..., \varphi^{m-1}, \varphi^m, ..., \varphi^b)$, where $\varphi^m, ..., \varphi^b$ is a solution of system (A2.1.20).

If η is another coordinate system in O and $\eta(x) = \eta(\chi(x))$, $x \in \partial\Omega$, then

$$\sigma_m(\chi(x), d\nu(x)) = \sigma_m(\eta(x), d\nu(x)).$$

So the mapping α does not depend on the local coordinates and is defined globally on $\partial\Omega$.

By construction, $\operatorname{Im}\alpha = \ker L_\gamma$.

It will now be shown that α is continuous.

Represent α in the form $\alpha = \alpha_{b-1}\alpha_{b-2}...\alpha_0$, where

$$\alpha_j : DC_{m-1} \to \ker \partial_\nu^j L :$$

$$(u, \partial_\nu u, ..., \partial_\nu^{m+j} u)\big|_{\partial\Omega} \to (u, \partial_\nu u, ..., \partial_\nu^{m+j} u, -(\Lambda(x))^{-1} \sum_{i=0}^{m+j-1} \Lambda_j^i(\partial_\nu^i u))\big|_{\partial\Omega}.$$

The image α_j coincides with the graph of mapping:

$$\overline{\alpha}_j : (u, \partial_\nu u, ..., \partial_\nu^{m+j} u)\big|_{\partial\Omega} \to (-(\Lambda(x))^{-1} \sum_{i=0}^{m+j-1} \Lambda_j^i(\partial_\nu^i u))\big|_{\partial\Omega}.$$

Therefore the continuity of the mapping $\overline{\alpha}_j$ is sufficiently verified (cf. [5, p. 373]).

However, the mapping $\overline{\alpha}_j$ is continuous, since the mappings $\Lambda_j^i : H^{b-i}(\partial\Omega) \to H^{b-m-j}(\partial\Omega)$ are continuous as differential operators of order l, and the operator of product on a smooth function is contiuous by the strength of the compactness of $\partial\Omega$.

§2. The difference potentials method

The scheme of the method of Ryaben'kii's difference potentials will now be studied in the application for the Lamé operator.

As it has already been mentioned, DPM utilizes the representation of the potential in the form (cf. (A2.1.9)):

$$p = \gamma(I - G_\Omega I_\Omega)\gamma_r^{-1}\,.$$

The role of the operator G is taken by the operators which are inverses to the operator of the boundary value problem in the auxiliary domain.

An approximation of the projector P is evaluated with aid of a replacement of the operators, entering into (A2.1.9), by their difference analogues.

Subsection 1^0 recalls the definition of a S.L.Sobolev space on a grid, subsections 2^0–7^0 describe discrete analogues of the operators, entering into (A2.1.9). Subsection 2^0 introduces the Lamé difference operator: Subsection 3^0 defines the Green difference operator. Subsection 4^0 deals with its particular realization – the difference fundamental solution. Subsection 5^0 establishes a property which is the discrete analogue of the property of locality of a differential operator. Subsection 6^0 defines the grid boundary. In subsection 7^0 the difference potential and the Calderon-Seely difference projector present. Subsection 8^0 formulates Reznik's theorem on the approximation of the continual potential by difference one. Subsection 9^0 decribes one of the methods of discretization of problem (A2.1.13):

$$\begin{cases} \varphi(x) - p\varphi(x) = 0\,, \\ \quad \tilde{l}\varphi(x) = f(x)\,. \end{cases}$$

In contrast to §1, the presentation here assumes to a large extent the character of a survey. A strict foundation of the method of difference potentials is given in V.S. Ryaben'kii's book [22].

1^0. **S.L.Sobolev grid space.** Let R_h^n be a cubic grid in R^n with step of length h. Denote by $U(\mathrm{R}_h^n)$ the manifold of all n–vectorial complex-valued functions, given on the grid R_h^n, and by $D(\mathrm{R}_h^n)$ its sub-manifold of functions u_h such that their support $\operatorname{supp} u_h$ is a bounded manifold.

For the functions $u_h \in D(\mathrm{R}_h^n)$, there are determined the direct and inverse Fourier transforms:

$$\tilde{u}(\xi) \equiv (Fu_h)(\xi) \equiv \sum_{x \in \mathrm{R}_h^n} u_h(x) \exp(-ix\xi)h^n\,,$$

$$u(x) \equiv (F^{-1}\tilde{u}_h)(x) \equiv \int_T \tilde{u}_h(\xi) \exp(i\xi z)\,d\xi\,,$$

where $\xi \in \mathbb{R}^n$, $T = \{\xi = (\xi_1, ..., \xi_n) \in \mathbb{R}^n | \; |\xi| \leq \pi h^{-1}\}$.

For any real s, determine the Sobolev space $H^s(\mathbb{R}_h^n)$ of grid functions as a subspace consisting of functions $u_h \in U(\mathbb{R}_h^n)$ such that

$$\|u_h\|_s \equiv \frac{1}{(2\pi)^n} \int\limits_T (\sigma_h(\eta))^s \exp(i\eta z)\, d\eta < \infty,$$

where $\sigma_h(\eta) \equiv 1 + 4h^{-2} \sum\limits_{i=1}^n \sin^2(\eta_i h/2)$.

In particular, for $s = 0$, the norm $\|u_h\|_0$ is equivalent to the norm

$$\|u_h\|_{L^2(\mathbb{R}^n)} = \sum\limits_{x \in \mathbb{R}_h^n} |u_h(x)|^2 h^n.$$

Let $\Omega \subset \mathbb{R}^n$ be a bounded domain with smooth boundary $\partial\Omega$. $\Omega_h = \Omega \cap \mathbb{R}_h^n$.

Denote by $\rho(\Omega_h)$ the restriction operator, comprising functions U_h, given on $\mathbb{R}^n$, and functions u_h, given on Ω_h, such that $U_h(x) = u_h(x)$ for any point $x \in \Omega_h$.

The functions u_h, given on Ω_h, belong to the space $H^s(\Omega_h)$, if one can find a function $v_h \in H^s(\mathbb{R}_h^n)$ such that $\rho(\Omega_h)v_h = u_h$. In this case, $\|u_h\|_{H^s(\Omega_h)} \equiv \inf \|v_h\|_s$, where the infimum is taken over all function v_h such that $\rho(\Omega_h)v_h = u_h$.

Note ([34]) that the Sobolev grid spaces form a scale of spaces, i.e., that $\|v_h\|_t \leq \|v_h\|_s$ for $s \geq t$, while product operator on the fin-function is bounded uniformly with respect to h for $0 \leq h \leq h^0$ as the operator from $H^s(\Omega_h)$ in $H^s(\Omega_h)$.

2^0. **The Lamé difference operator.** Let e_j be the ortho-normal base in $\mathbb{R}^n$.

Construct the Lamé operator difference scheme, obtained by replacement of derivatives by the difference expressions:

$$\frac{\partial^2 u_j(x)}{\partial x_i^2} \approx h^{-2}\left(u_j(x + he_i) - 2u_j(x) + u(x - he_i)\right),$$

$$\frac{\partial^2 u_j(x)}{\partial x_i \partial x_k} \approx (4h^2)^{-1} \sum_{s=0}^1 \sum_{t=0}^1 \left((-1)^{s+t} u_j(x + h(-1)^t e_i + (-1)^s e_k)\right).$$

Then the Lamé difference operator assumes the form

$$L_h u(x) = \frac{1}{h^2} \sum_{m \in M} a(x, m) u_j(x + mh), \qquad (A2.2.1)$$

where $x \in \mathrm{R}_h^n$, $u = (u_1, ..., u_n)$ is a vector–valued function of the replacement on the grid R_h^n, $a(m)$ is a $n \times n$–matrix and $M = \{m = (m_1, ..., m_n) \in Z^n| \, |m_j| \leq 1\}$ is the pattern. For example, for two–dimentional isotropic Lamé operator with λ and μ Lamé coefficients the matrix a is written:

$$a_{jj}(m) = \begin{cases} \lambda + 2\mu, & m \in \{(1,0), \, (0,1)\} \\ -4(\lambda + 2\mu), & m = (0,0) \\ 0, & \text{otherwise} \end{cases}$$

$$a_{ij}(m) = \begin{cases} \lambda + \mu, & m \in \{(1,1), \, (-1,-1)\} \\ -(\lambda + \mu), & m \in \{(1,-1), \, (-1,1)\} \quad \text{by } i \neq j \\ 0, & \text{otherwise}. \end{cases}$$

According to results of [35], the operator (A2.2.1), considered as an operator acting from $H^s(\mathrm{R}_h^n)$, in $H^{s-2}(\mathrm{R}_h^n)$, is bounded uniformly with respect to h for small h, i.e., $\|L_h\| \leq C$, where C does not depend on h.

3^0. **The Green difference operator.** Let the boundary value problem

$$\begin{cases} Lu(x) = f(x), & x \in \Omega' \\ l'u(x) = 0, & x \in \partial\Omega' \end{cases} \tag{A2.2.2}$$

be correctly soluble in the spaces $H^s(\Omega')$, $H^{s-2}(\Omega')$. Let G be the operator solving the auxiliary problem.

Let $\Omega_h^M = \{x \in \mathrm{R}_h^n | x = y + mh, \, y \in \Omega_h, \, m \in M\}$.

It will be assumed that for problem (A2.2.2) there has been constructed the stable difference scheme

$$\begin{cases} L_h u(x) = f(x), & x \in \Omega_h', \\ l'u(x) = 0, & x \in \Omega_h'^M/\Omega_h', \end{cases} \tag{A2.2.3}$$

where L_h is the Lamé difference operator (A2.2.1).

The operator G_h, comparing the grid function $f(x) \in H^{s-2}(\Omega_h')$ to the grid function $u(x) \in H^s(\Omega_h')$, which a solution of problem (A2.2.3) will be referred to as Green difference operator [22].

An example of a problem which satisfies the demands of this representation, is the Dirichlet problem for $\Omega' = \{x| 0 < x_j < 1\}$, $h = 1/N$, where N is a positive integer.

This problem is stable, by the strength of a priori stability estimates which may be obtained, for example, with the aid of the method of energy inequalities [26]:

$$\|u\|_{L_h^2(\Omega_h')} \leq C\|f\|_{L_h^2(\Omega')}, \tag{A2.2.4}$$

where $\|u\|_{L^2(\Omega'_h)} = \sum\limits_{x\in\Omega'} |u(x)|^2 h^n$.

According to [29], one can establish from the estimates (A2.2.4) the boundedness, uniform with respect to h, of the operator G_h, conceived as an operator from $H^{s-2}(\Omega'_h/(\mathrm{R}^n/\Omega)^M)$ into $H^s(\Omega'_h)$.

4^0. **The discrete fundamental solution**. The construction of the difference potential method employs, as a rule, a Green difference operator of some auxiliary problem, given in an enveloping domains. Nevertheless, a way of construction of the discrete fundamental solution for Lamé operator with constant coefficients will now be given.

Recall the definition of the symbol of the difference operator L_h:

$$\sigma_h(x,\xi) \equiv \exp(-ix\xi) \sum_{m\in M} h^{-2} a(m)\exp(i(x-mh)\xi),\ \xi\in\mathrm{R}^n.$$

The operator (A2.2.1) is homogeneous to degree 2 in h, whence the symbol can be presented in the form $\sigma_h(x,\xi) \equiv h^{-2}\sigma(\xi h)$.

Since the Lamé operator is elliptic, the matrix $\sigma(\eta)$ does not degenerate for any non-zero η. Besides, by definition, the function $\sigma(\xi h)$ is periodic in ξ with period $2\pi/h$.

The diffrence fundamental solution will now be defined by the expression

$$g_h(x,h) = h^2\, g((x-y)/h),\qquad (A2.2.5)$$

where

$$g(z) = \frac{1}{(2\pi)^n} \int_T \sigma^{-1}(\eta)\exp(i\eta z)\,d\eta,\qquad (A2.2.6)$$

$$T = \{\xi = (\xi_1,...,\xi_n)\in\mathrm{R}^n|\,|\xi|\le \pi h^{-1}\}.$$

It is readily seen by direct verification that one has the relations

$$L_h g_h(x,y) = h^{-n} I_h\, \delta_h(x-y).\qquad (A2.2.7)$$

Remark. The function $\sigma^{-1}(\eta)$ has a singularity $O(|\eta|^{-2})$ for $\eta\to 0$. For $n=3$, the singularity is integrable, while for $n=2$ one must verify the regularity of (A2.2.6). It is sufficient for this purpose to replace the function $\exp(i\eta z)$ by the function $\exp(i\eta z)-1$. •

Let D_h be the space of functions with bounded support in R^n_h.

On the strength of (A2.2.7), the convolution with a fundamental solution, acting on f from D_h according to the formula

$$G_h f(x) = \sum_{y\in\mathrm{R}^n_h} g_h(x,y)f(x)h^n,\ x\in\mathrm{R}^n$$

is the inverse to the operator L_h:

$$L_h G_h f(x) = f(x), \quad G_h L_h u(x) = u(x).$$

5^0. **The "locality" of difference operators**. The locality property of differential operators was essential for the construction of the potential (A2.1.9). For difference operators, it has been found that its following analogue of locality property holds:

Let $(\operatorname{supp} u)_h = \{x \in \mathrm{R}_h^n \,|\, u(x) \neq 0\}$ be the support of the function u.

Then $(\operatorname{supp} L_h u_h) \subset (\operatorname{supp} u)_h^M$.

The proof follows immediately from the definitions of L_h, $(\operatorname{supp} L_h u)_h$ and $(\operatorname{supp} u)_h^M$.

6^0. **The grid boundary.**

Definition [22]. The manifold $\Gamma_h = \Omega_h^M \cap (\mathrm{R}_h^n / \Omega_h)^M$ is called the grid boundary of a manifold Ω.

Example. Let $\Omega_h = \{x \in \mathrm{R}_h^2, x_2 \geq 0\}$. Then

$$\Omega_h^M = \{x \in \mathrm{R}_h^2, \ x_2 \geq -h\}, \ (\mathrm{R}_h^2 / \Omega_h)^M = \{x \in \mathrm{R}_h^2 \ x_2 \leq 0\},$$
$$\Gamma_h = \{x \in \mathrm{R}_h^2, \ x_2 = 0, \ x_2 = -h\},$$

i.e., the boundary Γ_h consists of the two one-dimentional layers $x_2 = 0$ and $x_2 = -h$. Such a multi-layer property is characteristic for grid boundaries also in the case of an arbitrary domain Ω. The principle and essence of the multi-layer property comprise the fact that a grid function on Γ_h is a discrete analogue of first order Cauchy data, and therefore it must be have "sufficiently many" points in order to allow approximation of derivatives along the normals of continual functions.

The space $Z(\Gamma_h)$ of functions, given on a grid boundary Γ_h, is the domain of definition of the difference potential. The difference potential will be defined in the next sub-section. At this stage, only a comment will be made regarding the operator γ_h^{-1}, mapping the functions from DC_1 into the function of $Z(\Gamma_h)$.

Formally, the operator γ_h^{-1} may be represented as a composition of the extension operator $\gamma_r^{-1} : DC_1 \to H^s(\mathrm{R}^n)$ and the operator γ_h of restriction into the grid boundary.

DPM deals with a discrete problem, it is sufficient to determine the operator γ_r^{-1} only within some subspace of the space DC_1. In that case, the design of the operator γ_r^{-1} may be greatly simplified.

Describe the construction of the extension of a continual smooth function $\varphi \in DC_1$ into the neighbourhood O of the boundary $\partial \Omega$.

Let the neighbourhoods O and O' of the boundary $\partial\Omega$ and the function $\psi(x) \in \Gamma(\mathbf{R}^n)$ be such that $O' \supset O$ and

$$\psi(x) = \begin{cases} 1, & x \in O \\ 0, & x \in \mathbf{R}^n/O' . \end{cases}$$

Define in the neighbourhood O of the boundary $\partial\Omega$ a mapping γ_r^{-1} in the following manner:

$$\gamma_r^{-1}(\varphi(x)) = \psi(x)(\varphi^0(x_0) + \varphi^1(x_0)(x - x_0)+$$

$$+\frac{1}{2!}\varphi^2(\varphi^0,\varphi^1,x_0)(x - x_0)^2 + \ldots + \frac{1}{b!}\varphi^b(\varphi^0,\varphi^1,x_0)(x - x_0)^b),$$

where φ^{2+j} is the solution of the algebraic system

$$\begin{cases} \partial_\nu^i u\big|_{\partial\Omega} = \varphi^i,\ 0 \leq i \leq 1 \\ \partial_\nu^i(Lu)\big|_{\partial\Omega} = 0,\ 0 \leq j \leq b - 2, \end{cases} \qquad (A2.2.8)$$

and x_0 is the point of $\partial\Omega$, closest to x.

The operator γ_r^{-1} is the right inverse to the operator γ, where, since φ is smooth, $\gamma_r^{-1}\varphi \subset H_L^s(\mathbf{R}^n)$.

One may take as operator γ_h the operator of taking the values at the points $x \in \Gamma_h$.

Comments. 1. The single-valued solubility of system $(A2.2.8)$ follows from the lemma of §1. 6^0.

2. The neighbourhood O is taken to be so small that for any point $x \in O$ there exists a unique point $x_0 \in \partial\Omega$ which is closest to x. Such a neighbourhood exists. In fact, let $E = \{(x,y) : x \in \partial\Omega,\ y \in \nu(x)\}$. The tangential space $T_x(E)$ to E at point x is $T_x(\partial\Omega) \oplus \nu(x)$. Consider the mapping $f : (x,y) \to x + y$. The mapping $f : (x,y) \to x + y$ is identical on $\partial\Omega$, its differential df is identical on $T_x(\partial\Omega)$ and on $\nu(x)$. Thus, df has a rank equal to n. By a theorem on non-explicit functions, there exists a neighbourhood $U \subset \partial\Omega$ of the manifold E such that the restriction f on U is mutually single-valued (for details chapter 4 of [30]).

7^0. **The difference potential.** Define the operator P_h by: $P_h = \Theta_h^M - \Theta_h^M G_h\Theta_h L_h$, where Θ_h and Θ_h^M are characteristic functions of the manifolds Ω_h and Ω_h^M, respectively.

Lemma. The operator P_h is a projector on the kernel of the operator $\Theta_h L_h$. The function $P_h u$ depends only on the restriction of u on Γ_h.

• The operator P_h is a projector, since $\Theta_h L_h$ does not depend on the position outside Ω_h^M and therefore $\Theta_h^M G_h \Theta_h L_h \Theta_h^M G_h L_h = \Theta_h^M G_h \Theta_h L_h$. Furthermore, let $u \in \operatorname{Im} P_h$, then $\Theta_h^M L_h u - \Theta_h^M L_h G_h \Theta_h L_h u = L_h \Theta_h^M u - \Theta_h L_h u = 0$ for any point $x \in \Omega_h$, by the locality property (5^0).

The inverse inclusion $\ker \Theta_h L_h \subset \operatorname{Im} P_h$ is obvious.

For the proof of the second assertion of the lemma, in view of the linearity, it is sufficient to establish that $P_h u = 0$ if $u(x) = 0$ on Γ_h.

Let $u(x) = 0$ on Γ_h. Present it in the form $u(x) = u_+(x) + u_-(x)$ so that $\Theta_h u_+(x) = 0$ and $\Theta_h^- u_-(x) = 0$, where Θ_h^- is the characteristic function of the manifold Ω_h'/Ω_h^M. Such an expansion is possible in view of the definition of the manifold Ω_h^M. But then $G_h \Theta_h L_h u_+ = \Theta_h^M u_+ = 0$, while for the function u_- one has: $\Theta_h^M G_h \Theta_h L_h u_- = \Theta_h^M G_h L_h u_- = \Theta_h^M u_- = 0$. Consequently, $P_h u = P_h(u_+ + u_-) = 0$. •

Note that in the terminology of §1. 2^0, one has thus proved: The restriction of a grid function on a grid boundary is a clear trace of the operator L_h.

It follows from the lemma just proved that there are defined operators $\pi_h : Z(\Gamma_h) \to \ker \Theta_h L_h$ and $p_h : Z(\Gamma_h) \to Z(\Gamma_h)$, acting according to the formula: $\pi_h = \Theta_h^M - \Theta_h^M G_h \Theta_h L_h \rho_h^{-1}$; $p_h = \rho_h \pi_h$, where $\rho_h : X_h \to Z(\Gamma_h)$ is the restriction operator of the grid function $u \in X_h$ on the grid boundary Γ_h, and $\rho_h^{-1} : Z(\Gamma_h) \to X_h$ is the operator of continuation of the function $u \in Z(\Gamma_h)$ by zero.

Definition. The operator π_h is a *difference potential* , and the operator p_h – a *(difference) Calderon-Seely projector*.

Note yet that the calculation of the action of the operator G_h is reduced to the solution of the auxiliary difference problem (A2.2.4) with a right hand side having an support belonging to Γ_h^M.

8^0. **The approximation of continual potentials by differences ones**. The metric connection between difference and continual potentials for elliptic operators has been studied in the work of A.A.Reznik [18–20].

Here a result of this work which relates to the approximation by difference potentials of a continual potential considered will be presented.

Theorem [19]. Let $\varphi = (\varphi^0, \varphi^1) \in C_\alpha(\partial\Omega) \oplus C_{\alpha-1}(\partial\Omega)$, where $C_\alpha(\partial\Omega)$ and $C_{\alpha-1}(\partial\Omega)$ is the space of Hölder functions on the boundary $\partial\Omega$, $\alpha = 4+\beta$, $0 < \beta < 1$, and $u \in C_\alpha(\Omega)$ be such that $\varphi^i = \partial_\nu^i u|_{\partial\Omega}$, $i = 0, 1$, $\tilde{\gamma}(Lu) = 0$, where $\tilde{\gamma} : C_{\alpha-2}(\Omega) \to \oplus_{j=0}^2 C_{\alpha-2-j}(\partial\Omega)$ is the operator of taking the trace of the functions from $C_{\alpha-2}(\Omega)$.

Let $\varphi_h = u|_{\Gamma_h}$. Then in any domain $\Omega_0 \subset \Omega'$ such that $\Omega \subset \Omega_0$ and

$\overline{\Omega}_0 \subset \Omega'$, for any ε, $0 < \varepsilon < 1$, holds the inequality

$$\left| \pi_h \varphi_h - \rho_h \pi \varphi \right|_{\Omega_0^h, 2+\varepsilon} \leq C h^{2-\varepsilon} \left(\max \left| \varphi^i \right|_{\Omega, \alpha-i} \right),$$

where ρ_h is the restriction operator on the grid R_h^n, $| \cdot |_{\Omega, \alpha-i}$ is the norm of the continual Hölder space, $| \cdot |_{\Omega_0^h, 2+\varepsilon}$ is the norm of the discrete Hölder space ($|v|_{\Omega_0^h, 2+\varepsilon} = \inf |\tilde{v}|_{\Omega_0, 2+\varepsilon}$, where the infinum is taken over all functions $\tilde{v} \in C_{2+\varepsilon}(\Omega_0)$ such that $\tilde{v}|_{\Omega_0^h} = v|_{\Omega_0^h}$).

Remark. The existence of functions $u \in C_\alpha(\Omega)$, which satisfy the conditions of the theorem, is proved essentially simmilar to the lemma of §1. 6^0, substituting the scale of the Hölder spaces for the scale of Sobolev spaces.

9^0. **The discrete approximating problem.** According to the scheme given in §1 the problem

$$\begin{cases} Lu(x) = 0, \; x \in \Omega, \\ lu(x) = f(x), \; x \in \partial\Omega, \end{cases}$$

where L is the Lamé operator and the operator l, of not larger than the first order, yields a boundary condition, is equivalent (in the sense of §1. 1^0) to the system

$$\begin{cases} \varphi(x) - p\varphi(x) = 0, \\ \tilde{l}\varphi(x) = f(x), \; x \in \partial\Omega, \end{cases} \tag{$A2.2.9$}$$

where

$$\varphi(x) = \gamma(u) = \begin{cases} u(x)\big|_{\partial\Omega}, \\ \partial_\nu u(x)\big|_{\partial\Omega}, \end{cases}$$

$p = \gamma\pi = \gamma(I - G_\Omega L_\Omega)\gamma_r^{-1}$ is a Calderon-Seely projector and $\tilde{l}$ is determined by the relation $\tilde{l}\gamma = l$.

Next, one method of discretization of problem $(A2.2.9)$ will be presented.

The replacement of the continual potential π, entering into the definition of the Calderon-Seely projector p, by the difference potential π_h, which approximates the potential π (§2. 8^0) lies at the basis of the transition from the continual problem $(A2.2.9)$ to the discrete problem.

Let $\partial\Omega_H = (x_1^H, ..., x_N^H)$ be the manifold of the points of the boundary $\partial\Omega$, regular in the sense that the distance $\rho(x_i^H, x_j^H)$ between any two points x_i^H and x_j^H of $\partial\Omega_H$ is larger than H and for any points $x \in \partial\Omega$ can be found a point $x_i^H \in \partial\Omega_H$ such that $\rho(x_i^H, x) < 2H$.

The domain of definition of the difference potential π_h is the space of the grid functions on the grid boundary Γ_h.

Define the extension operator $\gamma_h^{-1} : Z(\partial\Omega_H) \to Z(\Gamma_h)$ of the function on the grid boundary Γ_h by the relation: $\gamma_h^{-1} = r_h\gamma_r^{-1}R$, where R is the operator of filling out the grid functions $\varphi = (\varphi^0, \varphi^1) \in Z(\partial\Omega_H)$ into the functions $\psi = (\psi^0, \psi^1)$ such that $\psi^i \in H^{s-i-1/2}(\partial\Omega)$ (as a rule, R is a local spline), $\gamma_r^{-1} : DC_1 \to H^s(O)$ is an operator for which one may employ the continuation operator of the function ψ in some neighbourhood O of the boundary $\partial\Omega$, containing the Γ_h, decribed in §2. 6^0, $r_h : H^s(\Omega) \to Z(\Gamma_h)$ is the restriction operator on Γ_h.

Finally, define the operator $\gamma_h : Z(\Gamma_h) \to Z(\partial\Omega_H)$ by the relation: $\gamma_h = r_H\gamma\Re$, where $\Re$ is the filling out operator of the grid function on Γ_h into the functions of $H^s(O)$, $\gamma : H^s(O) \to DC_1$ is the operator of taking the Cauchy data and $r_h : DC_1 \to Z(\partial\Omega_H)$ is the restriction operator on $\partial\Omega_h$.

In terms of these operators, write down the discrete system of equations in terms of the unknown, discrete function $\varphi \in Z(\partial\Omega_H)$, approximating the continual problem (A2.2.9):

$$\begin{cases} \varphi - \gamma_h\pi_h\gamma_h^{-1}\varphi = 0 \\ l_H \equiv r_H\tilde{l}R\varphi = r_Hf, \ x \in \partial\Omega. \end{cases} \qquad (A2.2.10)$$

This system contains one and a half as many equations than there are unknowns, whence it is indeterminate, and, in general, does not have a solution. It is natural to take as generalized solution of system (A2.2.10) the element $\varphi \in Z(\partial\Omega_H)$ providing the minimum of square form

$$\left\| \varphi - \gamma_h p_h\gamma_h^{-1}\varphi \right\|^2_{Z(\partial\Omega_H)} + \left\| l_h\varphi - r_Hf \right\|^2_{\operatorname{Im} l_H}$$

in some discrete space norms $Z(\partial\Omega_H)$ and $\operatorname{Im} l_H$, corresponding to the norm of the spaces in which the initial continual problem was posed correctly. For example, the norm of the discrete Sobolev space $H^{s-1/2}(\partial\Omega_H)\oplus H^{s-3/2}(\partial\Omega_H)$ may be selected as norm in the space $Z(\partial\Omega_H)$. Simular, the norm of the discrete space $H^{s-1/2}(\partial\Omega_H)$, if the order of the operator l is zero, or $H^{s-3/2}(\partial\Omega_H)$, if the order of the operator l is one, may be selected as norm in the space $\operatorname{Im} l_H$.

In practice, it turns out to be sufficient to adopt the following norms of the spaces $Z(\partial\Omega_H)$ and $\operatorname{Im} l_H$:

$$\|\varphi\|^2_{Z(\partial\Omega_H)} = \sum_{j=1}^{n} \sum_{x\in\partial\Omega_{H'}} H'^{(n-1)}\Big((\varphi_j^0(x))^2 + \sum_{i=1}^{n-1}(\partial_h^i\varphi_j^0(x))^2 + (\varphi_j^1(x))^2\Big);$$

$$\|\psi\|_{\operatorname{Im} l_H}^2 = \begin{cases} \displaystyle\sum_{j=1}^{n} \sum_{x \in \partial\Omega} H'^{(n-1)}(\psi(x))^2\,, & \text{if } l \text{ is of order zero} \\[2ex] \displaystyle\sum_{j=1}^{n} \sum_{x \in \partial\Omega_{H'}} H'^{(n-1)}\big((\psi(x))^2 + \sum_{i=1}(\partial_h^i \psi(x))^2\big)\,, & \text{otherwise.} \end{cases}$$

Here ∂_h^i is the difference relation, approximating the derivative in the tangential direction x_i.

The problem of minimization is solved by any interative method, for example, the conjugate gradient method.

In DPM, other discretization schemes are also employed. In particular, an approximate solution of problem (A2.2.9) is sought as solution of the problem of minimization of the quadratic form

$$\|\xi - p\xi\|_{DC_1}^2 + \|l\xi - f\|_{\operatorname{Im} l}^2$$

on the manifold of functions ξ of the form $\sum c_k e^{(k)}(x)$, where $e^{(k)}(x)$ is the base in the space DC_1.

Different variants of the solution of problem (A2.2.9) by the method of the difference potential and their foundations are discussed in Ryaben'kii's book [22].

§3. The Sokhotskii-Plemelj theorems and the theorem on the involutions of the boundary operator in electro–elasticity

The derivation of the integral equations of the classical potential method for Laplace's equation rests on the following fundamental property of the potentials

$$W^{(1)}(\varphi, x) = -(4\pi)^{-1} \int_{\partial\Omega} \frac{1}{|x - y|}\, \varphi(y)\, d_y s$$

and

$$W^{(2)}(\varphi, x) = (4\pi)^{-1} \int_{\partial\Omega} \partial_\nu\left(\frac{1}{|x - y|}\right) \varphi(y)\, d_y s$$

of simple and double layers.

1. The limiting values $\big[W^{(i)}(\varphi, x)\big]^{\pm}$ of the simple– and double–layer potentials and the limiting values $\big[\partial_\nu W^{(i)}(\varphi, x)\big]^{\pm}$ of their normal derivatives, as x approaches the boundary $\partial\Omega$ from inside and outside of a domain along a non-tangential direction exist and are interrelated by

$$\big[\partial_\nu^{2\ j}{}^{j}W^{(i)}(\varphi, x)\big]^{-} - \big[\partial_\nu^{2-j}W^{(i)}(\varphi, x)\big]^{+} = \varphi(x)\delta_{ij}\,, \ x \in \partial\Omega,$$

$i = 1, 2$, $j = 1, 2$ (Lyapunov-Tauber theorem).

2. The direct value $W^{(2)}(\varphi, x)$ of the double–layer potential and the direct value of the normal derivative $\partial_\nu W^{(1)}(\varphi, x)$ of the simple–layer potential are defined. One has

$$\partial_\nu^{2-j} W^{(j)}(\varphi, x) = \frac{1}{2}\varphi(x) + \left[\partial_\nu^{2-j} W^{(j)}(\varphi, x)\right]^+, x \in \partial\Omega, j = 1, 2$$

(the real analogue of the Sokhotskii-Plemelj theorem).

The analogies between the mentioned properties remain true also for isotropic elastic potentials (cf., for example, chapter 6 of the present book) and for the potentials of some other operators of mathematical physics (the direct values $W^{(2)}(\varphi, x)$ and $T_\nu W^{(1)}(\varphi, x)$ for elastic potentials are defined in the sense of the principal value). The standard method of proof of these assertions is based on a study of the explicit expression for the kernel of the integral operators, defining the potentials.

The assumption that the mentioned facts can be widely generalized seems to be quite natural.

In the present section the analogy of the Lyapunov-Tauber and Sokhotskii-Plemelj theorems will be proved for the operator of electro-elasticity (and, in fact, for homogeneous, second order elliptic systems with constant coefficients). Note that the corresponding results have been presented in [17] without proof.

It is an essential fact that the proof is based on the definition of potentials in the form (A2.1.9) and that there is not required an explicit form of the fundamental solution or Green's function. Therefore these results may be employed in the construction of boundary integral equations also in those cases, when the fundamental solution or the Green's function of some canonical domain are given in the form of series, integrals or by any other algorithmic method (cf., for example, [3], where the fundamental solution of the Lamé operator for anisotropic media is represented in the form of series of spherical harmonics). Within the proposed approach the theorems of the Lyapunov-Tauber and Sokhotskii-Plemelj type are closely connected with the theorem on involution of the boundary operator, corresponding to the differential operator. This theorem is a generalization of the theorem on involution for the singular Cauchy operator which plays an important role in the Riemann-Hilbert problem.

Sections §3. 1^0–4^0 present an introduction to the boundary value problems of electro-elasticity for piezo-electric media, utilized later on. They are discussed in greater detail in [2,15,17].

§3. 5^0 defines the electro-elastic simple– and double–layer potentials in analogy with elastic potentials.

§3. 6^0 presents an alternative definition of the simple– and double– layer potentials and establishes its equivalence with the preceding definition.

§3. 7^0 proves a theorem on "jumps" of the potentials and their derivatives during passage through $\partial\Omega$. The theorem presented is a generalization of the Lyapunov-Tauber theorem.

§3. 8^0 gives a representation of the limiting values of potentials and their derivatives on $\partial\Omega$ in the form of a singular integro-differential operator.

§3. 9^0 establishes a theorem on involution of the derived integro-differential operator.

1^0. **The equation of electro-elasticity**. The equations of state of piezo-electric media in linear approximation have the form

$$\begin{cases} \sigma_{ij} = c_{ijkl}\, u_{k,l} + e_{kij}\, u_{0,k} \\ D_j = e_{jkl}\, u_{k,l} - \vartheta_{jk}\, u_{0,k} \end{cases}$$

where σ_{ij} is the mechanical stress tensor, D is the electric induction vector, u_0 is the electric potential, linked to the stress of the electric field by the relation $E_j = -\operatorname{grad} u_0$, $u = (u_1, ..., u_3)$ is the displacement vector, c_{ijkl} are the moduli of elasticity, measured for constant current, e_{mij} are the piezo-electric constants and ϑ_{jk} – the dielectric permeabilities.

The tensor c_{ijkl} is symmetric with respect to the first and second pairs of subscripts, the tensor e_{mij} is symmetric with respect to the latter.

The system of the linear equilibrium equations of electro-elasticity of piezo-electric media has the form:

$$Lu = 0, \qquad\qquad (A2.3.1)$$

where $L = \begin{cases} \sigma_{ij,j} \\ D_{j,j}\,. \end{cases}$

For physically meaningful values of the coefficients, the system $(A2.3.1)$ is an elliptic, homogeneous system of second order. In this context, as is readily seen, ellipticity of system $(A2.3.1)$ will occur if and only if the anisotropic Lamé operator $c_{ijkl}u_{k,lj}$ and the anisotropic Laplace operator $\vartheta_{jk}u_{0,kj}$ are ellitic.

The boundary conditions for system $(A2.3.1)$ subdivide into two groups - mechanical and electric conditions.

The mechanical conditions are the usual ones:

$$(l_M^0 u)_i \equiv u_i\big|_{\partial\Omega} \,, i = 1, 2, 3\,,$$

$$(l_M^1 u)_i \equiv \sigma_{ij}(u)\nu_j\big|_{\partial\Omega} \equiv c_{ijkl}\nu_j u_{l,k} + e_{kij}\nu_j u_{0,k}\,, i, j = 1, 2, 3\,,$$

where $\nu \equiv (\nu_1, \nu_2, \nu_3)$ is external normal vector. The electric conditions are:

$$l_\vartheta^0 u \equiv u_0\big|_{\partial\Omega}\,, \tag{A2.3.2}$$

$$l_\vartheta^1 u \equiv D_j(u)\nu_j\big|_{\partial\Omega} \equiv e_{jkl}\nu_j u_{l,k} - \vartheta_{jk}\nu_j u_{0,k}\,, i, j = 1, 2, 3\,. \tag{A2.3.3}$$

Condition (A2.3.3) describes the jumps of the normal derivative of the electric induction vector on the interface between two media, one of which is the air; condition (A2.3.2) gives the value of the potential on the electrode-bearing segment.

In the sequel, the boundary condition $l^i = (l_M^i, l_\vartheta^i)$ for $i = 0$ will be called Dirichlet condition, for $i = 1$ – Neumann condition; correspondingly, the boundary value problem $\{Lu = 0, \ l^i u = f\}$ will be called Dirichlet problem for $i = 0$ and Neumann problem for $i = 1$.

Proposition 1. The system $\{l^0, l^1\}$ is normal.

• Write the operator l^1 in the form: $l^1 = b\partial_\nu + b_1$, where $b = \|b_{ij}(x)\|$ is a 4×4 $(0 \le i, j \le 3)$ matrix, the elements $b_{ij}(x)$ of which are smooth functions and b_1 is a differential operator of first order which contains only tangential derivatives.

Since $b_1 \in L(H^{s-1/2}(\partial\Omega), H^{s-3/2}(\partial\Omega))$, and the product operator on a smooth function is likewise continuous in $H^{s-3/2}(\partial\Omega)$, then it will be sufficient for a proof of the normality of the system (l^0, l^1) to very the fact that the matrix b is invertible.

Let $(\tau_1, ..., \tau_{n-1}, \nu)$ be a coordinate system with centre at $x \in \partial\Omega$, and ν the external normal to $\partial\Omega$, $\{\tau_j\}$ the basis in the tangential plane to $\partial\Omega$ at the point x. Then $\partial_j = \nu_j\partial_\nu + \tau_{ij}\partial_{\tau j}$, whence

$$b = \begin{cases} c_{ijkl}\nu_l\nu_j + e_{kij}\nu_k\nu_j \\ e_{jkl}\nu_l\nu_j - \vartheta_{kij}\nu_k\nu_j\,, \end{cases}$$

which coincides with the symbol of the operator L.

Since the operator L is elliptic, the matrix b is non-degenerate.

Remark. If one sets the coefficients e_{ijk} equal to zero, system (A2.3.1) degenerates into the (anisotropic) Lamé system in terms of u_1, u_2, u_3 and the (anisotropic) Laplace equation in u_0. Thus, the system of electro-elastcity includes also the particular case of the Lamé system and the Laplace operator.

2^0. **Green's formula (reciprocity relation).** The operator L is formally self-conjugate with respect to the scalar product $(u, v) = \sum_{i=0}^{3} u_i v_i$.

Using relations of the form: $\sigma_{ij,j} u_i = (\sigma_{ij} u_i)_{,j} - \sigma_{ij} u_{i,j}$, and Stokes' formula:

$$\int_\Omega \operatorname{div} u_i \, d\omega = \int_{\partial\Omega} u_i \nu_i \, ds \, ,$$

Green's formula (the reciprocity relation) for system (A2.3.1) may be derived in the standard manner:

$$\int_\Omega (Luv - uLv) \, d\Omega = \int_{\partial\Omega} \big(\sigma_{ij}(u)v_i\nu_j - u_i\sigma_{ij}(v)\nu_j - \\ - D_i(u)v_0\nu_i + u_0 D_i(v)\nu_i \big) \, ds \, . \tag{A2.3.4}$$

Let $\varphi \in DC_1$. As has already been noted (§1. 10^0), if $\{l^j\}_{j=0}^1$ is a normal system, every element φ may be represented in a unique manner in the form (φ^0, φ^1), where $\varphi^j \in \operatorname{Im} l^j$. Green's formula (A2.3.4) determines on the space DC_1 the bilinear form

$$< \varphi, \psi > = \int_{\partial\Omega} \big(\varphi_i^1(y)\psi_i^0(y) - \varphi_j^0(y)\psi_j^0(y) \big) \, d_y s \, . \tag{A2.3.5}$$

Remark. Formula (A2.3.5) is of a much more general nature. In fact, for any elliptic operator, there exists a normal system, not the only one, $\{l^i\}_{j=0}^{m-1}$ which uniquely determines a normal system $\{\tilde{l}^j\}_{j=0}^{m-1}$ such that the skew-symmetric matrix form

$$< \varphi, \psi > = \int_{\partial\Omega} \sum_{j=0}^{m-1} (-1)^j \varphi^{m-1-j}(y)\psi^j(y) \, d_y s$$

is continuous and non-degenrate. Here $\varphi^j \in \operatorname{Im} l^j$, $\psi^j \in \operatorname{Im} \tilde{l}^j$.

3^0. **The kernels of the Dirichlet and Neumann problems.**

The kernel of the internal Dirichlet problem is trivial. The kernel of the internal Neumann problem consists of vectors u of the form

$$u_0 = c, \ u_1 = a_1 + b_1 x_3 - b_2 x_2, \ u_2 = a_2 + b_2 x_1 - b_3 x_3, \ u_3 = a_3 + b_3 x_2 - b_1 x_1 \, .$$

The kernels of the external Dirichlet and Neumann problems are trivial in the class of functions which satisfy the conditions:

$$u_i = O(|x|^{-1}), \operatorname{grad} u_i = O(|x|^{-2}), i = 0, 1, 2, 3. \qquad (A2.3.6)$$

• Let $\tilde{L}(u_1, u_2, u_3, u_0) \equiv L(u_1, u_2, u_3, -u_0)$.
Direct verification establishes the correctness of the identity

$$\int\limits_{\Omega} \tilde{L}u\, u = -(\mu(u) + \varepsilon(u)) + \int\limits_{\partial\Omega} \left(\tilde{\sigma}_{ij}(u)u_i\nu_j + \tilde{D}_j(u)u_0\nu_j\right) ds, \qquad (A2.3.7)$$

where

$\tilde{\sigma}_{ij}(u_1, u_2, u_3, u_0) \equiv \sigma_{ij}(u_1, u_2, u_3, -u_0),$
$\tilde{D}_j(u_1, u_2, u_3, u_0) \equiv D_j(u_1, u_2, u_3, -u_0),$
$\mu(u) = (1/2) \int\limits_{\Omega} c_{ijkl} u_{k,l} \varepsilon_{ij}\, dx$ is the energy of the mechanical deformation,
$\varepsilon(u) = (1/2) \int\limits_{\Omega} \vartheta_{ij}\, E_j E_i\, dx$ is the electric energy.

An analogous identity holds true for the external domain Ω (when the domain of integration is replaced by $\mathrm{R}^n/\overline{\Omega}$, and the external normal ν becomes internal). Conditions (A2.3.6) ensure that the quantities μ and ε are finite.

For physically meaningful values of the coeffcients, the quantity $\mu(u) + \varepsilon(u)$ is non-negative. The assertion of the proposition follows from this observation and identity (A2.3.7). •

4^0. **The fundamental solution of the operator of electro-elasticity.**

Let $a = \|a_{ijkl}\xi_i\xi_j\|_{k,l=0}^3$ be the symbolic matrix of the operator of electro-elasticity.

Then the function $g(z) = (2\pi)^{-3} \int a^{-1}(\xi)\exp(i\xi z)\, d\xi$ is the fundamental solution of the operator L.

The operator of electro-elasticity is a homogeneous, elliptic, second order operator. Hence (cf. for example, [29]), the function $g(z)$ can be represented in the form $g(z/|z|)|z|^{-1}$, where $g(\omega) \in \Gamma(\mathrm{R}^n/\{0\})$.

The convolution with the fundamental solution is defined on functions from $H_c^{s-m}(\mathrm{R}^n)$ and it is left–inverse to the operator L. Besides, $u = g * v$ satisfies the condition:

$$u(z) \leq c|z|^{-1}$$
$$l^1 u(z) \leq c|z|^{-2} \qquad (A2.3.8)$$

as $|z| \to \infty$.

On the other hand, if u satisfies the estimate (A2.3.8) and the equation $Lu = 0$, then it follows from Green's formula, written for a sphere S_R of radius R, containing the point x inside it,

$$u(x) = \int\limits_{S_R} g(y,x) l^1 u(y)\, d_y s - \int\limits_{S_R} l^1 g(y,x) u(y)\, d_y s = \int\limits_{S_R} o(R^2)\, d_y s$$

that $u(x) = 0$.

Thus the convolution with the fundamental solution satisfies the requirements of §1. 1^0, if the subspace of functions from $H^s_{loc}(\mathrm{R}^3)$, which satisfy inequality (A2.3.8), is considered as domain of definition of the operator L.

5^0. **Definition of the simple– and double–layer potentials.**

Let again $\Omega \subset \mathrm{R}^3$ be a bounded domain with the smooth boundary $\partial\Omega$.

By analogy with the elastic potential, introduce the concepts of simple– and double–layer potentials.

Definition 1. The quantity

$$W^{(1)}(\varphi, x) = - \int\limits_{\partial\Omega} l^0 g(y,x) \varphi(y)\, d_y s\,, \ x \in \mathrm{R}^n/\partial\Omega \qquad (A2.3.9)$$

is called a simple–layer potential with the vector density $\varphi = (\varphi_0, \varphi_1, \varphi_2, \varphi_3)$.

The quantity

$$W^{(2)}(\varphi, x) = \int\limits_{\partial\Omega} l^1 g(y,x) \varphi(y)\, d_y s\,, \ x \in \mathrm{R}^n/\partial\Omega \qquad (A2.3.10)$$

is called a double–layer with the vector density $\varphi = (\varphi_0, \varphi_1, \varphi_2, \varphi_3)$

6^0. **Equivalent definition of potentials.** Define the operators P^+ and P^-, acting on functions of $H^s_L(\mathrm{R}^n)$ by

$$P^+ u(x) = u(x) - \int\limits_{\Omega} g(x,y) Lu(y)\, dy = \int\limits_{\mathrm{R}^n/\overline{\Omega}} g(x,y) Lu(y)\, dy\,, x \in \mathrm{R}^n\,,$$

$$P^- u(x) = u(x) - \int\limits_{\mathrm{R}^n/\overline{\Omega}} g(x,y) Lu(y)\, dy = \int\limits_{\overline{\Omega}} g(x,y) Lu(y)\, dy\,, x \in \mathrm{R}^n\,,$$

Definition 2. Let the quantity

$$W^{(2-r)}(\varphi, x) = \begin{cases} P^+ u(x)\,, \ x \in \Omega \\ -P^- u(x)\,, \ x \in \mathrm{R}^n/\overline{\Omega}\,, \end{cases}$$

where $u \in H^s_{loc}(\mathbb{R}^n)$ such that

$$l^t u = \delta_{tr}\varphi \,, \, r, t = 0, 1$$
$$\tilde{\gamma}(Lu) = 0 \qquad\qquad (A2.3.11)$$

be called the potential of $(2 - r)$-th layer.

Lemma. The given definition is correct and agrees with definition 1.

• According to the lemma of §1. 6^0, for any first order Cauchy data there exist a function $u \in H^s(\mathbb{R}^n)$ which satisfies the second condition of (A2.3.11). Therefore the existence of a function u which satisfies both conditions (A2.3.11) follows from the fact that the system $\{l^0, l^1\}$ is normal (§3, 1^0).

Indentities (A2.3.9) and (A2.3.10) as well as the independence on the choice of the function $u \in H^s(\mathbb{R}^n)$, which satisfies conditions (A2.3.11) as well as Indentities (A2.2.11) follow from Green's formula, written for the domains Ω and $\mathbb{R}^n/\overline{\Omega}$ in the form:

$$P^+ u(x) = u(x) - \int_\Omega g(x,y) Lu(y)\, dy = \int_{\partial\Omega} l^1_\nu g(y,x) l^0 u(y)\, d_y s -$$
$$- \int_{\partial\Omega} l^0 g(y,x) l^1 u(y)\, d_y s \,, \quad x \in \Omega\,;$$

$$P^- u(x) = u(x) - \int_{\mathbb{R}^n/\overline{\Omega}} g(x,y) Lu(y)\, dy = -\left(\int_{\partial\Omega} l^1 g(y,x) u(y)\, d_y s - \right.$$
$$\left. - \int_{\partial\Omega} l^0 g(y,x) l^1 u(y)\, d_y s \right) \,, \quad x \in \mathbb{R}^n/\overline{\Omega}\,.$$

The minus sign on the right hand side of the second identity is due to the fact that the normal ν is external with respect to the domain $\mathbb{R}^n/\overline{\Omega}$. •

7^0. **The theorem on jumps in the limiting values of the simple– and double–layer potentials.** Let

$$\left[l^t W^{(2-r)}(\varphi, x_0) \right]^+ = \lim_{\substack{x \to x_0 \in \partial\Omega \\ x \in \Omega}} l^t W^{(2-r)}(\varphi, x)\,,$$

$$\left[l^t W^{(2-r)}(\varphi, x_0) \right]^- = \lim_{\substack{x \to x_0 \in \partial\Omega \\ x \in \mathbb{R}^n/\overline{\Omega}}} l^t W^{(2-r)}(\varphi, x)\,,$$

$r = 0, 1, \, t = 0, 1.$

Then the limiting values $\left[l^t W^{(2-r)}(\varphi, x_0)\right]^{\pm}$ exist and one has:

$$\left[l^t W^{(2-r)}(\varphi, x_0)\right]^{+} - \left[l^t W^{(2-r)}(\varphi, x_0)\right]^{-} = (-1)^t \delta_{rt}\varphi(x_0).$$

$\bullet$ Let $u \in H_L^s(\Omega)$ be such that $l^t u|_{\partial\Omega} = \delta_{rt}\varphi$, $r = 0, 1$; $t = 0, 1$. One has

$$\left[l^t W^{(2-r)}(\varphi, x_0)\right]^{+} - \left[l^t W^{(2-r)}(\varphi, x_0)\right]^{-} =$$

$$= \lim_{\substack{x_+ \to x_0 \in \partial\Omega \\ x_+ \in \Omega}} (-1)^t l^t P^+ u(x_+) + \lim_{\substack{x_- \to x_0 \in \partial\Omega \\ x_- \in \mathbb{R}^n/\bar\Omega}} (-1)^t l^t P^- u(x_-) =$$

$$= (-1)^t \left(l^t P^+ u(x_0) + l^t P^- u(x_0)\right) =$$

$$= (-1)^t l^t \left(P^+ u(x_0) + P^- u(x_0)\right) = (-1)^t l^t u(x_0) = (-1)^t \delta_{rt}\varphi(x_0).$$

The existence of $\lim P^t u^{\pm}(x_+)$ is ensured by the smoothness of $v^{\pm}(x) = P^{\pm} u(x)$, $x \in \mathbb{R}^n$: Since $v^{\pm}(x) \in H_c^s(\mathbb{R}^n)$ for $s > 5/2$, it follows from Sobolev embedding theorem that $v \in C^k(\mathbb{R}^n)$, where k–integer such that $0 \leq k < s - 3/2$ (cf., for example, [21]). $\bullet$

8^0. **Representation of limiting values of the potentials in singular integro-differential operators.**

According to the general reduction scheme, studied in §1, the initial equation $Lu = 0$ is equivalent to the condition $p\gamma u = \gamma u$, where p is a projector which acts in the first order Cauchy data space such that $\text{Im}\, p = \gamma(\ker L_\Omega)$.

The definition of §3. 6^0 and the theorem just proved determine the projector p as

$$p\psi(x_0) = \gamma(W^{(1)}(\psi^1, x_0)) + (W^{(2)}(\psi^0, x_0)), \qquad (A2.3.12)$$

where $\psi = (\psi^0, \psi^1) \in DC_1$ and γ is the operator of taking the first order trace, treated as operator of taking the limit $(\lim_{x \to x_0 \in \partial\Omega} l^t)_{t=0}^1$.

The theorem to proved next yields a description of the operator p in terms of operators which act in the space DC_1 without employing the extension into the domain.

In fact, this means a transition to an operator with the kernel obtained after execution of the limiting process.

Proposition 2. Let $x \in \partial\Omega$;

$$S_{01}\psi(x) = 2 \int\limits_{\partial\Omega} l^0 g(y, x)\psi(y)\, d_y s\,,$$

$$S_{00}\varphi(x) = -2 \int\limits_{\partial\Omega} l^1 g(y, x)\varphi(y)\, d_y s\,,$$

where the second integral is understood in the sense of a principal value.

Then

$$l^0 p^{\pm}(\varphi, \psi) = \left[W^{(2)}(\varphi, x) + W^{(1)}(\psi, x) \right]^{\pm} =$$

$$= \pm\frac{1}{2}\varphi(x) + \frac{1}{2}S_{01}\psi(x) + \frac{1}{2}S_{00}\varphi(x)\,.$$

Proof. Let $u \in H_L^s(\Omega)$ be such that $u|_{\partial\Omega} = \varphi$, $l^1 u|_{\partial\Omega} = \psi$, $u^{\pm} = P^{\pm}u$, where P^+ and P^- are the projectors defined in §3. 6^0.

Let $B_\varepsilon(x_0)$ be a ball of radius ε with centre at x_0, $S_\varepsilon = \partial B_\varepsilon(x_0)$ its boundary, $\Omega_\varepsilon^+ = \Omega \cup B_\varepsilon(x_0)$, $\Omega_\varepsilon^- = \Omega/\overline{B}_\varepsilon(x_0)$, $\sigma_\varepsilon = \partial\Omega \cap B_\varepsilon(x_0)$, $S_\varepsilon^- = S_\varepsilon \cap \Omega$, $S_\varepsilon^+ = S/S_\varepsilon^-$.

So, $x \in \partial\Omega$ is internal for Ω_ε^+ and external for Ω_ε^-.

In order to abbreviate the presentation, denote by $J(\omega; \varphi, \psi; x_0)$ the expression.

$$\int\limits_{\omega} \left(l^1 g(y, x_0)\varphi(y) - l^0 g(y, x_0)\psi(y) \right) d_y s\,.$$

Consider Green's formula for $u^+(x)$ with respect to the domain Ω_ε^+:

$$J(\partial\Omega/\sigma_\varepsilon; \varphi^+, \psi^+; x_0) + J(S_\varepsilon^+; \varphi^+, \psi^+; x_0) = u^+(x_0) - \int\limits_{\Omega_\varepsilon^+} g(y, x_0) L u^+(y)\, dy\,.$$

Since $Lu^+ \equiv 0$ in Ω, one must change the domain of integration Ω_ε^+ into $B_\varepsilon(x_0)$.

Using Green's formula with respect to the domain $B_\varepsilon(x_0)$, one obtains:

$$J(\partial\Omega/\sigma_\varepsilon; \varphi^+, \psi^+; x_0) = J(S_\varepsilon^-; \varphi^+, \psi^+; x_0)\,.$$

In an analogous manner, one finds from Green's formula, written for the function u^- with respect to the domain Ω_ε^-:

$$J(\partial\Omega/\sigma_\varepsilon; \varphi^-, \psi^-; x_0) = -J(S_\varepsilon^+; \varphi^-, \psi^-; x_0)\,.$$

Adding of the last two expressions yields

$$J(\partial\Omega/\sigma_\varepsilon;\varphi,\psi;x_0) = J(S_\varepsilon^-;\varphi^+,\psi^+;x_0) - J(S_\varepsilon^+;\varphi^-,\psi^-;x_0).$$

Analyse now the $\lim\limits_{\varepsilon\to 0} J(S_\varepsilon^\pm;\varphi^\pm,\psi^\pm;x_0)$.

First of all, note that $g(z)$ is an even function of z. In fact, let $F(g)$ be the Fourier transform of the function g. Since $L(\xi)$ is even and $L(\xi)Fg(\xi) = I$, $Fg(\xi)$ is even.

Besides, $Fg(\xi)$ is locally summable and belongs to the space S' of tempered distributions. Let $h(|\xi|)$ be a smooth function which is equal to 1 for $|\xi| < 1$ and equal to 0 for $|\xi| > 2$. Then $Fg_\varepsilon(\xi) \equiv h(\varepsilon|\xi|)Fg(\xi)$ converges in S' to $Fg(\xi) \in L^1(\mathbf{R}^n)$.

However, $g_\varepsilon(z)$ is an even function of z:

$$g_\varepsilon(z) = \int F(g_\varepsilon(\xi))\exp(iz\xi)\,d\xi\,;$$

$$g_\varepsilon(-z) = \int Fg_\varepsilon(\xi)\exp(-iz\xi)\,d\xi = \int\limits_\infty^\infty \tilde{g}_\varepsilon(-\xi)\exp(ix\xi)\,d\xi\,.$$

Furthermore, since the function $g(z)$ is even, it follows that also the function $l^1 g(z)|_{S_\varepsilon}$ is even. In order to verify this, write the operator l^1 in the form $\nu^j\partial_i a^{(i,j)}$, where $\nu^j = (x_j - y_j)/(|x-y|)$. With the aid of the Fourier transformation, reasoning as before, one finds that $\partial_i a^{(i,j)}g(z) = F^{-1}(\xi_i a^{(i,j)}Fg(\xi))$ is an odd function in z.

As a consequence of the smoothness of $\partial\Omega$, the surfaces S_ε tend to half spheres if ε tends to 0 (for more detail, cf. [13]). Therefore, as $\varepsilon \to 0$

$$J(S_\varepsilon^-;\varphi^+,\psi^+;x_0) - J(S_\varepsilon^+;\varphi^-,\psi^-;x_0) \to \frac{1}{2}J(S_\varepsilon;\varphi^+ - \varphi^-,\psi^+ - \psi^-;x_0).$$

According to Green's formula

$$J(S_\varepsilon;\varphi^+ - \varphi^-,\psi^+ - \psi^-;x_0) = \varphi^+(x_0) - \varphi^-(x_0) - \int\limits_{B_\varepsilon} g(y,x)Lu(y)\,dy\,.$$

The integral in the last formula tends to zero as $\varepsilon \to 0$, since

$$\int\limits_{B_\varepsilon} g(y-x)Lu(y)\,dy = \int\limits_0^\varepsilon r^1\left(\int\limits_{|\omega|=1} g(\omega)Lu(r\omega - x)\,d\omega\right)dr = O(\varepsilon)\,.$$

The proposition follows now from (A2.3.12) by application of the limiting process $\varepsilon \to 0$ taking into consideration the theorem of the preceding section.●

Proposition 3. Let $x_0 \in \partial\Omega$,

$$S_{10}\varphi(x_0) = -2 \int\limits_{\partial\Omega} l_x^1 g(y, x_0)\varphi(y)\, d_y s\,,$$

where the integral is understood in the sense of the principal value.

Then

$$\left[l^1 W^{(1)}(\varphi, x_0)\right]^{\pm} = \mp\frac{1}{2}\varphi(x_0) + S_{10}\varphi(x_0)\,.$$

Proof. 1. In component form, the kernel $h(y, x)$ of the double–layer potential is

$$h_l^m(y, x) = \nu^i(y)a_k^{(i,j)m}\frac{\partial}{\partial y_j}g_l^k(y, x)\,, \ \ x \in \Omega\,, \ y \in \partial\Omega\,.$$

Consider the operator with a kernel $h^*(y, x)$ such that

$$h_l^{*m}(y, x) = h_m^l(y, x)\,.$$

Then, as follows from proposition 2, there holds for an operator with the kernel h^* the relation

$$\int\limits_{\partial\Omega} h^*(y, x_0)\varphi(y)\, d_y s = -\frac{1}{2}\varphi(x_0) + \lim_{\substack{x \to x_0 \\ x \in \Omega}} \int\limits_{\partial\Omega} h^*(y, x_0)\varphi(y)\, d_y s\,.$$

2. Define beforehand an operator l^1 in the vicinity O of the boundary $\partial\Omega$, setting

$$l^1 u(x) = \nu^i(x_0)a_k^{(i,j)}\frac{\partial}{\partial x_j}u(x)\,,$$

where $x_0 \in \partial\Omega$ such that $(x - x_0) \in \nu(x_0)$ (cf. the remark in §2, 6^0).

Since $g(y, x) = g(y - x)$, then $\frac{\partial}{\partial y_j}g_l^k(y, x) = -\frac{\partial}{\partial x_j}g_l^k(y, x)$. Whence

$$h_l^{*m}(x, y) = -\nu^i(x)\frac{\partial}{\partial x_j}a_m^{(i,j)k}g_k^l(y, x) = (-l_x^1 g(x, y))_l^m\,.$$

3. Since the boundary $\partial\Omega$ is smooth, one has

$$\nu^i(y) - \nu^i(x) = \beta^{ij}(x, x - y)(y_i - x_i)\,,$$

where $\beta^{ij}(x, z)$ is a smooth function.

Taking into consideration the degree of the homogeneity of the fundamental solution, one finds

$$\lim_{x \to x_0} (l_x^1 W^{(1)}(x, \varphi) + W^{(2)}(x, \varphi)) = \lim_{x \to x_0} \int_{\partial\Omega} (-h^*(x, y) + h^*(y, x))\varphi(y)\, d_y s$$

$$= \lim_{x \to x_0} \int_{\partial\Omega} \beta^{ip}(y, x - y)(x - y)\nu^i(y)a_k^{(i,j)m} \partial_j g_l^k(y, x)\varphi_m(y)\, d_y s =$$

$$= \int_{\partial\Omega} (h^*(y, x_0) - h^*(x_0, y))\varphi(y)\, d_y s = 2(S_{00}\varphi(x_0) - S_{10}\varphi(x_0)),$$

whence, by proposition 2 proposition 3 is obtained •

According to the theorem of §3. 7^0

$$\left[l^1 W^{(2)}(\varphi, x_0)\right]^+ = \lim_{x \to x_0} l_x^1 \int_{\partial\Omega} l^1 g(y, x)\varphi(y)\, d_y s$$

exists and equals $\left[l^1 W^{(2)}(\varphi, x_0)\right]^-$.

It will now be shown that $\left[l^1 W^{(2)}(\varphi, x_0)\right]^+$ can be represented in the form of a singular, integro-differential operator with a zero term outside the integral.

Represent the operator l^1 in the form $l^1 = \alpha_k(x)\partial_k$. Let $v \in H_L^s(\mathbf{R}^n)$ be such that $l^0 v|_{\partial\Omega} = \varphi$, $l_\nu^1 v|_{\partial\Omega} = 0$. By Green's formula, one has

$$\int_{\partial\Omega} l_\nu^1 g(x, y)\varphi(y)\, d_y s = v(x) - \int_\Omega g(x, y)Lv(y)\, dy\,, \quad x \in \Omega\,.$$

Since $\partial_x g(x - y) = -\partial_y g(x - y)$, one finds, integrating by parts,

$$\partial_k \int_{\partial\Omega} l_\nu^1 g(x, y)\varphi(y)\, d_y s = \partial_k v(x) - \int_\Omega g(x, y)L\partial_{yk} v(y)\, dy\,.$$

Here it is taken account that $Lu|_{\partial\Omega} = 0$. Employing Green's formula for $\partial_k \nu(x)$, one obtains

$$\partial_k \int_{\partial\Omega} l_\nu^1 g(x, y)\varphi(y)\, d_y s = \int_{\partial\Omega} l_\nu^1 g(x, y)\partial_{yk} v(y)\, d_y s - \int_{\partial\Omega} g(x, y)l_\nu^1 \partial_{yk} v(y)\, d_y s$$

whence, as a consequence of proposition 2,

$$\lim_{x \to x_0 \in \partial\Omega} l_x^1 \int_{\partial\Omega} l_\nu^1 g(x,y)\varphi(y)\,d_y s = \frac{1}{2}(S_{01}\varphi)(x_0),$$

where

$$S_{01}(\varphi)(x_0) = 2\left(\int \alpha_k(x)l^1 g(y,x)\partial_k v(y)\,d_y s - \int \alpha_k(x)g(y,x)l^1\partial_k v(y)\,d_y s \right).$$

$$(A2.3.13)$$

Note still that by the condition $l^1 v = 0$, $Lv = 0$, $\partial_k v(y) = c_1 v(y)$, $l^1\partial_k v(y) = c_2 v(y)$, where c_j are differential operators of order j which contain only derivatives in tangential directions.

Remark. 1. Proposition 2 and 3 as well as formula (A2.3.13) remain true, if Green's function of the enveloping domain or parametrix is taken in place of the operator G. This is due to the fact that the difference of two parametrix is a smoothing operator.

2. Propositions 2 and 3 and (A2.3.13) also remain true in the case $n = 2$. The proof needs to change slightly, in order to confirm the character of the singularity of the fundamental solution $\log|x|$ which applies in this case.

9^0. **Theorem on involution.** It is known (chapter 2, §1. 5^0 of the present book) that the singular integral Cauchy operator

$$(S\varphi)(t) = \frac{1}{\pi i} \int \frac{\varphi(\tau)\,d\tau}{\tau - t}$$

is involution: $S^2 = I$. This fact can be generalized in the following manner.

Let $\varphi = (\varphi^0, \varphi^1) \in H^{s-1/2}(\partial\Omega) \oplus H^{s-3/2}(\partial\Omega)$, $s > 5/2$, $s \neq 1/2 +$ integer. Define the operator

$$S : H^{s-1/2}(\partial\Omega) \oplus H^{s-3/2}(\partial\Omega) \to H^{s-1/2}(\partial\Omega) \oplus H^{s-3/2}(\partial\Omega)$$

and the matrix $S = \|S_{tr}\|$, where the operators S_{tr} have been defined in propositions 2 and 3 and formula (A2.3.13).

Then $S^2 = I$. Moreover, $S\varphi = \varphi$ then and only then when φ are Cauchy data of the first order of the function u which satisfies the equation $Lu = 0$ in the domain Ω.

Proof. Let $u^r \in H_L^s(\mathbf{R}^n)$ be such that $l^t u^r = \delta_{tr}\varphi^r$, $r, t = 0, 1$. By the lemma of §3. 6^0

$$\left[l^{t-1}W^{(3-r)}(\varphi^r, x) \right]^{\pm} = l^t P^{\pm} u^r(x),$$

and, consequently, $(S_{tr}\varphi^r)(x) = l^t P^+ u^r(x) - l^t P^- u^r(x)$.

Let $u = u^0 + u^1$. Then $l^t u = \varphi^t$. Therefore

$$S\varphi = \gamma P^+ u - \gamma P^- u = \gamma P^+ \tilde{\gamma}_r^{-1} \alpha\gamma u - \gamma P^- \tilde{\gamma}_r^{-1} \alpha\gamma u = p^+ - p^- .$$

where $\gamma : H^s(\Omega) \to H^{s-1/2}(\partial\Omega) \oplus H^{s-3/2}(\partial\Omega)$ is the mapping determined by the relation $\gamma(u) = (l^0 u, l^1 u)$, α is the extension operator of the trace over the Cauchy data with the supplementary condition $\tilde{\gamma} L u = 0$ (cf. §1. 6^0), $\tilde{\gamma}$ is the operator of taking the trace and $\tilde{\gamma}_r^{-1}$ is its right inverse.

The operators p^+ and p^- are mutually supplementary (cf. theorem 3 of §1). Consequently

$$S^2 = (p^+ - p^-)(p^+ - p^-) = p^+ \mid p^- = I .\bullet$$

Remark. The method of the proof of propositions 1–3 can be transferred with minimum changes to the case of a homogeneous, strongly elliptic systems with constant coefficients. For this purpose, one need only a strengthened version of the lemma of §3. 6^0: let Green's formula for the system L be rewritten in the form

$$\sum_{r=0}^{m-1} \int_{\partial\Omega} \partial_\nu^{m-r-1} g(x,y) l^r u(y) \, d_y s - \int_\Omega g(x,y) L u(y) \, dy , \qquad (A2.3.14)$$

where $g(x,y)$ is the fundamental solution of an operator L, formally conjugate to the operator L^*. Define the potential $W^{(m-r)}(\varphi, x)$ of the $(m-r)$-th layer by the equality

$$W^{(m-r)}(\varphi, x) = \int_{\partial\Omega} \partial_\nu^{m-r-1} g(x,y) l^r u(y) \, d_y s .$$

Then

$$W^{(2-r)}(\varphi, x) = \begin{cases} P^+ u(x) \\ -P^- u(x) \end{cases} , x \in \mathrm{R}^n / \partial\Omega ,$$

where $u \in H^s_{loc}(\mathrm{R}^n)$ is such that $l^t u = \delta_{tr}\varphi$, $r, t = 0, m-1$ and $\tilde{\gamma}(Lu) = 0$.

The proof of the lemma in this case is analogous to the one given and it is based on the normality of the operator system $\{l^s\}_{s=0}^{m-1}$.

A generalization to the case of non-homogeneous, strongly elliptic systems is likewise not difficult. Using and asymptotic fundamental solution, it is readily shown that "jumps" determine the principal part of the operator L.

Finally, with the aid of the procedure of "freezing of coefficients" , without principal difficulties, a proof of the proposition may be extended to the case of variable coeffcients.

Theorem on involution, proved in the authors' paper [7] can be formulated in the follwoing manner.

Let $L = \sum\limits_{|k| \leq m} a_k(x)\partial^k$ be an elliptic differential operator of order m with a homogeneous symbol, acting in the bounded domain Ω with the smooth boundary $\partial\Omega$, $\{l^s\}_{s=0}^{m-1}$ be the normal system of operators entering into (A2.3.14). The operator

$$S = \|S_{tr}\| : \overset{m-1}{\underset{j=0}{\oplus}} H^{s-1/2-j}(\partial\Omega) \to \overset{m-1}{\underset{j=0}{\oplus}} H^{s-1/2-j}(\partial\Omega)$$

such that $S^2 = I$ is defined and continuous. And $S\varphi = \varphi$ then and only then when φ is Cauchy data of function $u \in H^s(\Omega)$ which satisfies the equation $Lu = 0$ in the domain Ω.

Under these conditions, S_{tr} is an integral operator of order $t - r$ which can be represented in the form

$$(S_{tr}\varphi)(x) = \int\limits_{\partial\Omega} l^{t-1}\partial_\nu^{m-r} g(x,y)\varphi^r(y)\,d_y s$$

for $t < r + 1$.

For $t > r$, an explicit expression of the operator S_{tr} in terms of the fundamental solution of the operator L^* can be written.

REFERENCES

Part 1

1. Arabadzhyan, L.G., Engibaryan, N.B., *Convolution equations and nonlinear functional equations.* (Russian), Itogi Nauki Tekh., Ser. Mat. Anal., Moscow, 22 (1984), 175–244.

2. Watson, G.N., *A Treatise on the Theory of Bessel Functions.* University Press, Cambridge, The Macmillan Company, New York, 1944, 804.

3. Vladimirov, V.S., *Wiener-Hopf equations on the half-axis in Nevanlinna and Smirnov algebras.* (Russian), Dokl. Akad. Nauk SSSR, 293 (1987), n. 2, 278–283.

4. Gakhov, F.D., *On the boundary-problem of Riemann.* (Russian), Mat. Sbornik, 2 (44), (1937) n. 4, 673–683.

5. Gakhov, F.D., Cherskii, Yu.I., *Equations of convolution type.* (Russian), Izdat. Nauka, Moscow, (1978), 295.

6. Gokhberg, I.Ts., Krupnik, N.Ya., *Introduction to the theory of one-dimensional singular integral equations operators.* (Russian), Izdat. Stiinca, Kishinev, 1973, 426.

7. Gokhberg, I.C., Feldman, I.A., *Convolution equations and projection for their solution.* (Russian), Izdat. Nauka, Moscow, 1971, 352.

8. Duduchava, R.V., *The integral equations of convolution with discontinuous presymbols, singular integral equations with motionless singularities and their applications to problems of mechanics.* (Russian), Tr. Tbilis. Mat. Inst. (1979), 60–135.

9. Kantorovich, L.V., *On the one method of approximate solution for differential equations in partial derivations.* (Russian), Dokl. Akad. Nauk SSSR, 2 (1934), n. 8-9, 532–536.

10. Kantorovich, L.V., Akilov G.P., *Functional analysis.* (Russian), Izdat. Nauka, Moscow, 1977, 742.

11. Kantorovich, L.V., Krylov V.I., *Approximate Methods of Higher Analysis.* (Russian), Gosudarstv. Izdat. Tehn.-Teor. Lit., Moscow-Leningrad, 1950, 695.

12. Kolmogorov, A.N., Fomin S.V., *Elements of the theory of functions and functional analysis.* (Russian), Izdat. Nauka, Moscow, 1981, 544.

13. Kohn, J.J., Nirenberg L., *Algebra of pseudodifferential operators. In: "Pseudodifferential operators".* (Russian), Izdat. Mir, Moscow, 1967, 9–62.

14. Krein, M.G., *Integral equations on the half-line with a kernel depending on the difference of the arguments.* (Russian), Uspehi Mat. Nauk, 13 (1958), n. 5, 3–120.

15. Markusevic, A.I., *Theory of Analytic Functions*. (Russian), Gosudarstv. Izdat. Tehn. Teor. Lit., Moscow-Leningrad, 1950, 703.

16. Markusevic, A.I., *"Mathematical encyclopaedia"*. (Russian), Izdat. "Sovetskaya encyclopaedia". 1–5, Moscow, 1974–1985.

17. Mikhlin, S.G., *Composition of double singular integrals*. (Russian), Dokl. Akad. Nauk SSSR, 2(11), n. 1(87), (1936), 3–6.

18. Mikhlin, S.G., *Singular integral equations with two independent variables*. (Russian), Mat. Sbornik, 1(43), n. 4, (1936), 535–552.

19. Mikhlin, S.G., *An addition to the paper "Singular integral equations with two independent variables"*. (Russian), Mat. Sbornik, 1(43), n. 6, (1936), 953–954.

20. Mikhlin, S.G., *Singular integral equations*. (Russian), Uspehi-Matem.-Nauk (N.S.) 3 (1948), n.3 (25), 29–112.

21. Mikhlin, S.G., *Integral Equations and their Applications*. (Russian), Gosudarstv. Izdat. Tehn. Teor. Lit., Moscow-Leningrad, 1949, 380.

22. Mikhlin, S.G., *Composition of multidimensional singular integrals*. (Russian), Vestnik Leningrad. Univ., (1955), n. 2, 25–41.

23. Mikhlin S.G. *On the theory of multidimensional singular integrals*. (Russian), Vestnik Leningrad. Univ. (1956), n. 1, 3–24.

24. Mikhlin, S.G., *Lectures on linear integral equations*. (Russian), Gosudarstv. Izdat. Fiz.-Mat. Lit., Moscow, 1959, 232.

25. Mikhlin, S.G., *Higher–dimensional singular integrals and integral equations*. (Russian), Gosudarstv. Izdat. Fiz.-Mat. Lit., Moscow, 1962, 254.

26. Mikhlin, S.G., *Numerical realization of variational methods*. (Russian), Izdat. Nauka, Moscow, 1966, 432.

27. Mikhlin, S.G., *Variational methods in mathematical physics*. (Russian), Izdat. Nauka, Moscow, 1970, 512.

28. Mikhlin, S.G., Smolickii, H.L., *Approximate methods of solution of differential and integral equations*. (Russian), Izdat. Nauka, Moscow, 1965, 383.

29. Moiseev, N.G., *On the factorization in close form*. (Russian), R.J. Math. 2, B101, 1987.

30. Muskhelishvili, N.I., *Singular integral equations*. (Russian), Izdat. Nauka, Moscow, 1968, 511.

31. Noble, B., *Methods based on the Wiener-Hopf technique for the solution of partial differential equations*. Pergamon Press, New York - London - Paris - Los Angeles, 1958, 246.

32. Perlin, P.I., *Numerical method for solving singular integral equations of basic three–dimensional elasticity theory problems*. (Russian), Izv. Akad. Nauk SSSR, Meh. Tverd. Tela, (1975), n. 3 109–111.

33. Petrovsky, I.G., *Lectures on the theory of integral equations.* (Russian), Gosudarstv. Izdat. Tehn.-Teor. Lit., Moscow, 1948.

34. Prössdorf, S., *Einige Klassen singulärer Gleichungen.* Akademie–Verlag Berlin, 1974, 353.

35. Riesz, F., *Über lineare Funktionalgleichungen*, Acta Mathematica, 41 (1918), 71–98.

36. Simonenko, I.B., *Certain general questions of the theory of the Riemann boundary value problem.* (Russian), Izv. Akad. Nauk SSSR, Ser. Mat. 32 (1968), 1138–1146.

37. Simonenko, I.B., Chin Ngok Min, *A local method in the theory of one-dimensional singular integral equations with piecewise continuous coefficients.* (Russian), Rostov. Gos. Univ., Rostov on Don, 1986, 59.

38. Slepyan, L.I., Yakovlev, Yu.S., *Integral transformations in nonstationary problems of mechanics.* (Russian), Izdat. "Sudostroenie", Leningrad, 1980, 344.

39. Smirnov, V.I., *A course of higher mathematics. Vol. 4.* (Russian), Gosudarstv. Izdat. Tehn. Teor. Lit., Moscow–Leningrad, 1951, 804.

40. Smirnov, V.I., *A course in higher mathematics. Vol. 5.* (Russian), Gosudarstv. Izdat. Fiz.-Mat. Lit., Moscow, 1959, 665.

41. Tricomi, F.G., *Integral equations.* Interscience Publishers, Ltd., New York – London, 1957, 238.

42. Uflyand, Ya.S., *The method of dual equations in probems of mathematical physics.* (Russian), Izdat. Nauka Leningrad. Otdel., Leningrad, 1977, 220.

43. Hvedelidze, B.V., *On the Puankare's boundary problem in the theory of logarithmical potential for multiconnected regions.* (Russian), Akad. Sci Georgia Soviet Rep., 1941, 2, n. 7, 571–578.

44. Hörmander, L., *Pseudodifferential operators. In: "Pseudodifferential operators".* (Russian), Izdat. Mir, Moscow, 1967, 63–67.

45. Hrapkov, A.A., *The problem of the elastic equilibrium of infinite wedge with axis-symmetrical incision in the top, solved in close form.* (Russian), Prikl. Mat. Mekh., 35, (1971), n. 6, 1062–1069.

46. Chebrikova, L.N., *Basic boundary problems for analytical functions.* (Russian), Kazan. Gos. Univ., Kazan, 1977, 328.

47. Anderssen, R.S., de Hoog, F.R., (ed.). *The numerical application of integral equations.* Sijthoff & Noordhoff, Alphen aan den Rijn. The Netherland, 1980, 259.

48. Carleman, T., *Sur la resolution de certaines equations integrales.* (French) Ark. Mat. Astr. Fys., 16, 1922, n. 26.

49. Delves, L.M., Walsh, J.E., (ed.), *Numerical solution of integral equations*. Clarendon Press, Oxford, 1974, 339.

50. Devinatz, A., Shinbrot, H., *General Wiener–Hopf operators*. Trans. Amer. Math. Soc., 1969, 145, 467–494.

51. Giraud, G., *Sur une classe generale d'equation a integrales principales*. C. R. Acad. Sc. Paris 202 (1936), 2124–2126.

52. Meister, E., Speck, F.-O., *Some multidimensional Wiener-Hopf equations with applications*. In: Trends in applications of pure mathematics to mechanics. Pitman, Boston, Mass. London (1979), 217–262.

53. Mikhlin, S.G., *Fehler in numerischen Prozessen*. Akademie–Verlag, Berlin 1985, 244.

54. Mikhlin, S.G., Prössdorf, S., *Singuläre Integraloperatoren*. Akademie–Verlag, Berlin, 1980, 514.

55. Pellegrini, V.J., *Unbounded general Wiener–Hopf operators*. Indiana Univ., Math. J., 21, (1971/1972), 85–90.

56. Seeley, R.T., *Singular integrals on compact manifolds*. Amer. J. Math., 81 (1959), n. 3, 658–690.

57. Shinbrot, M., *On singular integral equations*. J. Math. Mech., 13 (1964), 395–406.

Part 2

1. Alexandrov, V.M., Romalis, B.L., *Contact problems in mechanical engineering*. (Russian), Izdat. Mashinostroenie, Moscow, 1986, 174.

2. Antosik, P., Mikusinski, J., Sikorski, R., *Theory of distributions*. Panstw. wyd-wo nauk., 1973, 315.

3. Afyan, B.A., *Integral equations with fixed singularities in the theory of branching cracks*. (Russian), Akad. Nauk Armyan SSR Dokl., 79 (1984), n. 4, 177–181.

4. Afyan, B.A., Paukshto, M.V., *The method of conformal mappings in problems of elasticity theory for branched cracks*. (Russian), In: Studies in elasticity and plasticity, n. 5, Leningrad. Univ., Leningrad (1986), 7–12.

5. Banerjee, P.K., Butterfield, R., *Boundary element methods in engineering science*. McGraw–Hill Book Co. (UK) Ltd. London – New York, 1981, 452.

6. Brebbia, C.A., Walker, S., *Boundary Element Techniques in Engineering*. Newnes–Butterworths. London, 1980.

7. Buchuladze, T.V., Gegelia, T.G., *Development of the method in elasticity theory*. (Russian), Trudy Tbiliss. Mat. Inst. Razmadze Akad. Nauk Gruzin. SSR, 79, 1985, 226.

8. Vekua, N.P., *Systems of singular integral equations and certain boundary value problems.* (Russian), Izdat. Nauka, Moscow, 1970, 379.

9. Veryuzhskii, Yu.V., *The potential method useung for the theory of elasticity problems solution.* (Russian), Izdat. Naukova Dumka, Kiev, 1975.

10. Vishik, M.I., *On strong elliptical differential equations systems.* (Russian), Mat. Sbornik 29 (71), (1951), 615–676.

11. Vladimirov, V.S., *Equations of mathematical physics.* (Russian), Izdat. Nauka, Moscow, 1971, 512.

12. Vovkushevsky, A.V., Shoykhet, B.A., *The massive hydrotechnical edifices calculation with consideration of the seams opening.* (Russian), "Energoizdat", Moscow, 1981, 136.

13. Galin, L.A., *Contact problems of the theory of elasticity.* (Russian), Gosudarstv. Izdat. Tehn. Teor. Lit., Moscow-Leningrad, 1953, 264.

14. Hlavacek, I., Haslinger, J., Necas, J., Lovisek, J., *Solution of variational inequalities in mechanics.* (Russian), Izdat. Mir, Moscow, 1986, 272.

15. Goluzin, G.M., *Geometrical theory of functions of a complex variable.* (Russian), Izdat. Nauka, Moscow, 1966, 628.

16. Goldstein, R.V., *Plane crack of arbitrary aperture in an elastic medium.* (Russian), Izv. Akad. Nauk SSSR, Mekh. Solids, 14 (1979), n. 3, 111–126.

17. Gordadze, E.G., *On singular integrals with a Cauchy kernel.* (Russian), Soobsc. Akad. Nauk Gruzin. SSR, 37 (1965), 521–526.

18. Günter, N.M., *Theory of potential.* (Russian), Gosudarstv. Izdat. Tehn. Teor. Lit., Moscow-Leningrad, 1953, 415.

19. Johnson, K.L., *Contact mechanics.* Cambridge University Press, XI, 1985, 452.

20. Duduchava, R.V., *General singular integral equations and basic problems of plan elasticity theory.* (Russian), Tr. Tbilis. Mat. Inst., 82 (1986), 45–89.

21. Dynkin, E.M., Emelyanov, A.P., *A boundary integral equation of elasticity theory on a contour with a cusp.* (Russian), Vestnik Leningrad. Univ. Mat. Mekh. Astronom., 3 (1988), 95–96, 126.

22. Emelyanov, A.P., Paukshto, M.V., *Boundary integral equations in the problem of elastic peak-like inclusion.* (Russian), Izv. Akad. Nauk SSSR, Mekh. Solids, (1988), n. 5.

23. Zargaryan, S.S., *Integral equations of plane elasticity for multiply connected domains with corners.* (Russian), Izv. Akad. Nauk SSSR, Mekh. Tverd. Tela 17 (1982), n. 3, 87–98.

24. Zargaryan, S.S., Maz'ya, V.G., *Singularities of solutions of a system of equations of potential theory for Zaremba's problem.* (Russian), Vestnik Leningrad. Univ., Mat. Mekh. Astronom. (1983), 43–48, 126.

25. Zargaryan, S.S., Maz'ya, V.G., *The Asymptotic form of the solutions integral equations of potential theory in the neighbourhood of the corner points of a contour*. (Russian), Prikl. Mat. Mekh., 48 (1984), n. 1, 169–174.

26. Kolesnikov, Yu.V., Morozov, E.M., *Mechanics of the contact fracture*. (Russian), Izdat. Nauka, Moscow, 1989, 219.

27. Crouch, S.L., Starfield, A.M., *Boundary element methods in solid mechanics*. George Allen & Unwin, London – Boston, 1983, 322.

28. Kupradze, V.D.,

   a) *Potential–theoretic methods in the theory of elasticity*. (Russian), Gosudarstv. Izdat. Fiz. Mat. Lit., Moscow, 1963, 472.

   b) *More on the solution of boundary value problems of the theory of elasticity for inhomogeneous bodies with piecewise homogeneous structure*. (Russian), Trudy Tbilis. Univ., Mat. Mekh. Astronom. (1985), n. 18, 5–22.

29. Lazarev, M.I., *Solution of fundamental problems of the theory of elasticity for incompressible media*. (Russian), Prikl. Mat. Mekh., 44 (1980), n. 5, 867–874.

30. Lazarev, M.I., Matechin, N.A., *Compatibility conditions and different positings of the elasticity problems*. (Russian), Prepr. Akad. Sci. SSSR, 1988.

31. Leonov, G.Ya., *General problem of round punch pressing on the elastic half-space*. (Russian), Prikl. Mat. Mekh., 17 (1953), n. 1, 87–98.

32. Lekhnitskij, S.G., *Some cases of the plane problem of the theory of elasticity for anisotropic solid. In: "Experimental methods of the stress and strain definition in elastic and plastic fields"*. (Russian), ONTI NKTP, Moscow, (1935), 150–181.

33. Linkov, A.M., Mogilevskaya, S.G., *Finite–part integrals in problems of three–dimensional cracks*. (Russian), Prikl. Mat. Mekh., 50 (1986), n. 5, 844–850.

34. Lur'e, A.I., *Theory of elasticity*. (Russian), Izdat. Nauka, Moscow, 1970, 939.

35. Maz'ya, V.G., *In: "Annotaition of school–conf. reports in theory of elasticity."* (Russian), Telavi-Tbilisi, 1981, 55–56.

36. Matekhin, N.A., *Problems of elasticity theory for fields with conic point of boundary* (Russian), Ph. D. Diss., Leningrad, 1985, 135.

37. Matekhin, N.A., Morozov, N.F., Paukshto, M.V., *Some direct schemes of the potential method*. (Russian), Dokl. Akad. Nauk SSSR, 292 (1988) n. 2, 296–298

38. Maul, J., *Solution of plane problems of the coupled theory of elasticity for an isotropic inhomogeneous medium.* (Russian), Sakharth. SSR, Mekn. Akad. Math. Inst., 58 (1978), 150–167.

39. *Boundary–integral equation method: computational applications in applied mechanics.* Edited by T.A. Gluse, F.Y. Rizzo. The American Society of Mechanical Engincers, 1975.

40. Mikhlin, S.G., *Some cases of the theory of elasticity plane problem for inhomogeneous medium.* (Russian), Prikl. Mat. Mekh., 2 (1934), n. 1, 82–90.

41. Mikhlin, S.G., *Plane problem of the theory of elasticity.* (Russian), Trudy Seismol. Inst. Akad. Nauk SSSR, 1935, n. 65, 83.

42. Mikhlin, S.G., *Plane problem of the theory of elasticity for inhomogeneous medium.* (Russian), Trudy Seismol. Inst. Akad. Nauk SSSR, 1936, n. 66, 1–16.

43. Mikhlin, S.G., *Plane strain in the anisotropic medium.* (Russian), Trudy Seismol. Inst. Akad. Nauk SSSR, (1936), n. 16, 1–19.

44. Mikhlin, S.G., *Integral Equations and their Applications.* (Russian), Gosudarstv. Izdat. Tehn. Teor. Lit., Moscow–Leningrad, 1949, 380.

45. Mikhlin, S.G., *Higher–dimensional singular integrals and integral equations.* (Russian), Gosudarstv. Izdat. Fiz.-Mat. Lit., Moscow, 1962, 254.

46. Mikhlin, S.G., *Variational methods in mathematical physics.* (Russian), Izdat. Nauka, Moscow, 1970, 512.

47. Mikhlin, S.G., Morozov, N.F., Paukshto, M.V., *Boundary integral equations and problems in elasticity theory.* (Russian), Leningrad. Univ., Leningrad, 1986, 88.

48. Mishuris, G.S., *Integral transformations for the problems in multilayer elastic solids with slits.* (Russian), Ph. D. Diss., Leningrad, 1985, 174.

49. Mishuris, G.S., Paukshto, M.V., *On the antiplane shift of the crack in the multilayer elasticity medium.* (Russian), In: Oscillation and equilibrium of the mechanical systems. Leningrad, 215–221.

50. Movchan, A.B., Nazarov, S.A., *Asymptotic behavior of the stress-strained state near sharp inclusions.* (Russian), Dokl. Akad. Nauk SSSR, 290 (1986), n. 1,48–51.

51. Morozov, N.F., *Selected two–dimensional problems of elasticity theory.* (Russian), Leningrad. Univ., Leningrad, 1978, 182.

52. Morozov, N.F., *Mathematical problems of fracture theory.* (Russian), Izdat. Nauka, Moscow, 1984, 255.

53. Muskhelishvili, N.I., *Some basic problems of the mathematical theory of elasticity. Fundamental equation, plane theory of elasticity, torsion and bending.* Izdat. Nauka, Moscow, 1966, 707.

54. Muskhelishvili, N.I., *Singular integral equations.* (Russian), Izdat. Nauka, Moscow, 1968, 511.

55. Nowacky, W., *Theory of elasticity.* (Russian), Izdat. Mir, Moscow, 1975, 872.

56. Novozhilov, V.V., *Theory of elasticity.* (Russian), Izdat. Sudpromgiz, Leningrad, 1958.

57. Panasiuk, V.V., *Limited equilibrium of the brittle solids with cracks.* (Russian), Izdat. Naukova Dumka, Kiev, 1968, 246.

58. Parton, V.Z., Perlin, P.I., *Integral equations in elasticity theory.* (Russian), Izdat. Nauka, Moscow, 1977, 312.

59. Plamenevskii, B.A., Senichkin, V.N.,

  a) *Algebras of pseudodifferential operators with piecewise–smooth symbols.* (Russian), Dokl. Akad. Nauk SSSR, 298 (1988), n. 1, 40–44.

  b) *Representation of an algebra of pseudodifferential operators with multi-dimensional discontinuities in the symbols.* (Russian), Izv. Akad. Nauk SSSR, Ser. Mat., 51, (1987), n. 4, 833–859.

60. Paukshto, M.V., *N.I. Muskhelishvili's integral equations in the case of violation of conformality on the boundary.* (Russian), In: Studies in elasticity and plasticity. Leningrad. Univ., Leningrad, 1990, n. 16, 145–155.

61. Paukshto, M.V., Parfentyeva, O.B., *On the Kupradze method for the solution of the boundary integral equation in half-space with elasticity inclusion.* (Russian), VINITY, n. 6051–B89, 1989, 20.

62. Paukshto, M.V., Tovstik, P.E., *Conformal boundary elements for polygonal domain.* (Russian), "Effective methods for the solving of the boundary value problems of the solid body", Kharkov, 1989, 7.

63. Pobedrya, B.E.,

  a) *On the problem in stresses.* (Russian), Dokl. Akad. Nauk SSSR, 240 (1978), n. 3, 564–567.

  b) *Some general theorems of the mechanics of solids.* (Russian), Prikl. Mat. Mekh., 4 (1979), 531–541.

  c) *A new formulation of the solid mechanics problem in stresses.* (Russian), Dokl. Akad. Nauk SSSR, 253 (1980), n. 2, 295–297.

64. Rabotnov, Yu.N., *Mechanics of a deformable solid.* (Russian), Izdat. Nauka, Moscow, 1979, 744.

65. Radon, N., *On boundary problems for logarithmic potential.* (Russian), Uspekhi Math. Sci., 1.

66. *The evolution of the theory of contact problems in SSSR.* (Russian), Izdat. Nauka, Moscow, 1976, 493.

67. Ryzhov, E.V., Kolesnikov, Yu.V., Suslov, A.G., *Contact of solids under statical and dynamical loceding.* (Russian), Izdat. Naukova Dumka, Kiev,

1982, 172.

68. Timoshenko, S.P., Goodier, J.N., *Theory of elasticity.* McGraw–Hill Book Company, Inc., New–York, Toronto, London, 1951, 506.

69. Ugodchikov, A.G., Khutoryansky, N.M., *The boundary element method in the mechanics of a deformable solid body.* (Russian), Kazan. Gos. Univ., Kazan, 1986, 296.

70. Ungiadze, A.V., *The first fundamental problem for a piecewise homogeneous plane that contains a semi–infinite crack crossing the boundary at right angles.* (Russian), Trudy Tbiliss. Mat. Inst. Razmadze Akad. Nauk Gruzin. SSR, 81 (1986), 79–86.

71. Uflyand, Ya.S., *The method of dual equations in probems of mathematical physics.* (Russian), Izdat. Nauka, Leningrad. Otdel., Leningrad, 1977, 220.

72. Hay, M.V.,

 a) *Solution of some two-dimensional integral equations.* (Russian), Ukr. math. journal, 37 (1985), n. 6, 781–785.

 b) *Potential method for three-dimensional thermoelastic problems solids with space cracks.* (Russian), Dr. Sci. Diss, 1988, 377.

73. Hörmander, L., *On the division of distributions by polynomials.* (Russian), Ark. Mat., 3 (1958), 555–568.

74. Hörmander, L., *Linear partial differential operatiors.* (Russian), Izdat. Mir, Moscow, 1965, 379.

75. Zvang, V.A., Shevchenko, V.P., *Boundary integral equations in the theory of plates and shells.* (Russian), Don. Univ. Publ., Doneck, 1986, 100.

76. Sherman, D.I., *The theory of elasticity of static plane problems.* (Russian), Trudy Tbil. Mat. Inst., 2 (1937), 163–225.

77. Sherman, D.I., *The theory of elasticity of static plane problems for inhomogeneous media.* (Russian), Trudy seismol. Inst. Akad. Nauk SSSR, (1938), no 86.

78. Sherman, D.I., *The theory of elasticity of plane problems for an isotropic medium.* (Russian), Trudy seismol. Inst. Akad. Nauk SSSR, (1938), n. 86, 51–78.

79. Sherman, D.I., *On the solution of the theory of elasticity static plane problem under given external loading.* (Russian), Dokl. Akad. Nauk SSSR, (1940), n. 1, 25–28.

80. Sherman, D.I., *Plane strain for isotropic medium.* (Russian), Prikl. Mat. Mekh., 7 (1943), n. 4.

81. Sherman, D.l., *Rotation of the round cylinder reinforced by an elliptical rod.* Eng. sbornik, v. 21, Academizdat, Moscow, 1955.

82. Shoikhet, B.A., *An energetical identity.* (Russian), Prikl. Mat. Mekh, (1976), n. 2, 317–326.

83. Staerman, I.Ya., *The contact problems of the theory of elasticity.* (Russian), Gusudarstv. Izdat. Tehn. Teor. Lit., Moscow–Leningrad, 1949, 270.

84. Shubin, M.A., *Pseudodifferential operators and spectral theory.* (Russian), Izdat. Nauka, Moscow, 1978, 279.

85. Bielerbach, L., *Lehrbuch der Funktionentheorie.* Bd.II, B.G. Teubner-Verlag, Leipzig - Berlin, 1927, 366 p.

86. Cassisa, C., *Analytical problems originating in the study of the states of plane stress in an elastic cylinder.* Atti Accad. Naz. Lincei, Mem. Cl. Sci. Fis. Mat. Natur. Sez., VIII, 17 (1982), n. 1, 27.

87. Stephan, E.P., *A boundary integral equation method for three-dimensional crack problems.* Math. Methods Appl. Sci. 8 (1986), n. 4, 609–623.

88. Frank, F.C., Lawn, B.R., *On the theory of Hertrian fracture.* Proc. Roy. Soc. Lond, Ser. A, 299 (1967), 291–309.

89. Jentsch, L.,
   a) *Existenzsätze der Thermoelastostatik in stückweise homogener Körper.* Beiträge zur Anal. (1973), n. 5, 107–119.
   b) *Über Wärmespannungen in Körpern mit stückweise konstanten Lameschen Elastitätsmoduln.* Schriftenreihe, ZIMM, Berlin, 1972, n. 14.
   c) *Über ein Bimetallproblem der Ebene.* 257 (1985), 83–102.

90. Hertz, H., *Über die Berührung fester elastischer Körper.*, J. Reine Angew. Math., 92 (1982), 156–171.

91. Hayman, W.K., *A conformal mapping problem arising in elasticity.* Atti Accad. Naz. Lincei, Mem. Cl. Sci. Fis. Mat. Natur. Sez., 17 (1982), n. 2, 56.

92. Hsiao, G.C., Wendland, W.L., *On a boundary integral method for some exterior problems in elasticity.* Trudy Tbiliss. Univ., Mat. Mekh. Astronom (1985), n. 18, 31–60.

93. Kinoshita, N., Mura, T., *On boundary value problem of elasticity.* Res. Rept. Tae. Eng. Meiji Univ., (1956), 8, 193–198.

94. Lauricella, G., *Sur l'integration de l'equation relative a l'equilihre des plaques elastiques encastreis.* Acta Mathem., 32 (1909), 201–256.

95. Lawn, B.R., Wilshaw, T.R.. *Indentation fracture principles and application.* J. Mater. Sci, 10 (1975), n. 6, 1049–1081.

96. Lichtenstein, L., *Über die erste Randwertaufgabe der Elastizitätstheorie.* Math. Zeitschr, 20 (1924), 21–28.

97. Maul, J.,

a) *Mixed contact problems in plane elasticity.* Z. Anal. Anwendungen, Bd. 2 (1983), n. 3, 207-234; Bd. 2 (1983), n. 6, 481-509.

b) *Potential method for treating mixed problems in plane elasticity.* Adv. in Mech. 6 (1983), n. 3-4, 79-98.

98. Pham, T. L., *Journal de Mech.*, 6 (1967), n. 2, 211-242.

99. Scherman, D.J., *On the problem of plan strain in non-homogeneous media.* Pergamon Press, 1959, Heterogeneity in elasticity and plasticity, 3-20.

100. Smirnov, V.I., *Über die Ränderzuordnung bei konformer Abbildung.* Math. Ann., 107 (1933), 465-499.

101. Stephan, E.P., *A boundary integral equation method for three-dimensional crack problems.* Math. Methods Appl. Sci. 8 (1986), n. 4, 609-623.

102. Wendland, W.L., *Boundary element methods and their asymptotic convergence,* CISM, Courses and Lectures, 277, Springer Verlag, 1983, 135-216.

103. Yu Dehao, *A system of plane elasticity canonical integral equations and its application.* J. Comp. Math., 4 (1986), n. 3, 200-211.

Appendix I

1. Agranovich, M.S., Dynin, A.S., *General boundary problems for elliptical systems in multidimensional region.* (Russian), Dokl. Akad. Nauk SSSR, 146 (1962), n. 3, 151-154.

2. Beresansky, Yu.M., *An expansion of self-adjoint operators by eigenfunctions.* (Russian), Izdat. Naukova Dumka, Kiev, 1965, 798.

3. Mikhlin, S.G., *The probem of minimum of a quadratical functional.* (Russian), Gosudarstv.Izdat. Tehn. Teor. Lit., Moscow-Leningrad, 1952, 216.

4. Mikhlin, S.G., *Multidimensional singular integrals and integral equations.* (Russian), M. Phus. math. Publ. 1962, 254.

5. Mikhlin, S.G., *The spectrum of operators bundle of elasticity theory.* (Russian), Uspekhi Mat. Nauk, (1973), n. 3 (171), 43-82.

6. Smirnov, V.I., *Course of high mathematics.* (Russian), v. 4, Moscow-Leningrad, 1951, 804.

7. Treves, J.F. *Lectures on linear partial differential equations with constant coefficients.*

8. *Handbuch der Physik. Band VI. Mechanik der elastischen Körper.* Bearbeitet von G. Angenheister, A. Busemann, O. Foppl [u.a.], Redigiert von R. Crammel. Berlin. 1928.

9. Tricomi, F., *On linear second order partial differential equations of mixed type.* Transl. from Italian, Moscow-Leningrad, 1947, 192 p.

10. Edelstein, U.I., *On the "freezing" of deformations in the photoelasticity method.* (Russian), Vestnik Lenigrad. Univ., (1968), n. 7, 118–127.

11. Agmon, S., Douglis, A., Nirenberg, L., *Estimate near the boundary for solutions of partial differential equations satisfying general boundary conditions.* II. Comm. on Pure and Appl. Math., (1954), 17, 1, 35–92.

12. Cosserat, E. et F., *Sur les equations de theore de l'elasticite.* Comptes Rendus de l'Akad. d. Sci. Francaise, (1898), 126, 1089–1091.

13. Cosserat, E. et F., *Sur les fonctions potentielles de la theorie de l'elasticite.* Comptes Rendus de l'Akad. d Sci. Francaise, (1898), 126, 1129–1132.

14. Cosserat, E. et F., *Sur la deformation infiniment petite d'un ellipsoide elastique.* Comptes Rendus de l'Akad. d. Sci. Francaise, (1898), 127, 315–318.

15. Cosserat, E. et F., *Sur la solution des equations de l'elasticite dans le cas ou les valeurs des inconnues a la frontiere sont donnees.* Comptes Rendus de l'Akad. d. Sci. Francaise, (1901), 133, 145–147.

16. Cosserat, E. et F., *Sur une application des fonctions potentielles de la theorie de l'elasticite.* Comptes Rendus de l'Akad. d. Sci. Francaise, (1901), 133, 210–213.

17. Cosserat, E. et F., *Sur la deformation infiniment petite d'un corps elastique soumis a des forces donnees.* Comptes Rendus de l'Akad. d. Sci. Francaise, (1901), 133, 271–273.

18. Cosserat, E. et F., *Sur la deformation intiniment petite d'une enveloppe spherique elastique.* Comptes Rendus de l'Akad. d. Sci. Francaise, (1901), 133, 326–329.

19. Cosserat, E. et F., *Sur la deformation infiniment petite d'un ellipsoide soumis a des effoorts donnees a la frontiere.* Comptes Rendus de l'Akad. d. Sci. Francaise, (1901), 133, 361–364.

20. Cosserat, E. et F., *Sur un point critique particulier de la solution des equations de l'elasticite dans le cas ou les efforts sur la frontiere sont donnees.* Comptes Rendus de l'Akad. d. Sci. Francaise, (1901), 133, 382–384.

21. Lichtenstein, L., *Über die erste Randwertaufgabe der Elastizitätstheorie.* Math. Zeitschr., (1924), 20, 21–28.

Appendix II

1. Vainberg, B.R., *The radiation principle, limit absorbtion and limit amplitude in the general theory of partial differential equations.* (Russian), Uspekhi Mat. Nauk, 21 (1966), n.3 129; English transl. in Russian Mathematical Surveys, 21.

2. Korotkina, M.R., *Electro–magneto–elasticity* (Russian), Izdat. Nauka, Moscow, 1989.

3. Kuznetsov, S.V., *The fundamental solution of the Lame equations for anisotropic media.* (Russian), Izv. Akad. Nauk SSSR, Meh. Tverd. Tela, (1989), n. 4, 50–54; English transl. in Mechanics of Solids.

4. Kupradz,e V.D., *Potential method in the theory of elasticity.* London: Oldbourne Press, 1965.

5. Kupradze, V.D., Gegelia, T.G., Baschelejschwili, M.O., Burtchuladze T.V., *Three–dimensional problem of the mathematical theory of elasticity and thermoelasticity.* (Russian), Izdat. Nauka, Moscow, 1976.

6. Arkhangel'skii, A.V., Appendix to Russian edition of J.L. Kelley's *General topology.* (Russian), Izdat. Nauka, Moscow, 1981.

7. Lazarev, M.I., *Potentials of linear operators.* Dokl. Akad. Nauk SSSR, (1987) 292, n. 5, 1045–1047; English transl. in Soviet Mathematics. Doklady.

8. Lazarev, M.I., Chikin V.N., *Boundary equations for linear and quasi-linear elliptic operators.* (Russian), Preprint; NCBIAN, Puschino, 1989.

9. Lazarev, M.I., Chikin, V.N., *The boundary reduction of quasi–linear ellitptic operators.* (Russian), Thesis, Conference "Method diskretnykh vikhrei", Kharkov, 1987.

10. Lions, J.-L., Magenes E., *Problémes aux limites non homogénes et applications.* v. 1. Paris, Dunor, (1968).

11. Lopatinskii, Ya. B., *On a method of reduction of boundary value problems for systems of differential equations of elliptic type to regular integral equations.* (Russian), Ukrainskii matematicheskii journal, (1953) 5, 2, 123–151.

12. Mikhlin, S.G., *Variationsmethoden der mathematischen Physik.* Akademie–Verlag Berlin, 1962, 464.

13. Mikhlin, S.G., *Linear partial differential equations.* (Russian), Izdat. Vyss. Skola, Moscow, 1977, 431.

14. Mikhlin, S.G., *Higher–dimensional singular integrals and integral equations.* (Russian), Gosudarstv. Izdat. Fiz.-Mat. Lit., Moscow, 1962, 254.

15. Nowacki, W., *Efekty elektromagnetyczhe w stalychcialach odksztalcalnych.* (Polish), Warszawa, 1983.

16. Palamodov, V.P., *Linear differential operators with constant coefficients.* (Russian), Izdat. Nauka, Moscow, 1967.

17. Parton, V.Z., Kudryavtsev, B.A., *Electro–magnito–elasticity of piezoelectric and electricity conducting bodies.* (Russian), Izdat. Nauka, Moscow, 1988.

18. Reznik, A.A., *The approximation of surface potentials and the solution of boundary value problems.* (Russian), Candidate's didssertation, Moskov. Fiz.-Tehn.-Inst., Moscow, 1983.

19. Reznik, A.A., *The approximation of surface potentials of elliptic operators by difference potentials.* (Russian), Dokl. Akad. Nauk SSSR, (1982), 263, 6,; English transl. in Soviet Mathematics. Doklady.

20. Reznik, A.A., *The metric property of boundary potentials and Green operators of elliptic operators.* (Russian), Preprint 135, Keldysh Inst. Appl. Mat. Akad. Sci. SSSR, Moscow, 1988.

21. Rudin, W., *Functional Analysis.* N.-Y., McGRAW-HILL, 1973.

22. Ryaben'kii V.S., *The difference potentials method and its application to problems of continuum mechanics.* (Russian), Izdat. Nauka, Moscow, 1987.

23. Ryaben'kii, V.S., *Green's formula for systems of difference equations with constant coefficients.* Matematicheskie zametki, (1969), 5, 5; English transl. in Mathematical notes of the Akademy of Sciences of the SSSR.

24. Ryaben'kii, R.S., *Generalization of Calderon projections and boundary equations on the basis of the concept of the clear trace.* (Russian), Dokl. Akad. Nauk SSSR, 270(1983), n. 5, 288–292; English transl. in Soviet Mat. Dokl., (1983), 27, 600–604.

25. Ryaben'kii, V.S., *The general construction of Green difference formula.* (Russian), Preprint 15, Keldysh Inst. Appl. Mat. Akad. Sci. SSSR, Moscow, 1983.

26. Samarskii, A.A., Andreev, V.B., *Difference methods for elliptic equations.* (Russian), Izdat. Nauka, Moscow, 1976.

27. Fedorenko, R.P., *Iteration methods of solution of elliptic difference equations.* (Russian), Uspehi Mat. Nauk, (1973), 28, 2; English transl. in Russian Mathematical Surveys, 28.

28. Hormander, L., *Pseudo–differential operators and non–elliptic boundary problems.* Ann. of Math., 83, (1966), 129–209.

29. Hormander, L., *The Analysis of Linear Partial Differential Operators. I Distribution Theory and Fourier Analysis.* Springer–Verlag, 1983.

30. Hirsch, M.W., *Differential topology.* Springer–Verlag, 1976.

31. Calderon, A.P., *Boundary value poblem for elliptic equations. Soviet– American Symposium on partial differential equations. Novosibirsk.* (Russian), Gosudarstv. Izdat. Fiz.-Mat. Lit., Moscow, 1963, 303–304.

32. Seely, R.T., *Singular integrals and boundary value problems.* Amer. J. Mat., (1966), 88, 781–809.

33. Calderon, A.P., Zygmund, A., *Singular integral operators and differential equations*. Amer. J. Mat., (1957), 79, 901–921.
34. Frank, L.S., *Difference convolution operators*. (Russian), Dokl. Akad. Nauk SSSR, 181(1968), 2; English transl. in Soviet Mathematics. Doklady.
35. Frank, L.S., *Grid function spaces*. (Russian), Mat. Sbornik, 86, 2, (1971), 187–233; English transl. in Mathematics of the SSSR. Sbornik, (1971), 15.
36. Lazarev, M.I., Chikin, V.N., *Direct and indirect methods of the reduction of elliptic boundary value problems*. Computer mechanics of solids. Issue 3 (to be published).

INDEX